Wide Bandgap Semiconductors

As Moore's law approaches its physical limit, research results beyond Moore's law based on novel materials, structures and devices is providing a new development direction for the semiconductor industry. GaN, SiC and other wide bandgap semiconductor materials are the third generation of semiconductor materials, following silicon, germanium, GaAs and InP. They are the preferred materials for manufacturing solid-state light sources, power electronic devices, and microwave RF devices due to their superior frequency, efficiency, voltage resistance, temperature resistance, and radiation resistance. The study of new materials can meet national strategic requirements for energy conservation and emission reduction, intelligent manufacturing and information security among others. It has become the forefront of global semiconductor technology research and a new industrial focus, and has a significant impact on industrial growth.

The book series "Wide Bandgap Semiconductors" is dedicated to clarifying the fundamental frontiers, key scientific and technological issues of wide bandgap semiconductors. It provides a comprehensive and prospective discussion of the characteristics of new generation semiconductor materials, fundamental physical properties, device fabrication, international R&D trends, key scientific and technological problems encountered, major R&D results achieved, device applications, and development prospects. This series represents the latest findings in fundamental semiconductor theory research and has significant application value.

The book series "Wide Bandgap Semiconductors" brings together the research experience and achievements of more than twenty of China's leading experts in this field. It is a creative exploration of the "synergy between fundamental and applied disciplines". This series fills a gap in the publication of monographs in this field both domestically and internationally, and can serve as a resource for scientific and technical workers engaged in the research and development of wide bandgap semiconductor electronic materials and devices, as well as for enterprise engineers and technicians seeking to strengthen their fundamental knowledge and investigate the problems encountered in practical scientific research.

Tongbo Wei · Zhiqiang Liu · Jinmin Li

Quasi van der Waals Epitaxy of Nitride Semiconductor and Its Applications

Springer

Tongbo Wei
Institute of Semiconductors
Chinese Academy of Sciences
Beijing, China

Zhiqiang Liu
Institute of Semiconductors
Chinese Academy of Sciences
Beijing, China

Jinmin Li
Institute of Semiconductors
Chinese Academy of Sciences
Beijing, China

ISSN 2948-2607 ISSN 2948-2615 (electronic)
Wide Bandgap Semiconductors
ISBN 978-981-95-3468-5 ISBN 978-981-95-3469-2 (eBook)
https://doi.org/10.1007/978-981-95-3469-2

Jointly published with Xidian University Press
The print edition is not for sale in China (Mainland). Customers from China (Mainland) please order the print book from: Xidian University Press.
ISBN of the Co-Publisher's edition: 978-7-5606-6288-6

This Springer imprint is published by the registered company Springer Nature Singapore Pte Ltd.
The registered company address is: 152 Beach Road, #21-01/04 Gateway East, Singapore 189721, Singapore

Preface

As a typical representative of third-generation semiconductors, the forbidden bandwidths of group III nitrides AlN, GaN, InN and their alloys are continuously adjustable from 0.64 to 6.2 eV, and thus nitride-based optoelectronic devices can theoretically achieve a complete coverage of infrared to ultraviolet wavelength bands. Compared with silicon and gallium arsenide and other traditional semiconductor materials, nitride materials show a high breakdown field, high thermal conductivity, high electron saturation rate and higher radiation resistance, more suitable for the production of high-temperature, high-frequency, radiation-resistant and high-power devices. Therefore, the third generation of semiconductors in the field of optoelectronics and microelectronics compared to the first two generations of semiconductors has incomparable advantages.

However, due to the lack of a single crystal substrate, nitrides are mainly grown heterogeneously epitaxially on sapphire. Strong chemical bonds are created between the nitride epitaxial layer and sapphire, which can lead to crystal deformation and generate huge stresses. During the growth process, the stress energy accumulates with the thickness of the epitaxial layer, and when the thickness of the film exceeds a critical value, various dislocations or stacking faults are formed to release the stress. The lattice mismatch and thermal mismatch that exist between nitrides and sapphire result in dislocation densities as high as 10^9–10^{10} cm^{-2}, which seriously affects the device performance.

Since graphene was successfully prepared in 2004, the research on graphene has been developing rapidly, and these advances have also rapidly inspired people to study two-dimensional (2D) crystal materials. The absence of hanging bonds on the surface of 2D crystals can effectively reduce the interaction between the epitaxial layer and the substrate, which on the one hand can alleviate the high stress and high density of dislocations brought about by the lattice mismatch and thermal mismatch, and on the other hand, the weak van der Waals (vdW) force between the layers provides a new way for the mechanical stripping and transfer of the devices. Thus, its unique vdW epitaxial mode has become one of the important trends in the future development of nitride, the core material of third-generation semiconductors.

The authors have been engaged in the growth of nitride materials and research on light-emitting diodes, detectors and other devices for more than ten years and have lots of experience and solid accumulation in epitaxy, chip and device physics. Since 2015, the authors have cooperated with Academician Liu Zhongfan's team at Peking University to carry out in-depth research in the emerging direction of Quasi van der Waals epitaxy (QvdWE) nitrides on 2D crystals, and the results achieved are also due to the rich experience and strong technical reserves of Academician Liu's team in graphene growth. Based on the research results of the authors and their research team in recent years, this book introduces the theoretical calculations, material growth, device preparation and applications of QvdWE nitrides on 2D materials in detail with the goal of industrialization of third-generation semiconductors in combination with 2D materials, and the contents are both academic and practical. In addition to the research results of the authors and their co-workers, the examples in this book also draw on many useful references and other related works.

The book is divided into eight chapters. Starting from the definition and basic properties of 2D materials, the definition of QvdWE is introduced, and the principle and application of QvdWE nitrides on 2D materials are presented. Subsequently, the theory of 2D material/nitride QvdWE interface and bonding and nucleation are introduced, revealing the epitaxial growth mechanism of 2D/three-dimensional (3D) materials. Further, the graphene-assisted QvdWE technology and growth methods based on single crystal substrates and amorphous substrates are presented. Then the application of QvdWE nitride technology in device transfer and heat dissipation is introduced. Finally, the latest research and progress on 2D nitrides are also presented.

The content of the book is novel, theoretical and applied, can be used for reading and reference by scientific researchers, graduate students and corporate R&D personnel engaged in the field of third-generation semiconductor materials and semiconductor lighting and can also be used as a training material and training textbook for engineers in the industry, and hopefully, it can be helpful to relevant scientific and technological workers in the field of nitride research in the aspects of academic references and development trends.

This book is a summary of our recent years based on new nitride epitaxy and device preparation. However, due to the limited level, there are inevitably deficiencies. I hope our colleagues may offer more valuable suggestions! We would like to take this opportunity to express our special thanks to Academician Liu Zhongfan, Academician Hao Yue, Prof. Shen Bo, and many other scientists for their long-term guidance, as well as to our collaborators, Prof. Gao Peng, Prof. Sun Jingyu, Prof. Shenyuan Yang, and Prof. Chen Zhaolong for their help. Thanks to the support of the National Key Research and Development Program of China, the National Natural Science Foundation of China, and the Beijing Natural Science Foundation. Many colleagues from the lighting research and development Center of the Institute of Semiconductors, Chinese Academy of Sciences, have participated in the relevant research work. The authors would like to express their gratitude to Xi'an University of Electronic Science and Technology Press for their help in the publication of this book.

Beijing, China

Tongbo Wei
Zhiqiang Liu
Jinmin Li

About This Book

Based on the research results of the authors and their research team in recent years, this book aims at the industrial application of the combination of third-generation semiconductors and two-dimensional (2D) materials and describes the theoretical calculations, material growth, device preparation, and application of Quasi van der Waals (QvdWE) of nitrides on 2D materials in detail, with the contents combining academics and practicality. The book consists of 8 chapters, including the principles of 2D materials and QvdWE, theoretical calculations of interfaces, growth of QvdWE nitride materials assisted by 2D materials, QvdWE on large mismatch substrates, the application of 2D materials in flexible transfers and heat dissipation, and 2D nitrides and other current research frontiers.

The book is novel, progressive in sequence, theoretical and applied and can be used as a reference for researchers, graduate students and corporate R&D personnel engaged in the field of third-generation semiconductor materials and semiconductor lighting.

Contents

Chapter 1
Two-Dimensional Materials and the Principles and Applications of Quasi van der Waals Epitaxy

1.1 Two-Dimensional Materials and Their Synthesis Methods

Nanomaterials refer to materials whose dimensions reach the nanoscale (1–100 nm) in one-dimensional (1D), two-dimensional (2D), or three-dimensional (3D) directions. Based on their dimensionality, they are categorized into 1D, 2D, and 3D nanomaterials. Among these, 2D materials represent a significant class of substances characterized by a thickness reduced to the atomic layer limit in one dimension, while maintaining relatively larger dimensions in the other two directions [1].

In 2004, researchers Andre Geim and Konstantin Novoselov at the University of Manchester in the United Kingdom experimentally firstly demonstrated that it is possible to successfully isolate a single layer of carbon atoms [2], known as graphene, from graphite—a layered material—using a simple physical exfoliation method. Since then, both in theoretical research and practical applications, the study of 2D materials represented by graphene has garnered significant interest from scientists. Subsequently, some other 2D materials were successively separated, such as hexagonal boron nitride (h-BN), molybdenum disulfide (MoS_2), tungsten disulfide (WS_2), and transition metal carbides/nitrides/carbonitrides (Mxene) materials [3]. Compared with the nanomaterials of other dimensions, the advantages of 2D materials lie in their ability to be chemically modified, their structure and composition can be adjusted, and their electrical properties can be controlled by catalysis. They are also more conducive to electron transfer, which is beneficial to the improvement of electronic device performance. Furthermore, at the same time, their flexibility and transparency are high, and they have broad application prospects in wearable smart devices and flexible energy storage devices [4]. 2D materials have become a research hotspot in the fields of condensed matter physics, materials science, chemistry, and nanotechnology. At present, research on 2D materials focuses on preparation, characterization, modification, theoretical calculation, and application exploration. The performance of the same type of material, when derived from different materials

T. Wei et al., *Quasi van der Waals Epitaxy of Nitride Semiconductor and Its Applications*, Wide Bandgap Semiconductors,
https://doi.org/10.1007/978-981-95-3469-2_1

or through various preparation processes, may exhibit certain discrepancies. Great progress has been made in the fields of electronic/optoelectronic devices, catalysis, energy storage/conversion, sensing, and environmental management [5, 6].

2D materials can be divided into five major categories according to their composition and structure: elemental, non-metallic compounds, metallic compounds, salts, and organics. The typical materials in each category are shown in Fig. 1.1. The following provides a brief introduction to the composition and structural characteristics of various 2D materials.

1. Elemental materials include graphene, graphyne, black phosphorus, metals (Au, Ag, Pt, Pd, Rh, Ir, Ru) and newly emerged borophene, arsenene, germanene, silicene, bismuthene, etc.;

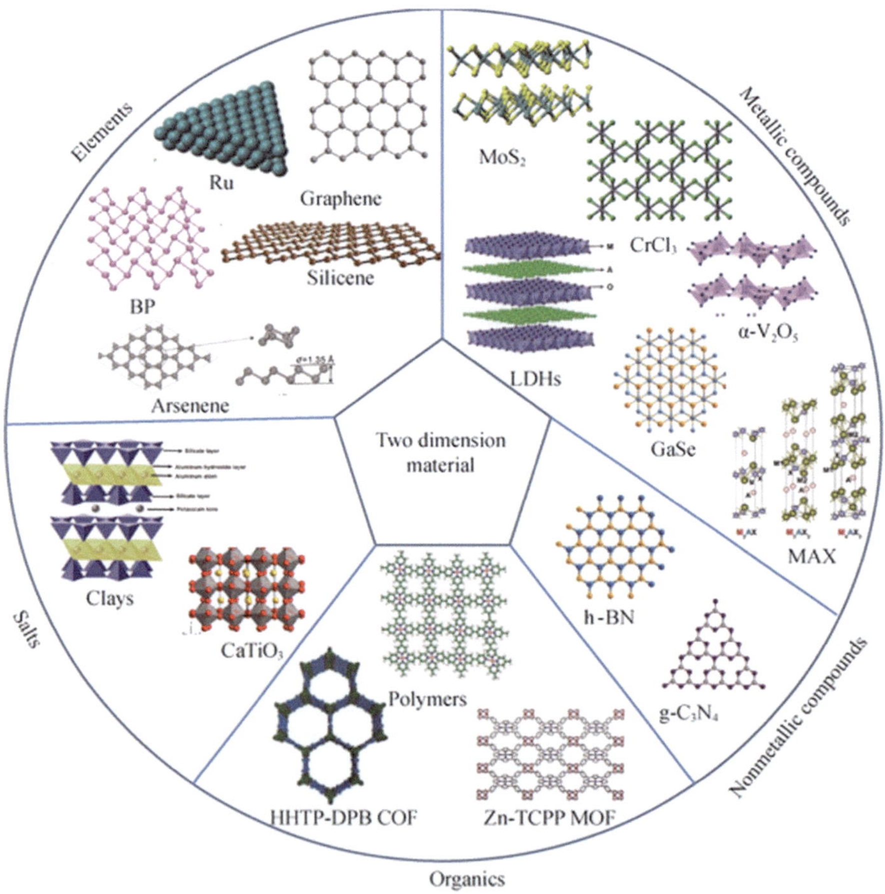

Fig. 1.1 Crystal structure classifiction of typical 2D materials [1]

2. Non-metallic compound materials include h-BN, graphite-like carbon nitride, boron carbon nitrogen, graphene oxide, halogenated graphene, and various graphene derivatives;
3. Metallic compound materials include transition metal dichalcogenides (TMDs), layered double hydroxides (LDHs), transition metal oxides (TMOs), transition metal carbon/nitrogen/carbonitrides (MXenes), metal phosphorus trisulfides (APX_3), metal halides, transition metal halogen oxides (MOX), III-VI group layered semiconductors (MX), etc.;
4. Salts that can yield 2D materials include inorganic perovskite compounds (AMX_3) and clay minerals (hydrated layered aluminosilicates);
5. Organic two-dimensional materials mainly include layered metal–organic framework compounds (MOFs), layered covalent organic framework compounds (COFs), and polymers.

1.1.1 Composition and Structure of Typical Two-Dimensional Materials

1. **Graphene**

As the most typical 2D material, graphene was first studied and prepared. In 2004, Andre Geim and Konstantin Novoselov obtained graphene from highly oriented pyrolytic graphite by mechanical exfoliation and demonstrated its unique and excellent electrical properties [2]. As shown in Fig. 1.2a, graphene is a hexagonal honeycomb structure composed of carbon atoms [7], with adjacent carbon atoms connected by σ bonds. The bond length may reach 0.142 nm with a bond angle of 120°, and can be stacked to form graphite by van der Waals (vdW) forces [8], with an interlayer spacing of about 0.335 nm.

Graphene's unique crystal structure enables many excellent characteristics, such as high mechanical strength, good ductility, excellent thermal conductivity, high electron mobility, and high specific surface area [9–13]. In terms of electrical properties, graphene demonstrates excellent conductivity, with electrons transmitting within the graphene sheet, and a mobility of up to 2×10^5 $cm^2/(V\ s)$, about 100 times

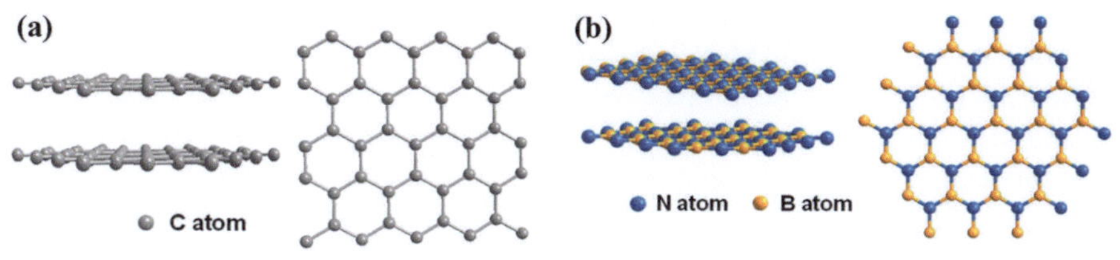

Fig. 1.2 Schematic diagram of the crystal structure of typical 2D materials of single-element and non-metallic compound types [3]. **a** Side view and top view of graphene. **b** Side view and top view of h-BN

that of silicon. In terms of thermal properties, the measured thermal conductivity of graphene is about 5000 Wm^{-1} K^{-1}, which is more than 10 times the thermal conductivity of copper at room temperature (400 $W{\cdot}m^{-1}$ K). As for optical property, a single layer of graphene can absorb 2.3% of visible and infrared light. Therefore, it shows many potential applications in fields such as sensors, polymer nanocomposites, optoelectronic functional materials, and controlled drug release.

2. **Hexagonal Boron Nitride**

h-BN, commonly known as "white graphene," is a graphene-like hexagonal structure covalently linked by boron and nitrogen atoms in equal proportions as shown in Fig. 1.2b. 2D h-BN has proven to be an ideal dielectric substrate of graphene for improved device quality due to its close lattice matching (~ 1.6%) with graphene, as well as dangling-bond free and atomically flat surface. In addition, h-BN occupies a special position, being an excellent electrical insulator with a wide band gap (5.97 eV) isostructural to graphene. It consists of alternating boron and nitrogen atoms arranged in a sp^2-bonded network within each layer, and the bond length of B-N is 1.44 Å while that of C–C in graphene is 1.42 Å. Hongxing Jiang from Texas Tech University pointed out that the generalized lattice mismatch between h-BN and wurtzite aluminum nitride (AlN) is only about 0.58%, and replacing highly insulating p-type AlN with p-type h-BN helps reduce contact resistance, and it has a lower acceptor ionization energy [14]. Paduano and Kobayashi et al. inferred that the generalized lattice mismatch between h-BN and sapphire does not exceed 5.5% [15, 16], which can have various uses and is key to advancing 2D optoelectronic devices [17]. However, there is still a gap between theoretically predicting 2D nitrides other than h-BN and experimentally realizing this structure.

3. **Transition Metal Dichalcogenides**

One of the most promising post-graphene nanomaterials is TMDs, which are compounds composed of transition metal elements and chalcogen elements (S, Se, or Te) [18]. The layers are connected by vdW forces, and each single layer contains three atomic layers, with the transition metal elements sandwiched between the chalcogen layers. Depending on the composition of different elements, TMDs have a variety of physical properties, such as: they can be used as insulating materials (like HfS_2), semiconductors (like MoS_2, WS_2), metals (like TiS_2, VS_2 and $NbSe_2$) etc. In addition, the band gap of TMDs depends on the number of layers of the material. For example, the band gap of MoS_2 can be increased from 1.2 eV of the bulk material to 1.8–1.9 eV of a single layer, and a single layer of MoS_2 assembling into a layered structure can enhance the catalytic performance of MoS_2 [19, 20]. Some materials can act as direct bandgap semiconductors in a single layer, and exhibit indirect bandgaps in double or thicker layers. Such TMDs materials are very suitable for optoelectronic applications [21, 22].

A notable feature of TMDs is that different crystal polymorphs can be formed according to different element coordination methods and layer stacking sequences. Most TMD materials have three phase characteristics: trigonal (1 T) also known

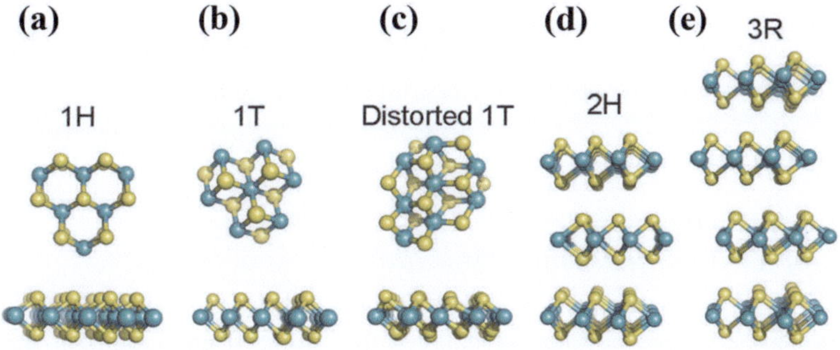

Fig. 1.3 Different polymorphs or phases of single-layer and stacked single layer TMDs: **a** 1 T phase, **b** ideal (a × a) 1 T phase, **c** distorted (2a × a) 1 T phase, **d** 2H phase and **e** 3R phase [23]

as metallic phase, hexagonal (2H), and rhombic (3R). These three phases can be transformed into each other under certain conditions. Taking MoS_2 as an example, it has four different crystal structures, namely 2H, 1 T, 1 T′, and 3R, according to different coordination models between Mo and S atoms or the stacking order between layers (refer to Fig. 1.3). The 2H structure has hexagonal close-packed symmetry and the atomic stacking order in the trigonal prismatic coordination (S-Mo-S′) ABA. Therefore, 2H-type MoS_2 is thermodynamically more stable. Additionally, the 1H structure is used to describe the single-layer structure of the 2H phase. The 1 T structure has octahedral coordination with square symmetry, and each layer has an atomic stacking sequence of (S-Mo-S′) ABC. The twisted 1 T (denoted as 1 T′) structure also has octahedral coordination, similar to the 1 T structure, but it contains an upper structure in each layer, such as tetramers, trimers, and sawtooth chains. It should be noted that ReS_2 and $ReSe_2$ have a naturally twisted 1 T′ structure. Compared with the 2H structure, the 3R structure has three layers per primitive unit, and the stacking order between the layers is different.

In recent years, the advantages of TMDs have been widely used in electrochemical energy storage and conversion, and integrated circuits (ICs). Because electrochemical reactions always occur at the surface or the interface, this 2D material has a large specific surface area and atomic exposure rate, providing abundant sites for electrostatic adsorption of ions, and providing conditions for the formation of high double-layer capacitance. Secondly, the ultra-thin structure can confine carriers within the 1 nm space of the interface, effectively suppressing the short channel effect of the transistors, thereby reducing equipment loss.

In addition to the above three commonly used 2D materials, the entire 2D material family also includes single-element silicene, germanene, stanene, black phosphorus, main group metal chalcogenides such as GaS, InSe, etc., and other 2D materials, graphitic carbon nitride (g-C_3N_4), graphene oxide (GO), metal phosphorus trisulfide compounds (such as $NiPS_3$), etc. These 2D materials have different band structures and electrical properties, covering a wide range of material types from superconductors, metals, semimetals, semiconductors to insulators, and have excellent optical, mechanical, thermal, magnetic, and chemical properties. They can build a more functional material system and are expected to be widely used in high-performance power electronic devices, optoelectronic devices, and the energy field.

1.1.2 Characteristics and Synthesis Methods of Two-Dimensional Materials

2D nanomaterials have unique advantages that differ from traditional materials [24, 25]:

First, in 2D single-layer materials, electrons are confined within the horizontal surface, making their quantum confinement effects more pronounced. This endows them with superior optoelectronic performance. Therefore, 2D materials, as ideal materials, have significant application value in the research of electronic and optoelectronic devices.

Second, each single layer of 2D materials has strong covalent bonds in the interior, while the forces between adjacent layers are weaker vdW forces. The next generation of semiconductor devices places great emphasis on the mechanical strength, flexibility, and optical transparency of materials. The in-plane covalent bonds and atomic layer thickness determine that 2D nanomaterials will become more promising candidate materials.

Third, due to their large planar dimensions and nanoscale thickness, 2D nanomaterials exhibit extremely high specific surface areas, which are beneficial for exposing surface atoms and providing more active sites. This has a significant impact on research related to surface area, such as catalysis and supercapacitors [26].

Fourth, the preparation method of ultra-thin 2D nanomaterials based on liquid-phase processing is simple. High-quality single crystalline thin films can be prepared by simple methods such as vacuum filtration, spin coating, drop coating, spray coating, and inkjet printing. This is very necessary for practical applications such as supercapacitors and solar cells.

Fifth, the atoms exposed on the surface of ultra-thin 2D nanomaterials can be used to regulate their characteristics through surface modification, doping, defects, strain, etc.

As mentioned above, the excellent performance of 2D materials has been proven to be a good candidate for the next generation of functional nanomaterials. Research hotspots of the 2D field focus on basic theoretical research and high-end applications, including the structure and properties of 2D materials, preparation methods, etc. The performance of devices based on graphene and other 2D materials has been deeply studied in the past few years, and some applications have gradually been industrialized as strong competitors of existing technologies. In general, great progress has been made in 2D materials in recent years, but problems such as low yield, poor structural controllability, unclear growth mechanism, poor stability, and insufficient performance diversity still need to be solved. The current research on the physical and chemical properties of 2D nanomaterials is still in the initial stage. Although theoretical studies have been conducted on the electronic properties of some 2D materials, other basic properties of these 2D material systems, including optical, mechanical, thermal, and chemical properties, have not yet been explored. The theoretical research on the formation mechanism of many 2D nanostructures and the relationship between their structure and performance is seriously insufficient, and experimental work on the controlled synthesis of different 2D morphologies and the explanation of the formation mechanism needs to be greatly strengthened.

2D materials also face many challenges in systematic and comprehensive research, including the exploration of the basic physical and chemical properties of single-layer, double-layer, and few-layer thin films; the engineering of heterojunction interface properties; and the research to change basic properties through doping, functionalization, electric field, and strain engineering. Based on these, it is expected to create new 2D materials, discover exotic properties, and enable 2D materials to be widely used in emerging technologies such as electronic circuits, spintronics, 2D light-emitting diodes (LED), nonlinear optical devices, THz generation and detection devices, supercapacitors, lithium-ion batteries, solar cells, chemical and biological sensors, and catalysts, etc. In addition, if 2D materials are stacked together as needed, new structures of double-layer or even multi-layer can be formed, which are maintained by vdW forces, such structures are called vdW heterostructures [6]. These artificial structures greatly enrich the properties of materials and can conveniently create artificial materials that do not exist in nature but have excellent performance. With the maturity of transfer technology and the continuous emergence of new ideas, it can be foreseen that more and more vdW systems will come into our view.

Due to the excellent performance and broad application prospects of 2D materials, the development of a simple, feasible, and reliable method for preparing 2D materials is of great significance for exploring their performance and applications. People are committed to exploring various synthesis methods to prepare 2D nanomaterials. Although the composition and crystal structure of different materials are different, they can all be divided into two types: layered and non-layered structure materials. Based on this, the current preparation methods are mainly divided into "top-down" and "bottom-up" methods [1]. The application premise of the "top-down" method is that the bulk material needs to have a 3D layered structure, using various driving forces to break the weak vdW interaction between the stacked layers, and peel the layered bulk crystals into single-layer or multi-layer nanosheets, such as mechanical exfoliation, liquid exfoliation, ion insertion and exfoliation, selective etching and exfoliation, etc. In comparison, the "bottom-up" method depends on the chemical reaction of the precursor under specific reaction conditions, such as physical vapor deposition (PVD) and chemical vapor deposition (CVD). In comparison with the top-down approaches, the bottom-up approaches based on the gas-phase vapor growth has been regarded as an efficient and controllable synthesis strategy for growth of high-quality large-area 2D materials, which are promising for meeting the requirements of industrial applications.

The method of mechanical exfoliation can obtain most of the 2D layered materials, but this method is inefficient. Usually, the lateral size of the exfoliated sample is small, and the thickness is not easy to control. Liquid phase exfoliation or CVD can prepare graphene and some transition metal chalcogenides, but it is difficult to control parameters such as the number of layers, edge morphology, defect density, doping concentration, etc. These methods need to be further optimized in the preparation of new 2D materials. In this section, we will focus on the commonly used synthesis and transfer methods of 2D materials used in subsequent chapters.

1. **Synthesis Methods of Two-dimensional Materials**

Top-down fabrication method:

The premise of the application of top-down method is that the bulk phase material needs to have a layered structure. Figure 1.4 shows various top-down fabrication methods of 2D materials. The micromechanical exfoliation method is to apply mechanical force through transparent tape to weaken the vdW interaction between the layers of large crystals, and obtain a few layers or single-layer 2D nanomaterials on the substrate as shown in Fig. 1.4a. This is one of the most effective methods to obtain electronic-grade high-quality single-crystal graphene with intrinsic physical properties. In 2004, Novoselov, Geim and colleagues successfully produced single-layer graphene from graphite using this method. Inspired by their work, many other types of single-layer or multi-layer nanosheets have been prepared, such as h-BN, MoS_2 and so on. The currently optimized micromechanical exfoliation method has been greatly promoted and applied to the exfoliation of other bulk materials. The lateral size of the obtained 2D materials can reach several to tens of micrometers. Because the preparation process does not introduce chemical substances and chemical reactions, the surface of the obtained 2D materials is very clean, maintaining the same crystal structure as its bulk crystal, and demonstrates good crystal quality, with almost no defects. However, this method is manual exfoliation, so there are disadvantages such as low yield, low preparation efficiency, and uncontrollable size, shape, and thickness of the prepared 2D materials. This method is mainly used for laboratory research on the physical properties of 2D materials and cannot be industrialized.

Besides the above method, applying mechanical forces such as sound waves or shear forces into the materials in the liquid phase can also disrupt the vdW forces between the layers of bulk materials to obtain 2D materials. For example, the ion insertion method is to insert small-radius cations (such as Li^+) into the layers of bulk materials to weaken the vdW forces between the layers as shown in Fig. 1.4c. The ion exchange-assisted liquid phase exfoliation method uses larger-radius organic or inorganic ions to replace the smaller-radius ions originally present between the layers of the bulk material Fig. 1.4d. The selective etching-assisted liquid phase exfoliation method uses etchants (hydrofluoric acid or fluoride/acid) to remove specific atomic layers. These methods have the potential to produce specific 2D materials efficiently and in large quantities, but the reaction process is complex and costly.

The oxidation-assisted liquid phase exfoliation method is also a method widely used to prepare graphene. It uses strong oxidants to oxidize graphite. The large number of functional groups introduced onto the graphene surface during the oxidation process will increase the interlayer distance, which is conducive to subsequent

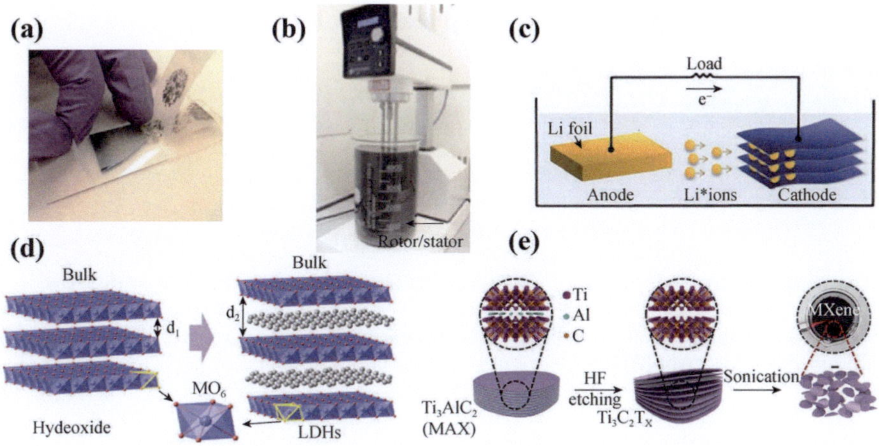

Fig. 1.4 Top-down methods: **a** Preparation of graphene by micromechanical cleavage; **b** Preparation of graphene by shear force-assisted liquid exfoliation; **c** Preparation of TMDs by ion intercalation-assisted liquid exfoliation; **d** Preparation by ion exchange-assisted liquid exfoliation; **e** Preparation of MXenes by selective etching-assisted liquid exfoliation [1]

exfoliation. This method can efficiently produce graphene, but the disadvantage is that it is difficult to promote and apply to the preparation of other materials, and it will introduce strong oxidants and increase experimental risks.

Bottom-up fabrication method:

In comparison to top-down, bottom-up method relies on the chemical reaction of the precursor under specific reaction conditions, and theoretically bottom-up method is more universal. The most commonly used method for preparing graphene, the CVD method, first appeared in the 1960s and was initially mainly used to prepare high-purity, high-performance solid films. The principle of graphene CVD preparation technology is followed: a carbon-containing gaseous substance is introduced into the furnace in a high-temperature and high-vacuum environment, using hydrogen gas as a reducing gas, to generate graphene. The equipment used to prepare graphene by the CVD method includes tube furnaces, microwave plasma CVD equipment, radio frequency CVD equipment, etc. Usually, the base metal foil is placed in the furnace, hydrogen and argon or nitrogen are introduced for heating to a stable temperature, the protective gas is stopped, and the carbon source (such as methane) gas is introduced. After the reaction is completed, the power is cut off, the methane gas is closed, and the protective gas is introduced again to purge the methane gas. Under the environment of the protective gas, until the equipment cools to room temperature, the metal foil is taken out, and the graphene on the metal foil can be obtained.

The metal foil acts as both a growth substrate and a catalyst in the CVD growth process of graphene. The formation of graphene by the cracking of hydrocarbon gases on the surface of the metal substrate is a complex catalytic reaction process. It mainly consists of three steps: the decomposition of carbon precursors, the nucleation stage

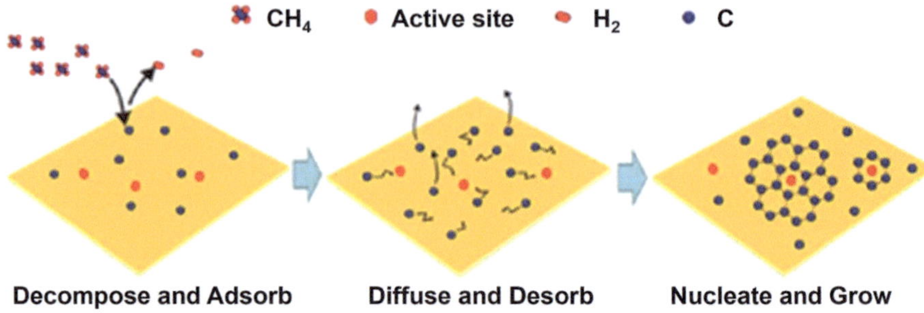

Fig. 1.5 Schematic illustration for the growth mechanisms of the CVD growth process of graphene with a methane precursor [27]

of graphene, and the gradual growth of graphene until they coalescence together, eventually connecting into a continuous graphene film, with the physical reaction process as shown in Fig. 1.5.

Since the first successful demonstration of CVD growth of monolayer (ML) graphene on Cu foils, CVD approaches have gradually become the dominant method for preparing various kinds of 2D materials including graphene, h-BN, metal disulfides (MX2, M=Mo, W, Ta, Cr, Nb, Re, etc., X=Se, S, Te), ternary compounds as well as 2D material-based heterostructures. The properties and applications of 2D materials are highly dependent on the thickness, domain size, geometric morphology and crystalline orientation, defect densities and dopant types, all of which can be controlled by optimizing the parameters in CVD growth. Accordingly, the strategies can be categorized to precursor design, space-confined growth, additive-assisted growth and substrate engineering [28].

Currently, CVD method has good controllability over material size, thickness, and composition, and can prepare large-area continuous thin films and single crystals. It is expected to prepare ultra-thin 2D materials for high-end electronic and optoelectronic industries, but it also has the disadvantage of relatively high production cost. Furthermore, many 2D crystals remain difficult to be directly grown on substrates via direct CVD growth owing to high melting points of their metal/metal oxide precursors as well as thermodynamics instabilities of metastable phases. Effective methods for directly growing large-scale high-quality 2D materials and their vdW heterostructures still need to be developed.

2. Transfer Methods of Two-Dimensional Materials

As the most typical 2D material, graphene prepared by the CVD method can usually grow on metal substrates such as copper and nickel foils, and they are the excellent conductors and heat conductors, which cannot directly test the electrical and thermal properties of graphene. In addition, the manufacture of devices must rely on silicon or other crystalline materials, so it is usually necessary to transfer operations to realize the application of graphene and other 2D materials. High precision, lossless transfer and accurate construction of 2D heterojunctions are essential for device applications.

The ideal graphene transfer technology should have the following characteristics: (1) the graphene structure is intact and undamaged after transfer; (2) no pollution to graphene (including doping); (3) the process is stable, reliable, and has high applicability. For graphene with only atomic-level or a few nanometers thickness, due to its low macro strength, it is very easy to damage during the transfer process. Therefore, the non-destructive separation from the initial substrate is the first problem that must be solved in the transfer process, and it is also the main problem currently faced.

Wet transfer is a common transfer method for 2D materials. According to the types of media layers used and the peeling methods, it can be divided into Polyvinyl Alcohol (PVA) adsorption transfer method, Polymethyl Methacrylate (PMMA) assisted transfer method, Poly (L-Lactic acid) (PLLA) rapid transfer method, small molecule doped Polystyrene (PS) transfer method, cellulose film transfer method, chemical etching transfer method, electrochemical peeling transfer method, sacrificial layer transfer method, metal-assisted peeling transfer method, etc. Dry transfer includes Polydi-Methylsiloxane (PDMS) peeling transfer method, vdWs interaction transfer method, etc. Transfers in non-atmospheric environments include inert atmosphere transfer method, vacuum transfer method, etc. [29]

Taking graphene as an example, the process of PMMA-assisted transfer is shown in Fig. 1.6. Graphene-on-BN devices are fabricated according to the procedure illustrated in Fig. 1.6d [30]. Graphene is exfoliated separately onto a polymer stack consisting of a water-soluble layer and PMMA, and the substrate is floated on the surface of a deionized water bath. Once the water-soluble polymer has dissolved, the silicon substrate sink to the bottom of the bath, leaving PMMA on top. Then the PMMA membrane is adhered to a glass transfer slide. By locating the position of graphene on the suspended polymer film, the graphene is aligned to the target BN and the two bring into contact. Once transferred, the PMMA is dissolved in acetone.

However, the wet chemical corrosion method to remove substrate also has certain limitations. For example, if the coated organic support layer is too thin, it is easy to

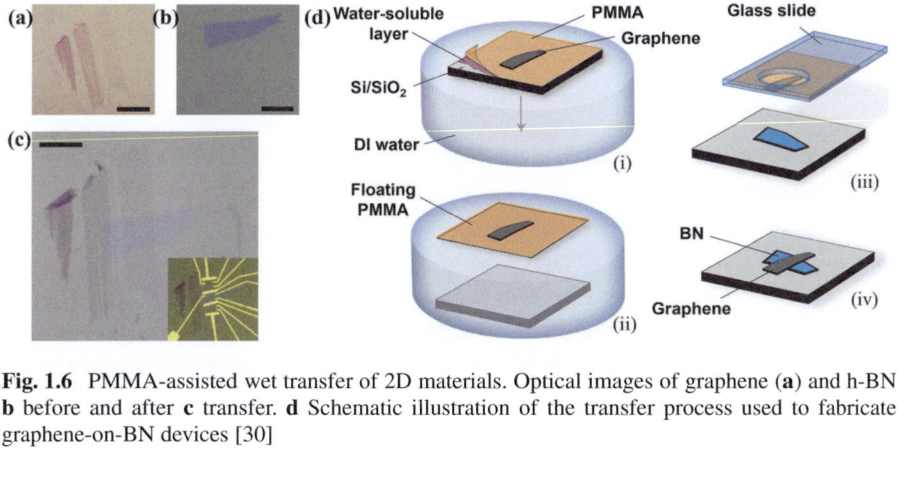

Fig. 1.6 PMMA-assisted wet transfer of 2D materials. Optical images of graphene (**a**) and h-BN **b** before and after **c** transfer. **d** Schematic illustration of the transfer process used to fabricate graphene-on-BN devices [30]

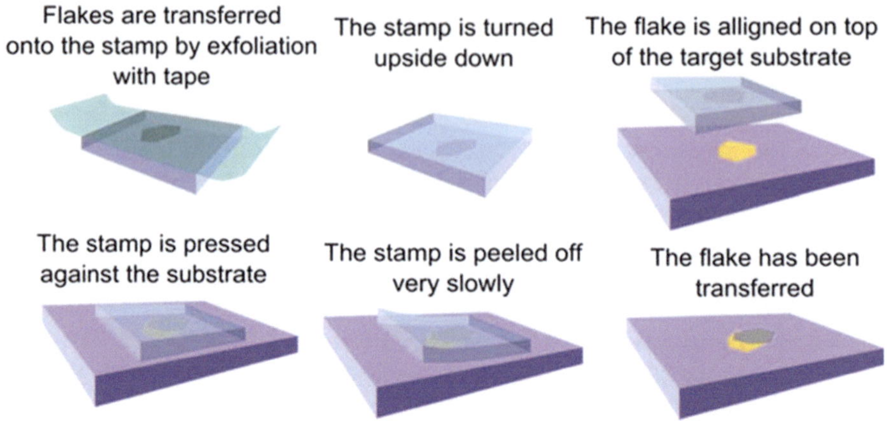

Fig. 1.7 Diagram of the steps involved in the preparation of the viscoelastic stamp and dry transfer of an atomically thin flake onto a user-defined location by PDMS [31]

cause film tearing during transfer, which is particularly unfavorable for the transfer of large-area graphene films. If the coated organic support layer is too thick, it has a certain strength, and the graphene and the target substrate cannot fully adhere. In addition, when the transfer medium is dissolved and removed, it will also easily cause the graphene film to be damaged.

Figure 1.7 shows the schematic process of PDMS dry transfer of 2D material. First, the 2D material is mechanically peeled off on stamp using PDMS. The viscoelastic PDMS with the sample is flipped, aligned with the target substrate, and the two are contacted. Then, the PDMS is peeled off from the substrate surface, and finally, the sample is successfully transferred to the target substrate. As the PDMS stamp is transparent, one can see the sample through it and thus it is possible to align the desired flake on the acceptor surface where one wants to transfer the flake at sub-micrometer resolution. Furthermore, in dry transfer, the metal substrate is not etched away and can be reused, greatly reducing the transfer cost. The quality of the graphene transferred to the polymer is very high, but there are still defects.

1.2 Heteroepitaxial Growth and Application of III-Group Nitrides

The semiconductor industry has gone through three stages to date. The first-generation semiconductor materials represented by silicon, and the second-generation semiconductor materials, gallium arsenide, have also been widely used. In recent years, third-generation semiconductor materials, such as gallium nitride

(GaN) and silicon carbide (SiC), zinc oxide (ZnO) and other wide-bandgap semiconductor materials, have attracted widespread attention from scientists and technologists around the world. Group III nitrides are typical materials of the third-generation semiconductors, including GaN, aluminium nitride (AlN), indium nitride (InN) and their alloys AlGaN, GaInN, AlInN, and AlGaInN. Group III nitride materials have excellent properties such as high breakdown voltage, high saturated electron velocity, and high stability. Their bandgap energy covers a wide spectrum range from near-infrared to deep ultraviolet, making them the most attractive materials in wide-bandgap semiconductors. They have broad application potential and huge market prospects in high-brightness LEDs, laser diodes (LDs), high-temperature high-power, radiation-resistant microwave devices, etc. and have become a hot field of scientific research.

The research on GaN thin film materials began in the 1960s. In 1969, Maruska et al. had deposited a large area of GaN thin film on sapphire using hydride vapor phase epitaxy (HVPE) [32]. However, due to the poor quality of the material and the difficulty of p-type doping, GaN materials were once considered to have no application prospects. Group III nitride semiconductors generally achieve their applications in optoelectronic devices, electronic devices, etc., through a multi-layer thin film structure within a few micrometers. Epitaxial growth is the most common way to achieve this multi-layer thin film structure. Epitaxy refers to the physical process of depositing and stacking one crystal material (called the epitaxial layer, epilayer) on another crystal (called the substrate, substrate), and the atoms of the epitaxial layer will be arranged in order according to the lattice parameters of the substrate. The epitaxial growth methods of group III nitrides mainly include: metal–organic chemical vapor deposition (MOCVD), HVPE, and molecular beam epitaxy (MBE). In the late 1980s, with the development of MOCVD technology, significant breakthroughs were made in the research of GaN. A series of key MOCVD growth technologies, such as heteroepitaxial buffer layer technology and p-type doping technology, solved the two major problems of growing GaN thin film materials on sapphire and GaN p-type doping, thus laying the foundation for the development of high-performance group III nitride devices. Benefiting from this, MOCVD technology has become the most mature and widely used epitaxial growth method for group III nitrides.

1.2.1 Heteroepitaxial Growth of III-Group Nitrides

Group III nitride materials have three crystal structures, namely wurtzite structure, zinc blende structure, and rock salt structure. Among them, the wurtzite structure is a thermodynamically stable structure. The zinc blende structure is a metastable structure, which can stably exist in thin films grown on cubic phase substrates. In comparison, the rock salt structure can be grown under very high pressure, and therefore is hard to obtain. As shown in Fig. 1.8, the wurtzite structure is a hexagonal structure stacked in the form of ABABAB… along the [0001] direction after two sets of hexagonal structures composed of Group III atomic layers and N atomic layers are

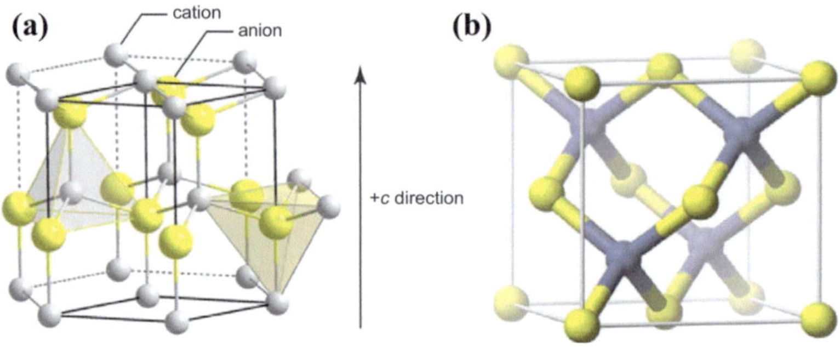

Fig. 1.8 Two typical crystal structures of Group III nitride materials. **a** Wurtzite structure. **b** Zinc blende structure [33]

nested by moving 5c/8 along the c-axis direction. However, the zinc blende structure is a cubic structure stacked in the form of ABCABC... along the [111] direction, which is formed by the displacement of two face-centered cubic lattices along the body diagonal by a quarter of the diagonal length.

Group III nitrides need to be epitaxially grown on substrates. Depending on the difference between the epitaxial layer and the substrate material, it can be divided into homoepitaxy and heteroepitaxy. When the epitaxial layer and the substrate are the same material, it is called homoepitaxy, otherwise it is called heteroepitaxy. Generally speaking, the first and second generation semiconductor materials can be obtained by homoepitaxy, which can be epitaxially grown on gallium arsenide (GaAs) and indium phosphide (InP) substrates. However, because the third-generation semiconductor GaN has special stability (melting point 2791 K, melting pressure 4.5 GPa), there is a lack of natural GaN bulk single crystal materials in nature, and it is very difficult to prepare GaN single crystals using Czochralski method for silicon single crystals. The homogeneous substrates of the third-generation semiconductor materials are relatively expensive, so heteroepitaxy growth of nitrides is mostly carried out on mismatched substrates.

At the atomic level, the sudden termination of the lattice at the surface of the sample will cause the outermost atoms to produce unpaired electrons, that is, unsaturated bonds, which are called dangling bonds in Fig. 1.9a [34]. When heteroepitaxy is performed, a strong chemical bond will be formed between the epitaxial layer and the substrate due to the interaction with these dangling bonds, prompting the atoms of the epitaxial layer to pair one by one with the atoms on the substrate. The atoms of the epitaxial layer are arranged according to the lattice orientation of the substrate, imitating the crystal symmetry of the substrate, and the lattice constant is also consistent with the substrate. In this way, the lattice constant of the epitaxial layer will change and no longer maintain its inherent parameters. The lattice mismatch is often used to describe the difference in size between the in-plane lattice constant of the epitaxial layer and the substrate, and the calculation formula is shown in Eq. 1.1.

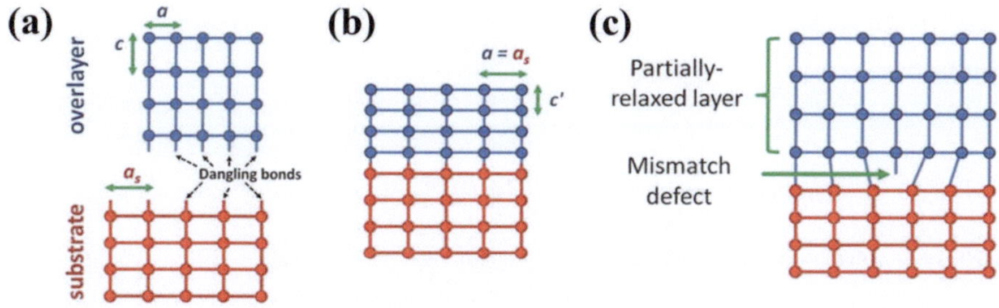

Fig. 1.9 Schematic diagram of lattice mismatch epitaxy. **a** Freestanding form. **b** Heteroepitaxy. **c** Mismatch defects in heteroepitaxy [34]

$$f = 100\% \times \frac{a_s - a}{a} \tag{1.1}$$

In the equation, a represents the in-plane lattice constant of the epitaxial layer material, a_s denotes the in-plane lattice constant of the substrate (the lattice constant after the epitaxial layer material changes).

During the heteroepitaxy process, when the lattice mismatch is relatively small, the first few atomic layers of the epitaxial layer will correspond one-to-one with the substrate, as shown in Fig. 1.9b, causing lattice deformation in the epitaxial layer, which leads to certain stress in the system. The stress will accumulate as the thickness of the epitaxial layer increases, but the ability of the epitaxial system to store stress is not unlimited. When the thickness of the thin film exceeds a certain critical value, it is energetically more favorable for the system to nucleate defects to relax the strain energy partially, as shown in Fig. 1.9c. The larger the lattice mismatch, the more defects formed, mostly in the form of penetrating dislocations, edge dislocations, stacking faults, etc. These dislocations may even extend into the active area of the multiple quantum wells (MQWs) through a certain thickness of the epitaxial layer, acting as non-radiative recombination sites or leakage current paths, severely affecting device performance [35]. Therefore, when performing heteroepitaxy, it is necessary to choose a substrate with lattice matching as much as possible to reduce the generation of such defects and improve the crystalline quality of the thin film.

Due to the lack of single crystal substrates in nature, the growth of Group III nitride materials is mainly achieved by heteroepitaxy on substrates such as sapphire, SiC, Si, as shown in Table 1.1. Sapphire is widely used due to its high stability, mature preparation technology, and low cost. However, as mentioned earlier, due to the mismatch of lattice constants and thermal expansion coefficients, heteroepitaxy based on sapphire will generate large stress and high dislocation density in the nitride epitaxial layer, resulting in unstable device performance. It usually requires the growth of AlN or GaN buffer layers at low temperatures to minimize mismatch [36, 37], thus increasing the cost. For example, the dislocation density of the GaN epitaxial layer obtained on sapphire using MOCVD technology is between 10^8 and $10^{10}/cm^2$, while the defect density allowed for high-stability LDs and deep ultraviolet

Table 1.1 Common substrates and parameters for GaN-based material epitaxy

Substrate	Sapphire (Al_2O_3)	Silicon(Si)	SiC	
			6H-SiC	3H-SiC
Crystal structure	Hexagonal	Cubic	Hexagonal	Cubic
Lattice constant (nm)	a = 0.4758 c = 1.2991	a = 0.5430	a = 0.308 c = 1.512	a = 0.4359
Mismatch with GaN Δa/a	16.2 (%)	17.1 (%)	3.5 (%)	3.56 (%)
Thermal expansion coefficient ($\times 10^{-6}$ K^{-1})	7.5 3.17	2.6	10.3	4.7
Thermal conductivity coefficient(W m^{-1} K^{-1})	46	150	490	490
Melting point (°C)	2030	1420	2700	2830
Stability	Good	Normal	Good	Good
Epitaxial GaN	Hexagonal	Cubic	Hexagonal	Cubic
Cost	Medium	Low	High	High

(DUV) LEDs is below 10^7 cm^{-2}. The thermal conductivity of sapphire is also very low, which severely affects the efficiency, reliability, and life of the devices, making heat dissipation a major problem for high-power devices. Si substrates with mature technology and low cost, can be integrated into silicon-based large-scale integrated circuits, but have a large thermal mismatch with GaN materials, and cracks are likely to appear after cooling. SiC substrate has good thermal conductivity and electrical conductivity, and the GaN grown SiC has a lower dislocation density. However, because SiC is very hard, it is difficult to polish the surface, and the price is too expensive. In addition, the p-type doping of nitrides, ohmic contacts, the acquisition of high-quality InGaN and AlGaN materials, and the integration and modularization of devices also face great challenges and many problems in the field of nitrides need to be solved.

1.2.2 Prospects of III-Group Nitrides

In the 1990s, Isamu Akasaki and Hiroshi Amano from Nagoya University in Japan, and Shuji Nakamura from Nichia Corporation successfully solved the problem of high-quality GaN thin film growth on sapphire by two-step growth. They also realized p-type doping of GaN using high temperature annealing, enabling the commercialization of high-efficiency nitride blue LEDs [38]. In 2014, the Nobel Prize in Physics was awarded to these three scientists in recognition of their invention of the blue LED, which brought about a new type of energy-saving light source. The breakthrough in GaN materials has led to a rapid development of devices based on nitride semiconductors, and also promoted the development of various other Group III nitrides and high-performance Group III nitride devices, such as LEDs, LDs, ultraviolet photodetectors (UV PD), and high electron mobility transistors (HEMT) for high-power electronic devices [39, 40].

The industrialization of optoelectronic devices based on the GaN material system has made significant progress worldwide. In particular, GaN-based LED devices, with their high efficiency, long life, wide spectrum, and intelligence, represent another revolution in lighting sources following incandescent and fluorescent lamps, providing an important solution to the increasingly severe energy and environmental problems. GaN-based LDs have important and wide applications in high-density information storage, laser display, underwater and fiber optic communications, biomedicine, and other civilian and military fields, and are leading products in the optoelectronics industry. GaN-based microwave power devices have significant implications in the field of military defense and can be used in phased array radars, electronic countermeasures, missiles, and radio communications. In the field of civilian commercial applications, they can be also used in wireless infrastructure (base stations), satellite communications, cable television, and power electronics. GaN-based detectors can be used in flame detection, missile warning, and other important applications. By adjusting the Al component, AlGaN-based solar-blind ultraviolet photodetectors can be obtained. GaN-based power electronic devices demonstrate nearly a thousand times lower conduction resistance and tens of times higher switching speed than traditional Si-based devices, greatly reducing power loss and saving electricity, and have broad application prospects in photovoltaic fields and electric/hybrid vehicles. In summary, group III-nitride devices have important application potential in new basic scientific fields, including the communication, control, sensing, information processing, electronics, software, and other technologies intersect [41], and thus have attracted great attention in recent years.

With the continuous development of flexible materials and devices, their applications span multiple fields such as aviation, consumer electronics, healthcare, robotics, and industrial automation. The products they can provide include flexible displays, chemical and biological sensors, flexible photovoltaics, flexible logic and storage, flexible batteries, wearable devices, electronic newspapers, electronic skin, and many other directions, which may change aspects of people's clothing, food, housing, and transportation in the future. The common growth methods for nitrides are MOCVD and MBE, and high-quality epitaxial nitride thin films can only be grown on lattice-matched single crystal substrates above 800 °C [37]. For large-area or flexible device applications, the devices need to be fabricated on glass, plastic, or metal substrates. However, glass substrates, as amorphous materials, cannot be used for the epitaxial growth of high-quality single crystal thin films, and plastic substrates do not tolerate high temperatures. General nitride semiconductors are grown on sapphire substrate using low-temperature AlN, low-temperature GaN, or AlON buffer layers. Due to the strong covalent bond between the buffer layer and the semiconductor, sp^3 Bonding, releasing nitride semiconductors from sapphire substrates presents certain difficulties, hence the need to peel off the nitride grown on sapphire or silicon substrates and transfer it to flexible substrates, or to find suitable buffer/release layers to achieve mechanical peeling of the nitride film from the substrate. Currently, laser lift-off (LLO) based on sapphire requires laser scanning and additional laser adjustments, or the silicon substrate is removed off by chemical etching, both of which require high energy consumption or time cost, and the yield is not high, making it difficult to apply on a large scale. With the increasing demand for foldable and wearable devices, combining nitride thin film devices with flexible substrates has become a research focus, and it is necessary to further overcome the lattice mismatch and thermal mismatch of heteroepitaxy, improve the crystal quality of nitride materials, reduce the cost of preparing flexible nitride devices, and thus new epitaxial technologies are needed to achieve this goal.

1.3 Principle and Application of Quasi van der Waals Epitaxy of Nitrides on Two-Dimensional Materials

1.3.1 Growth of Quasi van der Waals Epitaxy of Nitrides on Two-Dimensional Materials

The problems faced by III-nitride materials and devices urgently need to be solved, and the emergence of 2D materials has expanded the material range of traditional 3D material heterostructures. It triggers basic research on new coupling phenomena between 3D bulk materials and 2D atomic layers, aiming to exploit the unique interlayer van der Waals (vdW) forces of 2D materials, and such mixed heterostructures are crucial for material and functional device systems. The combination of crystal with different physical properties offers the capability of integrating them into functional systems.

In conventional epitaxy process, the crystalline layer is attached to the substrate by a strong chemical bond as shown in Fig. 1.10a [42]. Substrates used in conventional epitaxy possess a highly active surface with several dangling bonds. Due to the formation of the strong chemical bonds at the heterointerface, the in-plane lattice parameter of the epilayer should be modified to match that of the substrate. A high lattice mismatch between the epilayer and the substrate induces high interfacial strain and a number of mismatch-related defects, such as misfit dislocations and interface disorder. In 1984, in the process of studying the mechanism of heteroepitaxy, Koma et al. first demonstrated the Se/Te and $NbSe_2$/MoS_2 material systems [43], discovering that heteroepitaxy can also be connected by weaker vdW forces. In such systems, the vdW surface has no dangling bonds, and a vdW gap is generated at the heterointerface, producing a negligibly small strain. The epilayers in the van der Waals epitaxy (vdWE) have incommensurate in-plane lattices and a crystallographic orientation relationship with the vdW substrate and there is only weak interaction between the substrate and the epitaxial layer as shown in Fig. 1.10b. Furthermore, when conventional bulk crystals with dangling bonds are grown on a vdW substrate, Quasi vdWE (QvdWE) is possible due to the dangling-bond-free bottom substrate, as shown in Fig. 1.10c. The epitaxial layer will not completely arrange according to

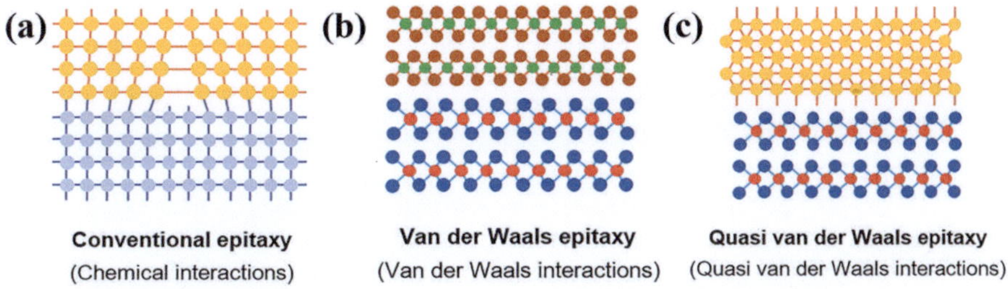

Fig. 1.10 Schematic illustration of **a** conventional epitaxy, **b** vdWE and **c** QvdWE [42]

the lattice constant of the substrate, but can maintain its inherent parameters, which can avoid the defect problems introduced by lattice mismatch.

However, before the appearance of graphene in 2004, there was less research on 2D materials. It is believed that due to thermodynamic limitations, as the number of material layers decreases, the melting point of 2D materials would rapidly decrease. Thus, it makes them more unstable, causing the thin film to become islands or dissolve. The emergence of graphene changes this traditional concept, and the latest theoretical research shows that wrinkles often appear in the preparation of 2D materials, which are usually regarded as 3D structures. This 3D warping will lead to an increase in strain energy and reduce the thermal vibration of the material, thereby reducing the total free energy to the lowest, allowing 2D materials to remain stable [44]. At the same time, similar to graphene, each single layer within 2D layered materials has strong covalent bonds, while the adjacent single layers are connected by weaker vdW forces, as shown in Fig. 1.11a. The so-called vdWE refers to material growth connected by vdW forces rather than by chemical bonds. Graphene and h-BN, as typical 2D atomic crystals, have a layered structure, with sp^2 covalent bonds between atoms within the layer and weak vdW bonds between layers. 2D/2D vdW heterostructures have been widely demonstrated using mechanical exfoliation and stacking processes.

When epitaxially growing 3D nitride thin films on 2D materials, that is, introducing a 2D material interlayer between the substrate and the epitaxial material, at this time, 3D covalent bonds and 2D vdW bonds coexist, which we can call QvdWE [45]. As shown in Fig. 1.11b, when 3D materials are epitaxially grown on 2D materials, there are also chemical interactions at the interface. However, compared with the chemical bonds between the 3D epitaxial layer and the 3D substrate, the chemical interactions between the 3D epitaxial layer and the 2D materials are about two orders of magnitude weaker. The vdW action helps with atomic alignment, and the structure arrangement between the epitaxial layer and the substrate crystal is consistent. Graphene and h-BN are typically used as a 2D substrate for the growth of nitride because they meet the temperature requirements for the growth conditions (up to 1000–1200 °C) of nitride due to their high melting temperature and chemical inertness [46, 47]. TMDs are also used as 2D substrates in the QvdWE of nitride. However, given the low thermal stability and high reactivity of TMDs relative to

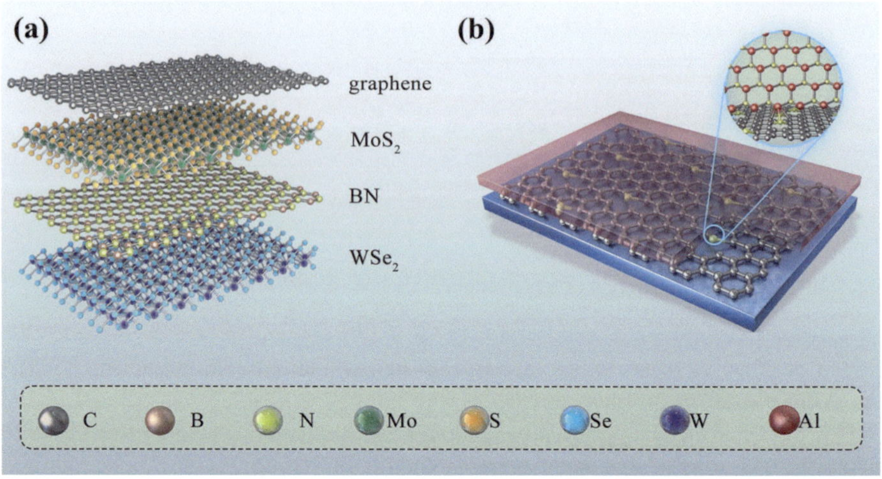

Fig. 1.11 Schematic growth mechanism of QvdWE nitride on 2D materials. **a** Schematic diagram of different 2D materials formed by vdW forces. **b** Schematic diagram of nitride growth on 2D materials through QvdWE [44]

graphene and h-BN, only a few experimental demonstrations of vdWE on TMDs have been reported [48].

Conventional heteroepitaxy of compound semiconductors relaxes the accumulated strain energy in the lattice mismatch film by forming mismatch dislocations through the introduction of a dislocation, as shown in Fig. 1.12a. However, QvdWE does not generate chemical bonds, and the 2D material surface has a lower energy barrier for interface movement of the epitaxial layer. The epitaxial layer grown on the 2D material can spontaneously relax before the accumulated elastic energy induced by lattice mismatch induces mismatch dislocations. This effect is more prominent in highly mismatched material systems (Fig. 1.12b). The weakened interface energy provides another way for strain relaxation during the heteroepitaxy process on the 2D material coated substrate, where the misfit strain can be immediately relaxed on the graphene surface [49]. QvdWE based on 2D materials opens a new door for the growth and flexible transfer of nitride thin films under large mismatch.

In traditional III-nitride heteroepitaxy, the chemical bonding of dangling bonds between the crystalline substrate and atomic building blocks is responsible for the lattice ordering in epitaxial growth, causing the epilayer to mimic the crystalline structure of the substrate. In contrast, the surface of the 2D material in QvdWE does not have dangling bonds. The initial substrate and epitaxial layer are combined by weak vdW forces between 2D materials, allowing the epitaxial layer and substrate to have a large lattice mismatch and different lattice symmetries. 2D materials like graphene here replace traditional low-temperature buffer layers, which can reduce the stress in the epitaxial layer. In addition, nitride QvdWE on 2D materials also has the following characteristics:

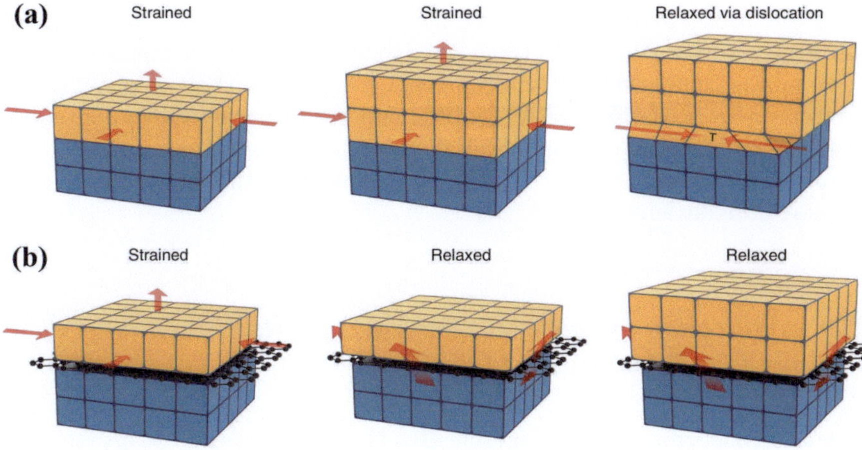

Fig. 1.12 InGaP grown on graphene/GaAs and GaAs substrate. **a** Schematic diagram of strain relaxation via the introduction of a dislocation. T indicates where a dislocation forms. **b** Schematic diagram of strain relaxation via spontaneous relaxation [49]

(1) Realizing the large-scale separation of the epitaxial layer and the substrate. Using the vdW force between the 2D material layers, the flexible peeling of the epitaxial film can be achieved through slip and other methods [50–52], without stripping equipment and chemical corrosives. The operation is also simple and fast, with good flatness, suitable for transfer to foreign substrates. Therefore, it is easy to achieve the transfer of GaN-based nitride devices to large-area, cheap, flexible substrates, and the substrate can be reused without the damage, improving economic efficiency.
(2) Avoiding the dependence on traditional single-crystal substrates. Thus, low-cost, large-size new substrates may be used. After using graphene or h-BN as the buffer layer, because it is QvdWE growth, 2D materials can shield the continuation of the substrate lattice, reducing the dependence on the substrate. Therefore, more diverse and cheaper substrates can be selected, such as quartz glass, metal substrates, etc. It not only promotes the GaN optoelectronics industry, but also promotes the advancement of crystal growth technology.
(3) Improving the heat dissipation performance of the device. 2D crystals, especially graphene, have good thermal conductivity. The thermal conductivity of graphene at room temperature is about 5000 $Wm^{-1} K^{-1}$, which is more than 10 times the thermal conductivity of copper. Using graphene as a buffer layer for epitaxial growth and device fabrication can alleviate the heat accumulation during the operation of the devices and play a good heat dissipation role.

In summary, 2D crystals represented by graphene, h-BN, and TMDs, with their unique crystal structure and electronic structure, provide a new direction and technical route for QvdWE nitride materials and device preparation. Exploring the method of low defect density nitride epitaxial growth on the transition layer of 2D crystals is expected to break through the limitations of substrate materials, develop flexible semiconductor devices and innovative applications. It will completely change the past nitride large mismatch heteroepitaxial growth method on sapphire, silicon and other single crystal substrates. QvdWE will become a disruptive preparation technology, contributing to basic physical research, and laying the foundation for constructing new semiconductor lighting, communication, energy, health and other fields of wide application.

1.3.2 Application of Quasi van der Waals Epitaxy Technology on Two-Dimensional Materials

Nitride semiconductor materials represented by GaN and AlN are gradually replacing the first-generation semiconductors (Si, Ge) and the second-generation semiconductor materials (GaAs, InP) in some specific fields, becoming the core support materials for the next-generation semiconductor industry. The QvdWE technology on 2D materials, due to its unique advantages, explores the integration methods of 2D/3D and 3D/2D, which can broaden its application fields to flexible wearable devices, etc. It may combine the characteristics of different materials to build high-performance devices, further exerting the excellent characteristics of nitride semiconductors. The main uses of QvdWE technology are 2D/3D, 3D/2D, 3D/2D/3D integration [53]. As shown in Fig. 1.13a, 2D/3D heterostructures are fabricated by transferring or growing 2D materials on 3D materials, particularly using vdWE technology to protect the original interface of 2D heterostructures on 3D, while fully relaxing lattice matching constraints. In traditional epitaxial processes, dangling bonds exist on the substrate surface, leading to chemical bonding of adsorbates. In qvdWE processes, adsorbates bind to the substrate surface through vdW forces, without transferring or sharing electrons. This allows the formation of heterostructures with large lattice mismatches, resulting in highly textured polycrystalline 2D materials on the substrate due to high strain energy. Compared to polycrystalline 3D materials, the presence of grain boundaries in 2D materials causes less damage to device performance, but many factors such as substrate type, precursor, temperature, and pressure must be precisely designed to control nucleation, agglomeration, film thickness, etc., to obtain high-quality heterostructures.

Unlike 2D/3D epitaxial growth, the transfer of 2D materials is not affected by these constraints, making it more flexible for the design of 2D/3D heterostructures. As shown in Fig. 1.13b, a controllable and scalable process is developed by wet transfer of grown 2D materials [54]. The freestanding polymer/2D material stack is subsequently scooped by the target 3D substrate from the solution and eventually the polymer handler is chemically dissolved. Although this process can be scaled up and stacked to form complex heterostructures, the direct contact of the 2D material surface with the wet chemical solution leads to a decrease in the quality of the 2D/3D

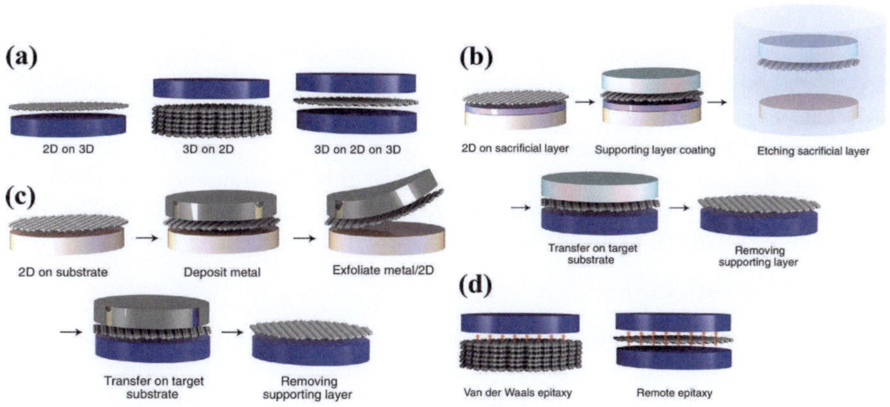

Fig. 1.13 3D/2D structures and fabrication methods. **a** Schematics of coupling structures. **b** Schematics of representative wet transfer of 2D materials using a supporting layer. **c** Schematics of representative quasi-dry transfer of 2D materials using a metal supporting layer. **d** Schematics of vdWE and RE or QvdWE [53]

interface. Further improvements have been made by adopting a metal-induced quasi-dry transfer process, where 2D materials are peeled off from the growth substrates using a metal handler in dry environment, and immediately bonded to the surface of the target 3D substrates. As such, metal-induced transfer can ensure the formation of clean 2D-on-3D interfaces as shown in Fig. 1.13c [55]. These techniques are applicable to a wide range of materials without thermal and chemical limitations, but processing residues such as polymers or metals are inevitably present on the surface of 2D materials.

Despite the maturity of 2D material transfer or vdWE processes in recent years, which have facilitated the preparation and research of 2D/3D heterostructures, the realization of 3D/2D heterostructures is still in an immature stage. The concept of vdWE has also inspired the epitaxial synthesis of strain-relaxed 3D materials on 2D materials in Fig. 1.13d. The growth of 3D nitride on 2D materials is usually called remote epitaxy (RE) or qvdWE according to the bonding state at the interface. However, the growth of planar single-crystal thin films of 3D materials remains challenging due to the poor wettability of the vdW surface of layered materials and the suppression of their nucleation. Although vdWE greatly relaxes the requirement of lattice matching, the weak interaction between substrate and epilayer can cause incommensurate short-range ordering in the epilayer. Current studies show that 3D overlayers can nucleate on 2D layered materials, but with only partial epilayer-substrate epitaxial alignment across the epitaxial films. Thus, it is hard to realize the large-size single-crystalline domain due to textured grain structure. In recent years, hot topics in qvdWE and RE research on 2D materials include interface theoretical calculations, growth nucleation control, dynamic process and new device development. The crystalline quality of nitride materials is gradually improved through studying growth mechanisms and processes, and developing flexible peeling and transfer methods to ultimately produce high-performance devices, including light-emitting devices, power devices, and piezoelectric devices. The qvdWE and RE of GaN and AlN materials on typical layered 2D materials such as graphene, h-BN, TMDs are widely investigated, laying the foundation for the realization of subsequent flexible semiconductor devices.

This section pays special attention to advanced growth technologies for obtaining high-quality 3D crystalline materials on 2D materials, and briefly introduces some applications of qvdWE of nitrides on graphene, h-BN and TMDs.

1. **Graphene buffer layer**

Since 2010, many groups have used various growth methods to prepare high-quality graphene, which has been used as a buffer layer to grow high-quality nitrides. This has also enabled the exfoliation of nitride devices, resulting in the preparation of semiconductor nitride devices with excellent performance [56]. The methods for preparing graphene buffer/release layers mainly include transfer printing (the most common of which include CVD synthesis and mechanical exfoliation to obtain graphene sheets), thermal decomposition, or direct synthesis onto the required substrate without transfer printing through low-temperature diffusion-assisted synthesis (Diffusion Assisted Synthesis, DAS) or atmospheric pressure chemical vapor deposition (Atmospheric Pressure Chemical Vapor Deposition, APCVD) [18]. This type of graphitized substrate can be directly used for III-nitride q qvdWE.

The use of graphene as a buffer layer for nitride growth has made great progress. The first report comes from Seoul National University. Kunook Chung et al. first achieved transferable devices using van der Waals forces of the graphene-layer structure. They fabricated transferable GaN-based LEDs on the oxygen-plasma-treated graphene, and high-density zinc oxide (ZnO) nanowalls were used as an intermediate layer to improve the quality of GaN films [57]. Subsequently, Jeehwan Kim et al. from IBM Watson Center demonstrated direct vdWE of high-quality single-crystalline GaN films on epitaxial graphene with low defectivity and surface roughness comparable with that grown on conventional SiC or sapphire substrates. The GaN film may be released and transferred onto arbitrary substrates [58]. D. H. Mun et al. from Korea Institute of Science and Technology prepared graphene multilayer buffer layer for GaN microstructures on sapphire substrates with different orientations [59], and Q. Zeng et al. from Institute of Semiconductors of Chinese Academy of Sciences grew high quality AlN thin films on graphene/sapphire without transfer fabrication process [60]. These early studies laid the foundation for later vdWE of nitrides and devices.

Thanks to the weak interaction between graphene and its epitaxial layer, the graphene buffer layer can be used for multiple growth and transfer cycles of nitride thin films. The upper layer of the LED can be easily transferred to foreign substrates such as metal or plastic to improve the thermal conductivity of the device or achieve the desired flexibility. For example, metal substrates have excellent heat dissipation capabilities, which are conducive to the construction of high-power LED devices, while plastic, glass and other substrates can realize the preparation of large-area, full-color inorganic LED displays and flexible devices. At present, graphene-based nitride epitaxy has been prepared to produce flexible vertical LEDs and flexible ultraviolet photodetectors, which can display excellent photoelectric performance [61].

2. **h-BN buffer layer**

As a nitride in 2D materials, BN provides a reference direction for the development of nitrides, and it is expected to realize the release and transfer of nitride semiconductor devices to large-area, flexible, and cost-effective substrates. BN appears to be the material of choice for emerging applications, including DUV optoelectronics, electron emitters and neutron detectors [62]. Although, BN exists in various crystalline forms, the hexagonal form is isostructural to graphite and is the stable phase of BN synthesized at any temperature and under normal pressure. Due to its close lattice match to graphite and its chemical inertness and resistance to oxidation, h-BN is considered an ideal template and barrier material for the construction of 2D heterostructures and quantum wells. Similar to graphene, h-BN is expected to possess rich new physics due to its layered structure with the exception that it is also complementary to graphene because of its wide bandgap.

As early as 1998, there was research on h-BN as a buffer layer/release layer for growing high-quality nitride semiconductor devices [63]. Subsequently, h-BN buffer layer for the growth of nitride materials and the mechanical transfer to prepare flexible devices had been developed, such as vertical InGaN/GaN LEDs, AlGaN/GaN HEMTs, flexible h-BN-based metal–semiconductor metal (MSM) photodetectors, AlGaN-based DUV LEDs, etc. [64–67] paving the way for the acquisition of transferable high-efficiency nitride devices with wafer size.

3. **TMDs biffer layer**

As early as 1999, A. Yamada from Tsukuba University used bulk MoS_2 as a substrate, and epitaxially grew GaN thin films on it using the MBE method [68]. Recent studies have shown that although MoS_2 decomposes during the high-temperature growth stage of nitride, it achieves the controlled growth of nitride on the substrate surface by affecting its initial nucleation state. The MoS_2 buffer may also improve the crystal quality [48], which is consistent with the role of other 2D materials previously. Currently, GaN epitaxy on TMDs/SiO_2/Si substrates has been realized on MoS_2, WS_2 single crystals [69], and continuous AlN films have also been prepared on WS_2/sapphire substrate to obtain DUV LEDs based on AlGaN [70].

P. Gupta et al. have conducted relevant research on a series of TMDs such as WS_2, MoS_2, WSe_2, $MoSe_2$, ReS_2 and $ReSe_2$ [48]. Figure 1.14 shows the relationship between the band gap and in-plane lattice parameters of different III-nitrides and TMDs. The lattice mismatch rates of WS_2 and MoS_2 with GaN are 1.0% and 0.8% respectively. Such small mismatches have opened up new research directions for the preparation of high-quality nitride thin films and high-performance devices.

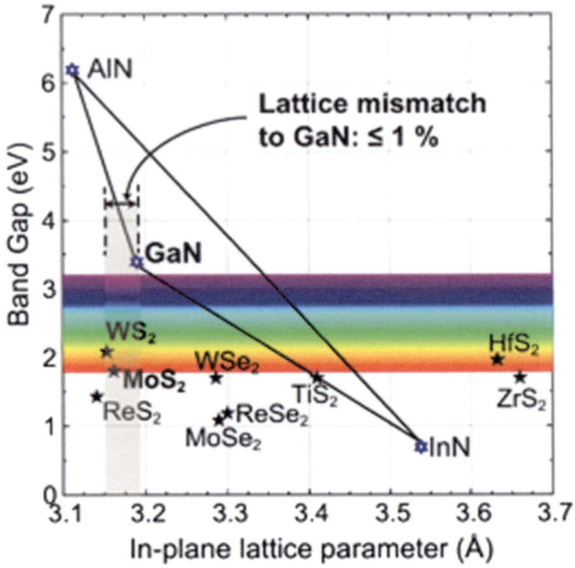

Fig. 1.14 Relationship between the band gap and in-plane lattice parameters of different III-nitrides and TMDs [48]

1.4 Summary

At the moment when the research on III-nitrides is booming, a key core problem that exists in both scientific research and industry is that the ubiquitous heteroepitaxy cannot achieve high-quality single crystal epitaxy. Epitaxial growth, a crystallographically oriented growth induced by the chemical bonding between crystalline substrate and atomic building blocks, has been a key technique in the nitride thin-film and heterostructure applications of semiconductors. However, the nitride epitaxial growth technique is limited by different lattice mismatch and thermal expansion coefficients of dissimilar crystals. With the continuous advancement of research and industry, more and more new substrates are needed. At present, the main method to solve the lattice mismatch problem existing in heteroepitaxy is the low-temperature growth buffer layer technology. With the booming development of 2D materials, using 2D materials as an insertion layer to cover different substrates, and using the QvdWE method to carry out the epitaxy of III-nitrides have become a promising research direction.

This chapter focuses on the development status of 2D materials and the principles and applications of QvdWE and RE, involving the definition, classification, physical and chemical properties of 2D materials, and various growth methods. Subsequently, the definition of QvdWE and RE is introduced, and the principles and applications of nitrides on 2D materials are introduced, which are expected to make the third-generation semiconductor materials an important direction for the industrial application of 2D crystals.

References

1. Zhu HW, Wang M (2017) Two-dimensional materials: Structure, preparation and properties. J Chin Ceram Soc 45:1043–1053
2. Novoselov KS, Geim AK, Morozov SV, Jiang D, Zhang Y, Dubonos SV, Grigorieva IV, Firsov AA (2004) Electric field effect in atomically thin carbon films. Science 306:666–669
3. Tan CL, Cao XH, Wu XJ, He QY, Yang J, Zhang X, Chen JZ, Zhao W, Han SK, Nam GH, Sindoro M, Zhang H (2017) Recent advances in ultrathin two-dimensional nanomaterials. Chem Rev 117:6225–6331
4. Zhang H (2015) Ultrathin two-dimensional nanomaterials. ACS Nano 9:9451–9469
5. Gao LF, Song ZQ, Sun ZH, Han DX (2018) Application and development of novel two-dimensional nanomaterials in electrochemistry. Chin J Appl Chem 35:247–258
6. Yu YJ, Zhang YB (2017) From two-dimensional materials to van der Waals heterostructures. Physics 4:5–13
7. Matthew JA, Vincent CT, Richard BK (2010) Honeycomb carbon: a review of graphene. Chem Rev 110:132–145
8. Balandin AA, Ghosh S, Bao W, Calizo I, Teweldebrhan D, Miao F, Lau CN (2008) Superior thermal conductivity of single-layer graphene. Nano Lett 8:902–907
9. Geim AK, Grigorieva IV (2013) van der Waals heterostructures. Nature 499:419–425
10. Zhang YB, Tan YW, Stormer HL, Kim P (2005) Experimental observation of the quantum hall effect and berry's phase in graphene. Nature 438:201–204
11. Stoller MD, Park S, Zhu Y, An J, Ruoff RS (2008) Graphene-based ultracapacitors. Nano Lett 8:3498–3502
12. Nair RR, Blake P, Grigorenko AN, Novoselov KS, Booth TJ, Stauber T, Peres NMR, Geim AK (2008) Fine structure constant defines visual transparency of graphene. Science 320:1308–1308
13. Lee C, Wei XD, Kysar JW, Hone J (2008) Measurement of the elastic properties and intrinsic strength of monolayer graphene. Science 321:385–388
14. Jiang HX, Lin JY (2017) Basic properties of h-BN epilayers. ECS J Solid State Sci Technol 6:Q3012
15. Hiroki M, Kumakura K, Kobayashi Y, Akasaka T, Makimoto T, Yamamoto H (2014) Suppression of self-Heating effect in AlGaN/GaN high electron mobility transistors by substrate-transfer technology using h-BN. Appl Phys Lett 105:193509
16. Paduano Q, Snure M, Siegel G, Thomson D, Look D (2016) Growth and characteristics of AlGaN/GaN heterostructures on sp^2-bonded BN by metal-organic chemical vapor deposition. J Mater Res 31:2204–2213
17. Yu JD, Wang L, Hao ZB, Luo Y, Sun CZ, Wang J, Han YJ, Xiong B, Li HT (2020) van der Waals epitaxy of III-nitride semiconductors based on 2D materials for flexible applications. Adv Mater 32:1903407
18. Chen Y, Fan ZX, Zhang ZC, Niu WX, Li CL, Yang NL, Chen B, Zhang H (2018) Two-dimensional metal nanomaterials: synthesis, properties, and applications. Chem Rev 118:6409–6455
19. Martella C, Mennucci C, Lamperti A, Cappelluti E, de Mongeot FB, Molle A (2018) Designer shape anisotropy on transition-metal-dichalcogenide nanosheets. Adv Mater 30:1705615
20. Guo BJ, Yu K, Li HL, Qi RJ, Zhang YY, Song HL, Tang Z, Zhu ZQ, Chen MW (2017) Coral-shaped MoS_2 decorated with graphene quantum dots performing as a highly active electrocatalyst for hydrogen evolution reaction. ACS Appl Mater Interfaces 9:3653–3660
21. Deng J, Li H, Wang S, Ding D, Chen M, Liu C, Tian Z, Novoselov KS, Ma C, Deng D, Bao X (2017) Multiscale structural and electronic control of molybdenum disulfide foam for highly efficient hydrogen production. Nat Commun 8:14430
22. LaMountain T, Lenferink EJ, Chen YJ, Stanev TK, Stern NP (2018) Environmental engineering of transition metal dichalcogenide optoelectronics. Front Phys 13:138114
23. Voiry D, Mohite A, Chhowalla M (2015) Phase engineering of transition metal dichalcogenides. Chem Soc Rev 44:2702–2712

24. Geim AK, Novoselov KS (2007) The rise of graphene. Nat Mater 6:183–191
25. Zhao BY (2019) Preparation and theoretical study of two-dimensional GaN-based materials by CVD. Xi'an University of Technology, Xi'an, pp 2–3
26. Tang C, Zhong L, Zhang BS, Wang HF, Zhang Q (2018) 3D mesoporous van der Waals heterostructures for trifunctional energy electrocatalysis. Adv Mater 30:1705110
27. Yu JX, Li J, Zhang WF, Chang HX (2015) Synthesis of high quality two-dimensional materials via chemical vapor deposition. Chem Sci 6:6705–6716
28. Cheng C, Chen W, Chen Y, Chen Y, Chen Y, Ding F, Fan C, Fan HJ, Fan Z, Gong C, Gong Y, He Q, Hong X, Hu S, Hu W, Huang W, Huang Y, Ji W, Li D, Li L-J, Li Q, Lin L, Ling C, Liu M, Liu N, Liu Z, Loh KP, Ma J, Miao F, Peng H, Shao M, Song L, Su S, Sun S, Tan C, Tang Z, Wang D, Wang H, Wang J, Wang X, Wang X, Wee ATS, Wei Z, Wu Y, Wu Z-S, Xiong J, Xiong Q, Xu W, Yin P, Zeng H, Zeng Z, Zhai T, Zhang H, Zhang H, Zhang Q, Zhang T, Zhang X, Zhao L-D, Zhao M, Zhao W, Zhao Y, Zhou K-G, Zhou X, Zhou Y, Zhu H, Zhang H, Liu Z (2021) Recent progress on two-dimensional materials. Acta Phys-Chim Sin 37:2108017
29. Liao JY, Wu JX, Dang CH, Xie LM (2021) Methods of transferring two-dimensional materials. Acta Physica Sinica 70:227–243
30. Dean CR, Young AF, Meric I, Lee C, Wang L, Sorgenfrei S, Watanabe K, Taniguchi T, Kim P, Shepard KL, Hone J (2010) Boron nitride substrates for high-quality graphene electronics. Nat Nanotech 5:722–726
31. Castellanos-Gomez A, Buscema M, Molenaar R, Singh, Janssen L, van der Zant HSJ, Steele GA (2014) Deterministic transfer of two-dimensional materials by all-dry viscoelastic stamping. 2D Mater 1:011002
32. Maruska HP, Tietjen JJ (1969) The preparation and properties of vapor-deposited single-crystalline GaN. Appl Phys Lett 15:327–329
33. Bhat I (2019) 3-physical properties of gallium nitride and related III-V nitrides. In: Baliga BJ (ed) Wide bandgap semiconductor power devices. Woodhead Publishing, pp 43–77
34. Utama MI, Zhang Q, Zhang J, Yuan YW, Belarre FJ, Arbiol J, Xiong QH (2013) Recent developments and future directions in the growth of nanostructures by van der Waals epitaxy. Nanoscale 5:3570–3588
35. Chen ZL, Gao P, Liu, Z (2020) Graphene-based LED: from principle to devices. Acta Phys-Chim Sin 36:1907004
36. Narukawa Y, Ichikawa M, Sanga D, Sano M, Mukai T (2010) White light emitting diodes with super-high luminous efficacy. J Phys D: Appl Phys 43:354002
37. Kim C, Robinson IK, Myoung J, Shim KH, Kim K (1999) Buffer layer strain transfer in AlN/GaN near critical thickness. J Appl Phys 85:4040–4044
38. Akasaki I (2007) Key inventions in the history of nitride-based blue LED and LD. J Cryst Growth 300:2–10
39. Ishida M, Ueda T, Tanaka T, Ueda D (2013) GaN on Si technologies for power switching devices. IEEE Trans Electron Devices 60:3053–3059
40. Chen W, Wong KY, Chen KJ (2009) Single-chip boost converter using monolithically integrated AlGaN/GaN lateral field-effect rectifier and normally-off HEMT. IEEE Electron Device Lett 30:430–432
41. Monemar B, Pozina G (2000) Group III-nitride based hetero and quantum structures. Prog Quantum Electron 24:239–290
42. Ryu H, Park H, Kim JH, Ren F, Kim J, Lee GH, Pearton SJ (2022) Two-dimensional material templates for van der Waals epitaxy, remote epitaxy, and intercalation growth. Appl Phys Rev 9:031305
43. Koma A, Sunouchi K, Miyajima T (1984) Fabrication and characterization of heterostructures with subnanometer thickness. Microelectron Eng 2:129–136
44. Liang DD, Wei TB, Wang JX, Li JM (2020) Quasi van der Waals epitaxy nitride materials and devices on two dimension materials. Nano Energy 69:104463
45. Alaskar Y, Arafin S, Wickramaratne D, Zurbuchen MA, He L, McKay J, Lin Q, Goorsky MS, Lake RK, Wang KL (2014) Towards van der Waals epitaxial growth of GaAs on Si using a graphene buffer layer. Adv Funct Mater 24:6629–6638

46. Kim J, Bayram C, Park H, Cheng CW, Dimitrakopoulos C, Ott JA, Reuter KB, Bedell SW, Sadana DK (2014) Principle of direct van der Waals epitaxy of single-crystalline films on epitaxial graphene. Nat Commun 5:4836
47. Kobayashi Y, Kumakura K, Akasaka T, Makimoto T (2012) Layered boron nitride as a release layer for mechanical transfer of GaN-based devices. Nature 484:223–227
48. Gupta P, Rahman AA, Subramanian S, Gupta S, Thamizhavel A, Orlova T, Rouvimov S, Vishwanath S, Protasenko V, Laskar MR, Xing HG, Jena D, Bhattacharya A (2016) Layered transition metal dichalcogenides: promising near-lattice-matched substrates for GaN growth. Sci Rep 6:23708
49. Bae SH, Lu K, Han Y, Kim S, Qiao K, Choi C, Nie Y, Kim H, Kum HS, Chen P, Kong W, Kang BS, Kim C, Lee J, Baek Y, Shim J, Park J, Joo M, Muller DA, Lee K, Kim J (2020) Graphene-assisted spontaneous relaxation towards dislocation-free heteroepitaxy. Nat Nanotechnol 15:272–276
50. Garcia JM, Wurstbauer U, Levy A, Pfeiffer LN, Pinczuk A, Plaut AS, Wang L, Dean CR, Buizza R, Van Der Zande AM, Hone J, Watanabe K, Taniguchi T (2012) Graphene growth on h-BN by molecular beam epitaxy. Solid State Commun 152:975–978
51. Shi Y, Zhou W, Lu AY, Fang W, Lee YH, Hsu AL, Kim SM, Kim KK, Yang HY, Li LJ, Idrobo JC, Kong J (2012) van der Waals epitaxy of MoS_2 layers using graphene as growth templates. Nano Lett 12:2784–2791
52. Fernandez-Garrido S, Ramsteiner M, Gao G, Galves LA, Sharma B, Corfdir P, Calabrese G, de Souza Schiaber Z, Pfuller C, Trampert A, Lopes JMJ, Brandt O, Geelhaar L (2017) Molecular beam epitaxy of GaN nanowires on epitaxial graphene. Nano Lett 17:5213–5221
53. Bae SH, Kum H, Kong W, Kim Y, Choi C, Lee B, Lin P, Park Y, Kim J (2019) Integration of bulk materials with two-dimensional materials for physical coupling and applications. Nat Mater 18:550–560
54. Kim KS, Zhao Y, Jang H, Lee SY, Kim JM, Kim KS, Ahn JH, Kim P, Choi JY, Hong BY (2009) Large-scale pattern growth of graphene films for stretchable transparent electrodes. Nature 457:706–710
55. Desai SB, Madhvapathy SR, Amani M, Kiriya D, Hettick M, Tosun M, Zhou YZ, Dubey M, Ager JW, Chrzan D, Javey A (2016) Gold-mediated exfoliation of ultralarge optoelectronically-perfect monolayers. Adv Mater 28:4053–4058
56. Chen RF, Song YJ, He R, Wang JX, Li JM, Wei TB (2024) III-nitride semiconductor membrane electronics and optoelectronics for heterogeneous integration. Prog Quantum Electron 98:100536
57. Chung K, Lee CH, Yi GC (2010) Transferable GaN layers grown on ZnO-coated graphene layers for optoelectronic devices. Science 330:655–657
58. Kim J, Bayram C, Park H, Wei CW, Dimitrakopoulos C, Ott JA, Reuter KB, Bedell SW, Sadana DK (2014) Principle of direct van der Waals epitaxy of single-crystalline films on epitaxial graphene. Nat Commun 5:4836
59. Mun D-H, Bae H, Bae S, Lee H, Ha J-S, Lee S (2014) Stress Relaxation of GaN microstructures on a graphene-buffered Al_2O_3 substrate. Phys Status Solidi (RRL) 8:341–344
60. Zeng Q, Chen ZL, Zhao Y, Wei TB, Chen X, Zhang Y, Yuan GD, Li JM (2016) Graphene-assisted growth of high quality AlN by metalorganic chemical vapor deposition. Jpn J Appl Phys 55:085501
61. Song WR, Chen Q, Yang KL, Liang M, Yi XY, Wang JX, Li JM, Liu ZQ (2023) Recent advances in mechanically transferable III-nitride based on 2D buffer strategy. Adv Funct Mater 33:2209880
62. Maity A, Grenadier SJ, Li J, Lin JY, Jiang HX (2021) Hexagonal boron nitride: epitaxial growth and device applications. Prog Quantum Electron 76:100302
63. Boo JH, Rohr C, Ho W (1998) MOCVD of BN and GaN thin films on silicon: new attempt of GaN growth with BN buffer Layer. J Cryst Growth 189:439–444
64. Makimoto T, Kumakura K, Kobayashi Y, Akasaka T, Yamamoto H (2012) A vertical InGaN/GaN light-emitting diode fabricated on a flexible substrate by a mechanical transfer method using BN. Appl Phys Express 5:072102

65. Chung K, Oh H, Jo J, Lee K, Kim M, Yi GC (2017) Transferable single-crystal GaN thin films grown on chemical vapor-deposited hexagonal BN sheets. NPG Asia Mater 9:e410
66. Wu QQ, Guo YN, Sundaram S, Yan JC, Zhang L, Wei TB, Wei XC, Wang JX, Ougazzaden A, Li JM (2019) Exfoliation of AlN film using two-dimensional multilayer hexagonal BN for deep-ultraviolet light-emitting diodes. Appl Phys Express 12:015505
67. Glavin NR, Chabak KD, Heller ER, Moore EA, Prusnick TA, Maruyama B, Walker DE Jr, Dorsey DL, Paduano Q, Snure M (2017) Flexible gallium nitride for high-performance, strainable radio-frequency devices. Adv Mater 29:1701838
68. Yamada A, Ho KP, Maruyama T, Akimoto K (1999) Molecular beam epitaxy of GaN on a substrate of MoS_2 layered compound. Appl Phys A 69:89–92
69. Zhang Y, Chang TR, Zhou B, Cui YT, Yan H, Liu ZK, Schmitt F, Lee J, Moore R, Chen YL, Lin H, Jeng HT, Mo SK, Hussain Z, Bansil A, Shen ZX (2014) Direct observation of the transition from indirect to direct bandgap in atomically thin epitaxial $MoSe_2$. Nat Nanotechnol 9:111–115
70. Yin Y, Ren F, Wang Y, Liu Z, Ao J, Liang M, Wei T, Yuan G, Ou H, Yan J, Yi X, Wang J, Li J (2018) Direct van der Waals epitaxy of crack-free AlN thin film on epitaxial WS_2. Materials 11:2464

Chapter 2
Theoretical Calculations of Quasi van der Waals Epitaxial Interfaces in Two-Dimensional Material-Based Nitrides

Graphene consists of a single layer of carbon atoms bonded by sp^2-hybridization and arranged in a regular hexagonal honeycomb lattice structure [1, 2]. The absence of dangling bonds makes the graphene surface chemically inert to any foreign atoms, which makes it challenging to grow three-dimensional semiconductor materials on its surface by conventional epitaxial methods. One possible solution to this problem is to use Quasi van der Waals epitaxy [3]. Compared to conventional epitaxy with strong chemical bonding, the bonding mechanism in vdWE is different so that the grown material does not necessarily have to match the graphite substrate lattice and strain-induced interfacial defects do not occur. Due to the increasing research work on the use of graphene as a substrate material for epitaxial growth, it is necessary to study the theoretical calculation of the interface of nitride Quasi van der Waals epitaxy.

2.1 Adsorption and Migration of Nitride Metal Atoms on Two-Dimensional Materials

Density Functional Theory calculations (DFTs) have been used to study the adsorption sites and adsorption energies of different atoms and molecules on the graphene surface [4]. In this subsection, the adsorption orientation (i.e., the positions of adsorbed atoms with respect to each other and with respect to graphene) as well as the adsorption kinetics of different kinds of atoms on graphene are discussed. Finally, the adsorption of nitrides on graphene is provided more accurately based on the physical/chemical properties of the nitride metal elements.

T. Wei et al., *Quasi van der Waals Epitaxy of Nitride Semiconductor and Its Applications*, Wide Bandgap Semiconductors,
https://doi.org/10.1007/978-981-95-3469-2_2

2.1.1 Adsorption and Migration of Atoms on Graphene

Trends in electronic and structural properties include: stability (binding energy), most stable position (bridge, null, top) (Fig. 2.1), adsorbed atom heights, migration energy (barrier energy for adsorbed atoms to move freely on graphene), Fermi energy shifts (from pristine graphene), graphene deformation/aberration, magnetization, and electron-energy bandgaps at the Fermi energy level (E_g) and charge transfers (from atoms to graphene). Classify materials as metals, semiconductors and insulators. The E_g threshold for semiconductors and insulators is set by convention and is usually blue ($\approx$ 3.1 eV, 1 eV = 96.4869 kJ/mol).

Kazuki Nakada and Hiroshi Ishii [6] used a first-principle band computation technique based on DFT. They used VASP, [7–9] which was a high-precision first-principle computational code using the PAW method. They used LDA as an exchange–correlation term with a cutoff energy of 500 eV, and all calculations were nonmagnetic. The cells of the graphene layers have a 3 × 3 structure. The lattice constants of graphene use values optimized by calculation. The distance between graphene layers is about 14.7 Å, and the distance between adsorbed atoms is about 7.3 Å. The final potential is constructed self-consistently from the eigenstates in the non-commensurable Brillouin zone at 24 sampled k points. To obtain the final band structure, they chose 68 sampled k-points in the IBZ. For the calculation of atoms at certain positions, the coordinates of their positions parallel to the surface were fixed and their coordinates perpendicular to the surface were completely relaxed. One atom at the edge of the 3 × 3 structure of the graphene sheet was fixed while relaxing to other carbon atoms on the sheet. The calculations were performed on the three adsorption points H6, B and T shown in Fig. 2.2. They calculated the adsorption

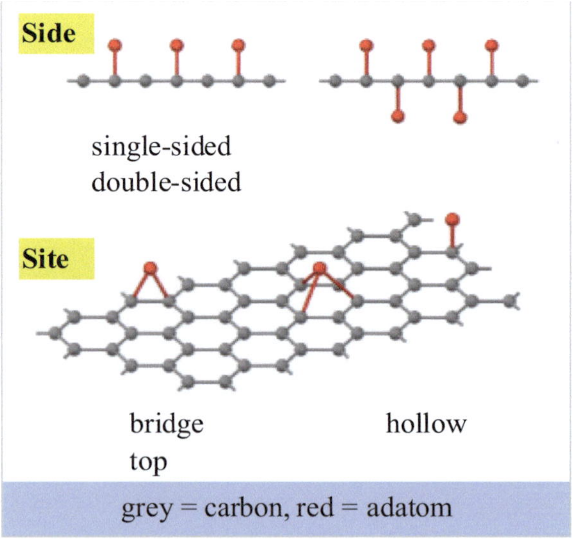

Fig. 2.1 Lateral schematic of elemental adsorption on graphene (single side, double side) and adsorption sites (bridge, null, top) [5]

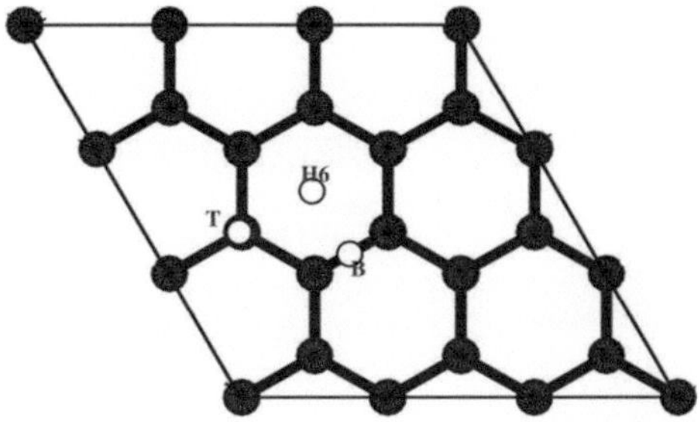

Fig. 2.2 Three adsorption sites B, H6 and T of a 3 × 3 graphene layer [6]

energy according to Eq:

$$E_{bond} = (E_{graphene} + E_{(adatom)} - E_{total}) \quad (2.1)$$

where E_{bond} was the binding energy of an adsorbed atom adsorbed onto a graphene sheet, $E_{graphene}$ was the total energy of a layer of graphene, and E_{adatom} was the total energy of an atom isolated as an atom. They considered almost all elements in the periodic table except lanthanides and rare gases as atoms and performed calculations from H to B.

Figure 2.3 showed the results of the calculations they did for the non-magnetic state. It showed the most stable adsorption sites and bond energies of various atoms when adsorbed at the three adsorption sites. In the figure, the most stable site for each adsorbed atom was indicated by color. The green, red and yellow boxes indicated that the most stable adsorption sites were the B, H6 or T sites, respectively. The values in Fig. 2.3 indicated the magnitude of the adsorption energy when each adsorbed atom was adsorbed to the most stable adsorption site. The results showed that for the nitride metal elements (Ga, Al, In, etc.), the most stable site was the H6 site, and the adsorption energy decreased from Al to Ga to In in the order. For H, F, Cl, Br and I, where the number of valence electrons of adsorbed atoms was 1, the T site was the most stable adsorption site. In addition, for the transition metal elements, the adsorption energy of each adsorbed atom was very large, and the binding energy tended to increase with the increase of the number of d electrons. For the nonmetallic elements C, N, O, the adsorption energies were very large and the difference in adsorption energies between the sites was also large (higher than 3.0 eV), which suggested that it was easy to firmly adhere to the B site. Figure 2.4 showed a table of bonding distances between graphene and adsorbed atoms. ("Bonding distance" refered to the average value of the distance between the position of the graphene layer and the adsorbed atoms.) If the bonding distance between the adsorbed atoms and graphene was large, the bonding energy tended to decrease. In the case of large bonding distances, the adsorbed atoms showed physisorption-like bonding. When the bonding distance was short, the bonding energy tended to increase. In this case, the bonding was characterized like chemisorption.

The adsorption energy is the energy with which an adsorbed atom adsorbs, while the migration energy is the susceptibility of an adsorbed atom to migrate on a surface.

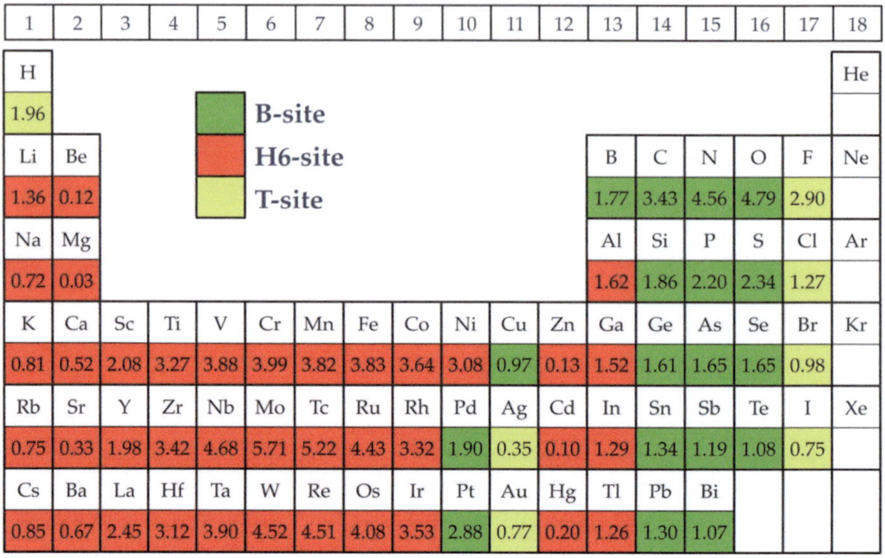

Fig. 2.3 Most stable sites and bond energies upon adsorption of adsorbed atoms: green for B site, red for H6 site, yellow for T site [6]

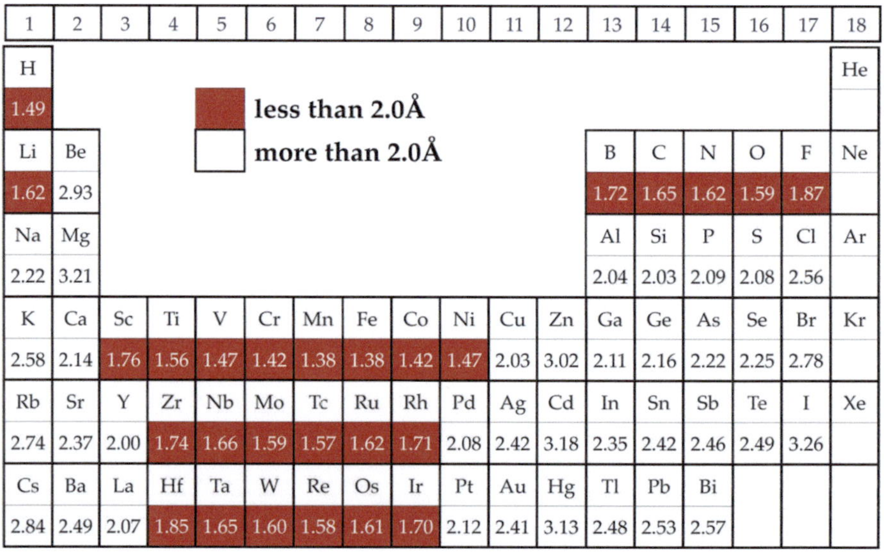

Fig. 2.4 Distances of adsorbed atoms from graphene at the most stable adsorption site: red fill shows distances less than 2 Å, white fill shows distances greater than 2 Å [6]

For the growth of semiconductor materials and devices on graphene or the design of nanodevices using graphene, the migration energy is important for the discussion of adsorption of adherents on the surface. When discussing the surface growth, it is necessary to discuss the migration energy. Nakada and Ishii further estimated the minimum limit of the migration energy of adsorbed atoms based on the difference of adsorption energy. Figure 2.5 showed the migration energies of the most stable sites. They found that the type of adsorption can be categorized into two types: fixed adsorption sites and unfixed adsorption sites. The migration barrier energy of the type with immobile adsorption sites was less than 0.5 eV. 0.5 eV corresponded roughly to the threshold energy of atomic migration at room temperature. In order to estimate the threshold energy, it was possible to deal with large system sizes of more than one million atoms per unit surface, which was comparable to the actual device size at the nanoscale. In thermodynamic equilibrium, the migration rate R was $R = R_0\exp(-E/k_BT)$, where R_0 was the preconditioner, k_B was the Boltzmann constant, and T was the substrate temperature. The migration energy required for an atom to start moving through a large system size of more than one million atoms per unit surface was about 0.5 eV. For example, Yoshihiro Ito and Kenji Shiraishi reported kMC simulations of MBE considering the electron counting model [10, 11]. They used parameters such as jump energy barriers obtained by first-principles calculations. The transition metals Ti, V, Cr, Mn, Fe, Co, Nb, Mo, Tc, Ru, Ta, W, Re and Os had atomic species with stabilization sites of H6 and very high migration potentials. On the other hand, nitride metals such as Ga, Al, and In had very low migration energies. At the same time, the migration energy decreased gradually from Al to Ga to In. For N and O elements, the migration energy was very large, and the stabilizing site was the B site.

These calculations above were performed in the nonmagnetic case without corrections (e.g., vdW and dipole corrections), but were still valuable for providing a picture of elemental adsorption on graphene.

Chan Kam-tai et al. [12] calculated some metals (Li, Na, K, Ca, Al, Ga, In, Sn, Ti, Fe, Pd, Au) on adsorbed graphene supercells using DFT. Their calculations were performed in first-principles DFT under the generalized gradient approximation (GGA) of Perdew, Burke, and Ernzerhof (PBE) [13], including spin polarization, vdW forces, and dipole corrections. VASP was used to perform all calculations [7–9]. Ion cores were modeled using a projector enhanced wave (PAW) potential [14, 15]. The valence electron wave function used a plane wave basis set with a plane wave energy of up to 500 eV. All parameters in the calculations were chosen so that the total energy converged to 0.01 eV. These more accurate calculations supported the results of Nakada and Ishii in terms of the most stable sites, binding energies, atomic heights, and mobility energies, except for the charge transfer values.

They used a calculated graphene lattice constant of 2.47 Å and let the x and y directions be parallel and the z direction perpendicular to the graphene plane. The adsorbed atom-graphene system was modeled using a metal adsorbed atom in a 4×4 hexagonal graphene supercell in Fig. 2.6. The metal adsorbed atom arrangement was referred to as a 4×4 layer. The setup corresponded to a coverage of 1 atom per 32 C atoms. The in-plane lattice constant was 9.88 Å, which was also the distance between neighboring atoms. They used a super-unit length of 15 Å in the z direction.

1	2	3	4	5	6	7	8	9	10	11	12	13	14	15	16	17	18
H																	He
0.60																	
Li	Be											B	C	N	O	F	Ne
0.30	0.02											0.12	0.25	1.00	1.02	0.45	
Na	Mg											Al	Si	P	S	Cl	Ar
0.13	0.02											0.05	0.05	0.45	0.46	0.02	
K	Ca	Sc	Ti	V	Cr	Mn	Fe	Co	Ni	Cu	Zn	Ga	Ge	As	Se	Br	Kr
0.12	0.07	0.34	0.61	1.05	1.45	1.26	0.97	0.77	0.40	0.03	0.02	0.03	0.07	0.20	0.23	0.00	
Rb	Sr	Y	Zr	Nb	Mo	Tc	Ru	Rh	Pd	Ag	Cd	In	Sn	Sb	Te	I	Xe
0.09	0.04	0.12	0.39	0.83	1.47	1.40	0.96	0.39	0.06	0.01	0.01	0.02	0.03	0.03	0.09	0.00	
Cs	Ba	La	Hf	Ta	W	Re	Os	Ir	Pt	Au	Hg	Tl	Pb	Bi			
0.10	0.05	0.18	0.23	0.60	1.17	1.23	0.75	0.15	0.19	0.03	0.01	0.00	0.01	0.00			

more than 0.50 eV

under 0.50 eV more than 0.30 eV

less than 0.30 eV

Fig. 2.5 Migration energy of the most stable sites: dark green > 0.5 eV, light green < 0.5 eV but > 0.3 eV [6]

Calculations using 4 × 4 coveraged approximate the interaction of isolated atoms with graphene. Although the adsorbed atom-adsorbed atom interactions were not negligible, the distances between adsorbed atoms were sufficiently large that the overlap of the electronic states of neighboring adsorbed atoms was small. For K, in the absence of graphene, the difference between the total energy of the 4 × 4 metal layer of K and the total energy of the isolated K atoms (described below) was less than 0.01 eV. Of all the atoms considered, K had the largest atomic radius (2.20 Å) [16]. The adsorbed atom-graphene system lacked inversion symmetry, so its net electric dipole moment was perpendicular to the surface. In order to eliminate spurious dipole interactions between periodic images along the z-direction, they made corrections to the local electrostatic potential and the total energy [17, 18]. These corrections were necessary to obtain the correct vacuum energy levels, which they used to determine the work function. They found corrections to the total energy up to 0.2 eV.

Calculations were carried out on isolated 4 × 4 graphene, isolated 4 × 4 adsorbed atom layers and 4 × 4 adsorbed atom-graphene systems using hexagonal supercells of the same size. The Brillouin zone was sampled with a 9 × 9 × 1 1 Γ-centered k-point grid, and the electronic energy levels were occupied by Gaussian drag tails with $\sigma = 0.05$ eV. They considered the binding of adsorbed atoms at three highly symmetric positions: the hollow (H) position at the center of the hexagon, the bridge (B) position at the midpoint of the carbon–carbon bond, and the top (T) position directly above the carbon atom in Fig. 2.7. For each adsorption point of the adsorbed atom-graphene system, the adsorbed atoms relaxed along the z-direction and the C ions in the graphene relaxed in all directions until the force on the ions was less than

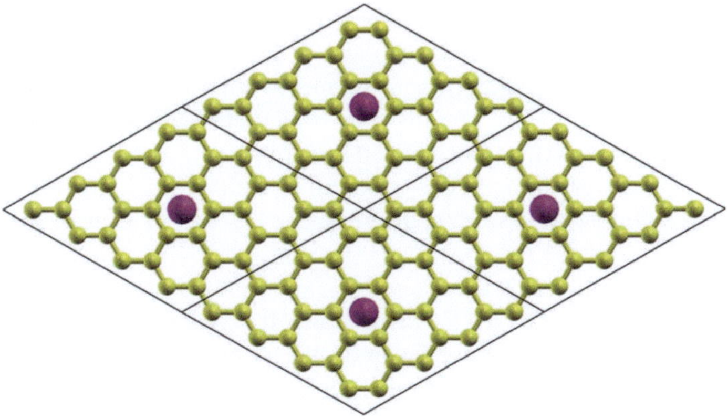

Fig. 2.6 Adsorbed atoms at hollow positions on the surface of a one-cycle 4 × 4 arrangement of graphene [12]

0.05 eV/Å. The dimensions of the super-unit were kept constant in all calculations. In order to calculate the adsorption energy, the total energy of an isolated atom was also required, which can be calculated from a single atom in a cubic supercell of length 14 Å. In the case of a cubic supercell of length 14 Å, the energy of an isolated atom was calculated from the total energy of an isolated atom in a cubic supercell. In this case, only the Γ point in the Brillouin zone was sampled.

In their work, the adsorption energy was defined as

$$\Delta E = E_{ag} - E_a - E_g \tag{2.2}$$

where E_{ag} was the total energy of each adsorbed atom in a 4 × 4 adsorbed atom layer on graphene, E_a was the total energy of an isolated atom, and E_g was the total energy of each 4 × 4 supercell (containing 32 C atoms) of isolated graphene.

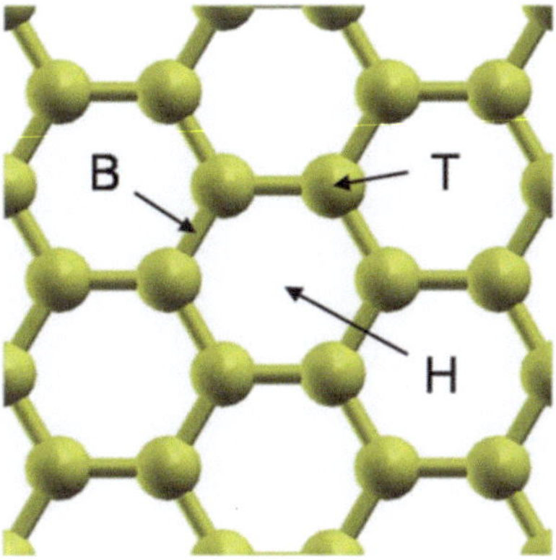

Fig. 2.7 Three adsorption sites considered for the calculation: hollow (H), bridge (B) and bridge (T) [12]

The adsorption geometry was obtained from the positions of the atoms after relaxation. The adsorbed atom height (h) was defined as the difference between the z-coordinate of the adsorbed atom and the mean value of the z-coordinate of the C atom of the graphene layer. The distance ($d_{AC)}$ between the adsorbed atom and its nearest carbon atom (s) was also calculated. In some cases, the distortion of the graphene layer was very severe. They quantified the distortions by calculating the maximum deviation of the C atom in the graphene layer from its average position in the z direction.

Of the three adsorption sites considered, the site with the largest adsorption energy (smallest total energy) was called the favored site. It was assumed that the most probable diffusion pathway between the favored sites was through the highly symmetric site with the second largest adsorption energy. The energy difference between these two sites was called the diffusion potential (E_{diff}). A more accurate determination of the diffusion barrier would require either calculating the entire potential energy surface or using the microshifted elastic band method. For all atoms bound to the H or T site (B site), the next largest adsorption energy was the B site (T site). The adsorption energies and structural properties of the three sites considered were summarized in Fig. 2.7. Metal adsorption atoms of I-III groups were all most strongly bound to the H site. In III-group, the adsorption energy decreased monotonically with increasing atomic number. However, the alkali metals did not follow this trend. Based on their calculated adsorption sites, the adsorption energies and heights agreed reasonably well quantitatively with previous calculations for bases [19–22] and for In [23]. Between previous experimental and theoretical work, the nature of the bonding of Al to graphite was not yet in complete agreement [24–28], but in their calculations, Al fitted the consistent picture found across the different elements. For alkali metals, the ratio of adsorption energy to bulk binding energy was higher, which suggested that alkali metals can form 2D layers on the graphene surface rather than 3D layers. On the other hand, the $\Delta E/E_c$ ratios of III-group elements were lower than those of alkali metals, so Al and In formed islands or 3D clusters on graphite [24, 28, 29].

There is a general consensus, both experimental and theoretical, that for dispersed alkali metal atoms, the bonding to graphite is mainly ionic [30]. For I-III group elements, the adsorption energies and heights of adsorption at the B sites are similar to those at the T sites, while the H sites have larger adsorption energies and smaller adsorption heights. For ionically bonded adsorbed atoms, the equilibrium height is determined by the balance between electrostatic attraction and short-range electron repulsion between the relatively charged adsorbed atoms and the surface. Since the electron density of graphene is lower at the H-site than at the B-site or T-site, the adsorbed atoms are stabilized closer to the surface. Thus, ionic bonding seems to favor adsorption at the H-site because, in addition to having higher coordination as described above, the adsorbed atoms are also closer to the sheet, which reduces the electrostatic energy [31]. The diffusion barrier of the adsorbed atoms of I-III group through the B-site is about 0.1 eV, but the size of Li is significantly smaller than that of the other atoms. The trend of the adsorption height or diffusion barrier is closely related to the atomic radius in I or III groups. For ionic bonding, the height increases as the radius of the adsorbed atoms increases, and the corrugation of graphene decreases. The elements in I-III groups do not distort the graphene sheets much even at the B or T sites where the distortion is the largest. As a result, the C–C bonds close to the atoms retain their sp^2 character and are not significantly rehybridized with any atomic orbitals. This result provides further support for the ionic bonding diagram (Tabe 2.1).

The 3d transition metal Ti has the largest adsorption energy of the adsorbed atoms considered, while Fe has a more moderate adsorption energy. Like the elements of I-III group, Ti and Fe are biased towards the H-site. However, the diffusion barrier is much larger than that of I-III group, about 0.5 eV. In addition, the H-site adsorption height of Fe is small compared to the other adsorbed atoms studied, and Ti and Fe are covalently bonded to graphene. The covalent bonds are oriented and the bond formation depends on the coordination of the adsorbed atoms. Therefore, it is reasonable that the adsorption energies strongly depend on the adsorption sites. Their calculations of the binding energies and geometrical configurations are in reasonable agreement with previous calculations for Ti and Fe on graphene [32–34].

2.1.2 Adsorption and Migration of Nitride Semiconductor Metal Atoms on Graphene

Mazid Munshi and Helge Weman [4] discussed different atomic arrangements of semiconductor atoms on graphene. They concluded that the possible adsorption sites for semiconductor atoms on graphene can be identified as (1) above the center of the hexagonal carbon ring (H-site), (2) above the bridge between carbon atoms (B-site) and (3) on top of the carbon atoms (T-site) as shown in Fig. 2.8. As early as the pioneering work of Kei Hirum et al., it was known that semiconductor NWs grew mostly along the [111] direction in the case of cubic semiconductors and along the [0001] direction in the case of hexagonal semiconductors [35, 36]. Thus, the atoms in the (111) plane of the cubic semiconductor (the (0001) plane of the hexagonal semiconductor) have hexagonal symmetry with respect to the carbon atoms in graphene. Recent work by Ami Munshi et al. had shown that depending on the preferred adsorption sites and the type of semiconductor atoms, different lattice mismatches of semiconductors and graphene will lead to different atomic arrangements [37]. Figure 2.8a–d showed the locations of such arrangements when the atoms were adsorbed on both H and B sites (Fig. 2.8a, b, d) as well as on H or B (Fig. 2.8c). As can be seen from the calculations in the previous section, the adsorption energies of most semiconductor atoms were lower at the T-site and were therefore not discussed here [6]. The lattice constants calculated from the four different atomic arrangements in Fig. 2.8a–d were plotted together with the band gap versus lattice constant diagrams for some conventional semiconductors. Figure 2.8 represented generalized plots of lattice mismatch conditions for the realization of non-vdWE growth in the (111) ((0001)) direction for cubic (hexagonal) semiconductors on graphene. In Fig. 2.8e, the band gap energies of III-V semiconductors (as well as Si and ZnO) corresponded

Table 2.1 Energetic and structural properties of sites H, B, T for 12 atoms [12]

Atom	Site	ΔE (eV)	Ec (eV)	$\Delta E/Ec$	$E^{max}_{a} - \Delta E$ (eV)	h (Å)	dAC (Å)	dGC (10^{-1} Å)
Li	*H*	1.096	1.630	0.672		1.71	2.23	0.0
	B	0.773			0.322	1.88	2.09	0.2
	T	0.754			0.342	1.89	2.02	0.3
	H	0.462	1.113	0.415		2.28	2.70	0.1
Na	*B*	0.393			0.069	2.44	2.59	0.1
	T	0.389			0.074	2.49	2.54	0.1
	H	0.802	0.934	0.859		2.60	2.99	0.1
K	*B*	0.739			0.063	2.67	2.85	0.3
	T	0.733			0.069	2.67	2.77	0.2
	H	0.632	1.840	0.343		2.29	2.72	0.1
Ca	*B*	0.484			0.148	2.33	2.53	0.2
	T	0.478			0.154	2.34	2.46	0.2
	H	1.042	3.390	0.307		2.13	2.56	0.1
Al	*B*	0.927			0.115	2.22	2.33	0.1
	T	0.911			0.131	2.22	2.24	0.2
	H	0.858	2.810	0.305		2.20	2.63	0.0
Ga	*B*	0.762			0.096	2.30	2.41	0.0
	T	0.749			0.109	2.31	2.33	0.1
	H	0.690	2.520	0.274		2.45	2.83	0.1
In	*B*	0.622			0.069	2.53	2.63	0.0
	T	0.614			0.077	2.55	2.56	0.1
	H	0.114			0.142	3.19	3.48	0.1
Sn	*B*	0.253			0.003	2.79	2.81	0.8
	T	0.256	3.140	0.082		2.82	2.75	0.7
	H	1.869	4.850	0.385		1.80	2.33	0.2
Ti	*B*	1.301			0.568	2.05	2.27	0.3
	T	1.301			0.568	2.00	2.18	0.3
	H	0.748	4.280	0.175		1.53	2.11	0.1
Fe	*B*	0.231			0.517	2.22	2.35	0.0
	T	0.149			0.599	2.18	2.13	0.5
	H	0.852			0.230	2.03	2.46	0.3
Pd	*B*	1.081	3.890	0.278		2.21	2.18	1.4
	T	1.044			0.038	2.21	2.10	1.1
	H	0.085			0.011	3.53	3.80	0.0
Au	*B*	0.089			0.007	3.06	3.11	0.4
	T	0.096	3.810	0.025		2.69	2.55	1.4

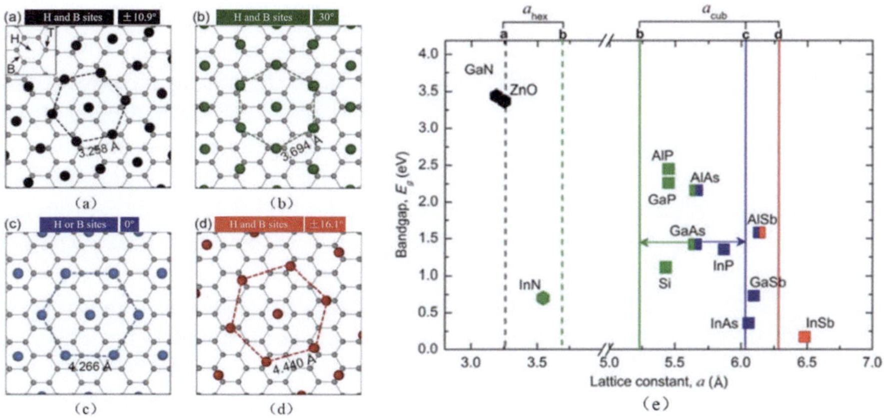

Fig. 2.8 Modeling of adsorption sites and lattice constants of common semiconductors on graphene. Relative orientation and arrangement of semiconductor atoms adsorbed on (**a**, **b**, **d**) H and B sites and (**c**) H or B sites. **e** Generalized modeling of the semiconductor band gap with lattice constants for lattice-matched atomic arrangements on graphene shown in **a** (black vertical line), **b** (green vertical line), **c** (blue vertical line), and **d** (red vertical line) [4]

to their lattice constants. The vertical solid (dashed) color lines depicted the lattice constants of an ideal crystal, which had an exact lattice match with graphene for cubic (hexagonal) crystals with four different atomic arrangements (Fig. 2.8a–d). This figure showed the great possibilities of epitaxial growth of semiconductors on graphite substrates.

Hongliang Chang et al. [38] demonstrated first-principles calculations about the role of graphene in lowering the surface migration barriers and facilitating the lateral migration of metal-aluminum adsorbed atoms and realized graphene-assisted fast QvdWE of high-quality AlN thin films on NPSS substrates. They concluded that the initial growth of AlN was closely related to the adsorption and diffusion of Al adsorbed atoms on the substrate. It was well known that the mobility (diffusivity) of the adsorbed atoms was related to the diffusion barrier through an Arrhenius-type exponential law.

$$r = v \exp\left(-E_{diff}/kt\right) \tag{2.3}$$

where E_{diff}, k and T were the diffusion barrier, Boltzmann's constant and temperature respectively [39].

In order to elucidate the growth and migration processes of AlN during epitaxial growth on MOCVD, they performed first-principles calculations using the Vienna ab-initio simulation program package (VASP) [9] to study the adsorption energies and diffusion barriers of Al atoms. The exchange–correlation function was adopted from the diffusive wave potential [14] and the generalized gradient approximation of Perdew, Burke and Ernzerhof [13]. The energy cutoff for the plane wave expansion was 400 eV, using the Becke88 optimization function including the vdW interaction

[40]. The diffusion barriers were calculated using a microprojectile elastic band method based on the theory of leapfrog sttes [41]. The Al terminated $Al_2O_3(0001)$ 2 × 2 surface was simulated by an eight-layer flat plate with the bottom layer passivated by pseudo-hydrogen atoms. During relaxation, the passivated pseudo-hydrogen atoms and the three underlying layers of Al_2O_3 remained fixed. The adsorption and diffusion of Al atoms on graphene were simulated on 4 × 4 graphene. The Brillouin zone was sampled as 4 × 4 × 1 and 9 × 9 × 1 for the two systems [42]. The atomic coordinates were fully relaxed until the Hermann-Feynman force was less than 0.02 eV/Å per atom.

In Al_2O_3, two Al atoms are not coplanar and they are called low and high Al atoms. According to previous studies, the most stable surface (0001) has only one low-Al atom, which is tightly bound to the O atom under the surface [43]. The two adsorption sites of the Al adsorbed atoms were considered to be the high Al and empty sites with adsorption energies of 2.69 and 1.67 eV, respectively, as shown in Fig. 2.9. Their calculations indicated that the diffusion of the surface Al atoms from the low-Al sites to the other sites was very unfavorable. Therefore, the possible diffusion path of Al atoms should be between the high Al and vacancy sites with a diffusion barrier of 1.02 eV in Fig. 2.9a. The large diffusion barrier suggested that Al atoms had difficulty diffusing on the $Al_2O_3(0001)$ surface even at 1200 °C, and thus they preferred 3D longitudinal growth and formed relatively dispersed island-like structures rather than macronuclei, which was in agreement with the experimental observation. However, on the graphene surface, the calculated diffusion barriers between different sites (H, B and T sites) were less than 0.1 eV, much smaller than the diffusion barriers on Al_2O_3 [44], as shown in Fig. 2.9b. At a growth temperature of 1200 °C, this diffusion barrier was even smaller than kT, which meaned that Al atoms can diffuse almost freely on graphene. In addition, the reactivity of Gr was greatly improved by introducing defects after plasma treatment, which was beneficial for AlN nucleation [45]. The strong binding of Al atoms to the defect sites and the free diffusion in the non-defective regions ensured the efficient nucleation and fast growth of the AlN layer.

2.2 Atomic Configurations of Nitride Interfaces on Two-Dimensional Materials

From the theoretical calculations in the previous subsection, it can be seen that the lack of chemical activity on the surface of the pristine 2D materials and the low adsorption energies of atoms such as Al, Ga, and In on the surface of the 2D materials make it difficult to directly grow nitride films directly on the 2D materials. This means that the surface chemical activity of 2D materials should be improved to realize the growth of nitrides [46]. Usually, the surfaces of 2D materials have some steps and boundaries, while the size of crystal domains varies from tens of nanometers to several micrometers. Nucleation usually occurs at places where the

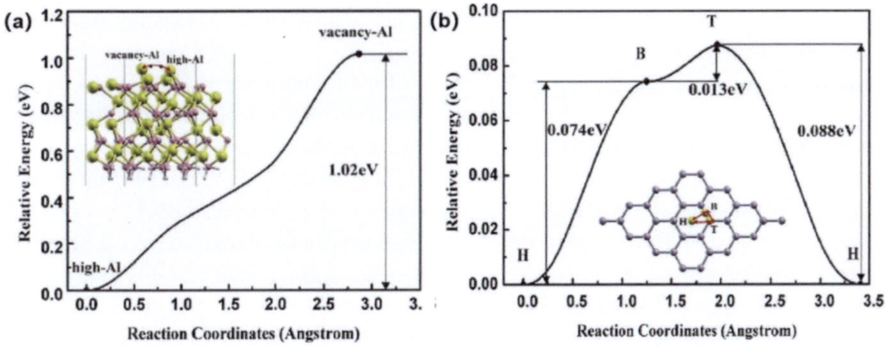

Fig. 2.9 Diffusion barriers of Al atoms on Al_2O_3 and graphene surfaces. **a** Diffusion barrier for Al atoms diffusing from high Al to vacancies on the $Al_2O_3(0001)$ surface. The inset showed the adsorption geometry at both sites. The two small black dots indicated the intermediate positions of Al atoms along the diffusion path. **b** Diffusion barriers between three adsorption sites of Al atoms on the Gr surface: the H site (at the top of the center of the hexagonal ring), the B site (at the top of the center of the C–C bond), and the T site (on a C atom). The yellow, pink and gray balls represent Al, O and C atoms, respectively [38]

surface state changes, such as steps and ridges. However, for epitaxial growth only on steps and boundaries, it is difficult to obtain high-quality nitride films. For this reason, many scholars have opted for plasma treatment to introduce defects and form dangling bonds to improve the ability to adsorb Al, Ga, In and N atoms.

2.2.1 Atomic Configurations of Nitride Interfaces on Two-Dimensional Materials

Zhaolong Chen et al. [47] investigated the nucleation process of AlN on graphene. The ML graphene structure was confirmed by high-resolution transmission electron microscopy (HRTEM) in Fig. 2.10a. After N_2 plasma treatment, the defect density (n_D) in graphene was changed from $n_D = 2.13 \times 10^{11} cm^{-2}$ to $n_D = 3.23 \times 10^{11} cm^{-2}$. From the x-ray photoelectron spectroscopy (XPS) spectra of Fig. 2.10d, the new N-sp^3C peak ($\approx$ 286.5 eV) verified the doping of nitrogen [48]. The doped nitrogen was identified as pyrrole N, with a content of 5.6% in Fig. 2.10e. DFT calculations were performed to verify the effect of introduced pyrrole N on AlN nucleation. The adsorption energies of Al and N atoms in the hollow, bridge and top sites of pristine graphene showed that Al atoms bound more strongly to graphene than N atoms, with adsorption energies of ca. 1.1 eV. The adsorption energies of Al atoms on pyrrole N in plasma-treated graphene were greatly increased to 5.9–8.6 eV, which suggested that graphene's reactivity towards AlN nucleation was greatly enhanced as shown in Fig. 2.10b. The adsorption of Al atoms can partially recover the effect of the introduced pyrrole N on the AlN nucleation. Atom adsorption partially restored the honeycomb structure in the vicinity of the N defects, but induced significant structural deformation and sp^3-like properties of the neighboring C atoms. The interfacial bonding characteristics of AlN/Gr were further analyzed by DFT total energy calculations. As shown in Fig. 2.10c, the apparent electron accumulation around the pyrrole N atoms indicated the formation of Al-N bonds. Notably, the formation energy (defined as the total energy difference between AlN and plasma-treated graphene before and after bonding) was reduced by more than 0.28 eVÅ^{-2} compared to pristine graphene. Pyrrole N in plasma-treated graphene tended to have a graphitic N conformation during AlN growth at high temperatures (1200 °C), [49] and it also showed a decrease in formation energy of about 0.004 eVÅ^{-2}. They also found that the intensity of $2p_z$ orbitals (π-orbitals) occupied by C atoms in the vicinity of N atoms was significantly reduced as shown by the projected density of states (PDOS) of $2p_z$ orbitals on different C atoms in Fig. 2.10f. This suggested a hybrid transition from sp^2 to sp^3 for C atoms in the vicinity of the N defect and electronic coupling between the C π and N π orbitals. Thus, the N_2 plasma treatment greatly facilitated AlN nucleation through the formation of Al-N bonds.

Yue Qi et al. [50] also investigated the role of graphene in strain relaxation during the growth of AlN on sapphire and studied the interfacial properties of AlN/graphene/sapphire. They extracted the bonding energy (BE) of AlN/sapphire and the internal stresses in the AlN films by DFT calculations. Figure 2.11a, d showed the computational models of AlN/sapphire and AlN/graphene/sapphire, respectively. From the 3D charge density top views of Fig. 2.11b, e, it can be observed that there was a significant charge redistribution after graphene insertion. At the AlN/sapphire interface, there was an obvious charge exchange between N in AlN and Al in sapphire (Fig. 2.11b, c). After graphene insertion, charge exchange occured between all interfaces of graphene-AlN and graphene-sapphire (Fig. 2.11e, f), and strong interactions (− 2.51 eV for graphene-sapphire, and − 6.20 eV for graphene-AlN) (Fig. 2.11g) leaded to deformation of the graphene layer, which can explain the strong compressive strain introduced in graphene. For AlN/sapphire, the calculated binding energies were − 10.84 eV for AlN/sapphire and − 2.04 eV for AlN/graphene/sapphire in Fig. 2.11g. Weakened AlN/sapphire interactions in AlN/graphene/sapphire were largely adapted to their lattice and thermal mismatch. As a result, the compressive strains in AlN were substantially released. The calculated internal stresses in AlN were reduced from 16.0 GPa in AlN/sapphire to 14.9 GPa in AlN/graphene/sapphire in Fig. 2.11h, i.

Besides graphene, h-BN is also a popular 2D material for growing nitrides. Qingqing Wu et al. [51] analyzed the growth mechanism of AlN on h-BN. ML h-BN on copper foil was first grown by low-pressure chemical vapor deposition (LPCVD) and then transferred onto a sapphire substrate. Next, the h-BN on sapphire was treated with oxygen plasma at an RF power of 100 W for 3 min. As a result, after oxygen plasma treatment, the surface of h-BN formed the chemical state of N–B–O as shown in the XPS spectra in Fig. 2.12a–e. Here, the sp^2 hybridization orbitals of B were transformed into sp^3 hybridization orbitals, which allowed the dangling bonds to be used for subsequent AlN film growth.

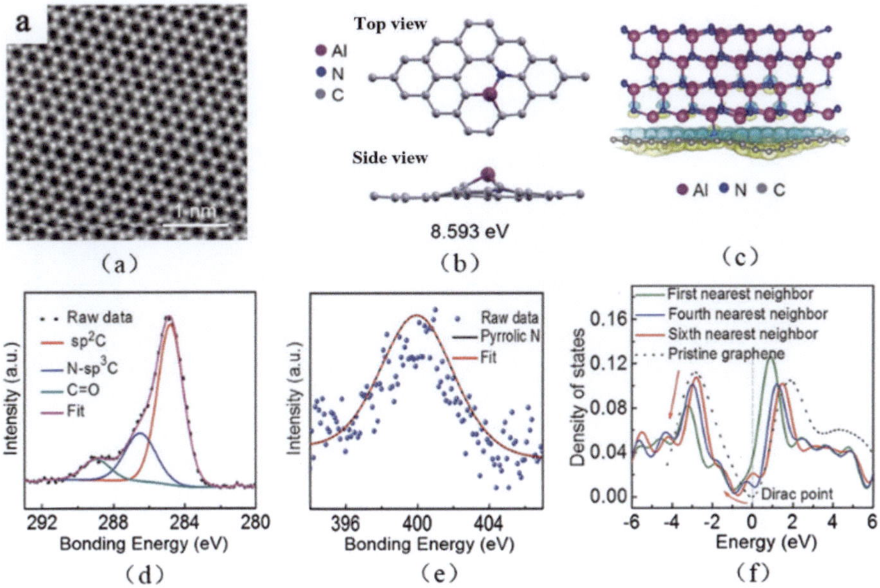

Fig. 2.10 Model of Al atom bonding on graphene. **a** Typical atomically resolved z-contrast image of graphene. **b** DFT calculation of the bonding energy of an aluminum atom on pyrrole-based N as 8.593 eV. **c** Isosurfaces of the model structure and electron density difference of AlN atoms formed through Al-N bonding on N_2 plasma-treated graphene. The blue and yellow isosurfaces correspond to charge depletion and accumulation, respectively. **d** C 1 s XPS of graphene films treated with N_2 plasma, showing that graphene had the characteristic signals of C-sp^2 peak (284.8 eV), N-sp^3 peak (286.5 eV), and broad O–C–O peak (288.9 eV). **e** N 1 s XPS spectrum of graphene film treated with N_2 plasma, showing the characteristic signal of pyrrole N feature (400.1 eV). **f** Calculated density of states (PDOS) for 3 C atoms to N atoms at different distances from the 2p$_z$ orbitals (i.e.: first nearest neighbor, second nearest neighbor, sixth nearest neighbor) [47]

2.2.2 Interfacial Atomic Configurations of Other III-V Compounds on 2D Materials

YoungJoon Hong et al. [52] utilized MOCVD to grow InAs nanostructures on both sides of suspended single-layer graphene (S-SLG). They used the DFT method for first-principle total energy, vdW binding energy and electronic structure calculations. They investigated the crystal growth orientation of InAs nanostructures grown by vdW-MOCVD by cross-section high angle annular dark field (HAADF) STEM and energy dispersive spectroscopy (EDX) analysis. Figure 2.13a showed atomic resolution HAADF-STEM tomography images and EDX atomic column mapping images recorded for the InAs nanowires, which indicated that the nanowires were composed of zinc-blende/Wurzitecite (ZB/WZ) mixtures, similar to their microstructure. The nanowires were grown by conventional catalyst-free and metal catalyst-assisted MOCVD [53, 54]. Despite the presence of high-density stacking defects, the nanowires maintained the same polarization direction $[\bar{1}\bar{1}\bar{1}]_{(ZB)}$ or $[00\ \bar{1}]_{(WZ)}$

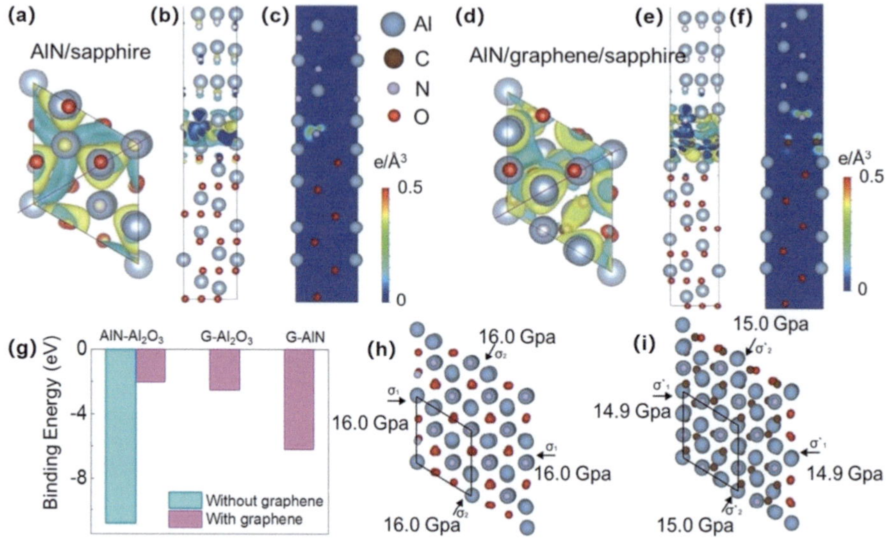

Fig. 2.11 DFT calculations of AlN/Al_2O_3 bonding energies and internal stresses in AlN. **a, d** Computational models for AlN/sapphire and AlN/graphene/sapphire, respectively. The black parallelogram indicates the computational unit. **b, e** Front view of the 3D charge density corresponding to (**a**, **d**), respectively. The yellow and blue regions represented the electron gain and loss, respectively. (**c**, **f**) Charge density cross sections along the purple line in (**a**, **b**), respectively. (Contours in $e\text{Å}^{-3}$). **g** Calculated BE for AlN/sapphire and AlN/graphene/sapphire in AlN/sapphire were − 10.84 and − 2.04 eV, graphene-AlN was − 6.20 eV, and graphene-sapphire was − 2.51 eV, respectively. **h, i** Calculated values of internal stresses in AlN without (**g**) and with (**h**) graphene intercalation (16.0 and 14.9 GPa)

growth direction along the entire length. Similarly, single-crystal InAs nano-islands exhibited the same preferred crystal growth direction along $[\bar{1}\bar{1}\bar{1}]_{(ZB)}$ in Fig. 2.13b. It should be noted that the nano-islands were usually structurally single-crystalline ZBs with a significantly lower density of stacking defects or polymorphism, presumably due to the size-dependent thermodynamic stability of the compound semiconductors [55], e.g., III-V nanowires grown by covalent epitaxy [56, 57] and II-VI nanostructures grown by vdW epitaxy [58]. Cross-sectional STEM studies showed the formation of InAs nanostructures along the exposure direction perpendicular to the SLG surface. This suggested that the InAs/S-SLG/InAs bi-heterostructures showed polarity reversal at the vdW hetero-interface as shown in Fig. 2.13c.

Considering the extent of semiconductor/graphene lattice mismatch [37], InAs/SLG must satisfy one of the three possible binding sites for In and As atoms on graphene: (i) on the hollow (H), which was the center of the carbon honeycomb; (ii) on the top (T), which was directly above the carbon atoms; or (iii) at the midpoint of the carbon–carbon bonding bridges (B) in Fig. 2.13d. This suggested that the adsorption energy difference between In and As atoms played a crucial role in determining the crystal growth direction during the initial growth phase of InAs. According to the atomic resolution STEM analysis, In atoms seemed to precede As

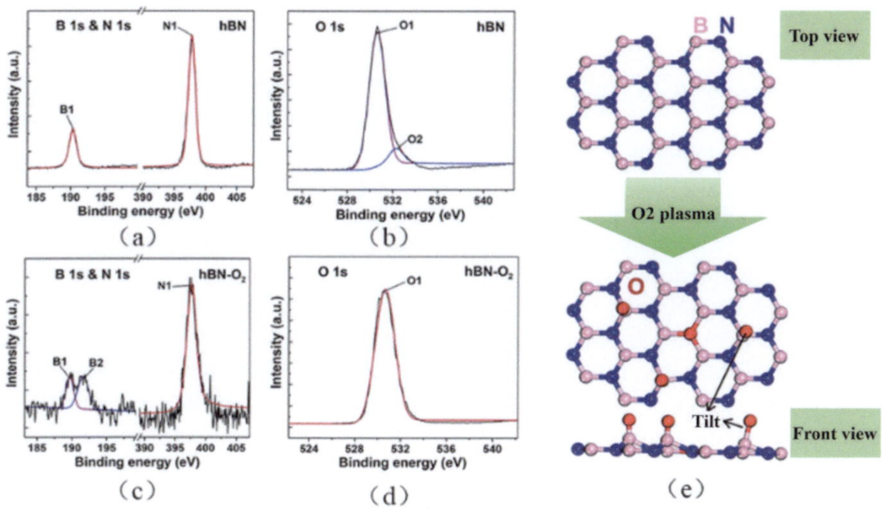

Fig. 2.12 AlN bonding model on h-BN. **a–d** B 1 s, N 1 s and O 1 s XPS peaks of h-BN and h-BN-O_2 on sapphire. **e** Schematic representation of h-BN atomically connected structures before and after O_2 plasma treatment [51]

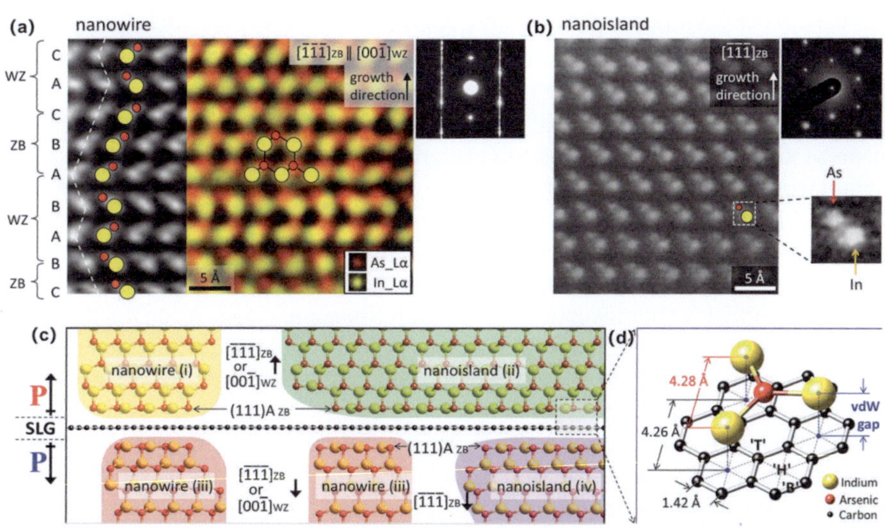

Fig. 2.13 Preferred crystal growth orientation of InAs on graphene substrates. **a** Cross-sectional atomic resolution HAADF-STEM filtered image and EDX atomic column mapping image of InAs nanowires (yellow and red colors correspond to In and As, respectively). **b** HAADF-STEM images of InAs nano-islands. (the insets in (**a**) and (**b**) were in SAED mode, indicating that all images were taken along a [1 $\bar{1}$ 0] ZB-type projection. **c** Atomic epitaxy schematic, depicting the epitaxial InAs/SLG/InAs double heterostructure. **d** Schematic of the atomic structure of the vdW epitaxial InAs/graphene heterostructure. The three possible configurations of In (or As) atoms at the top of the carbon atom and on the carbon–carbon bond bridge at the honeycomb hollow are denoted as H, T and B, respectively [52]

atoms and into the SLG surface. According to the calculations of Kevin Chan et al. [12] presented in the previous section, the occupation of the H site by In adsorbed atoms was favorable for a substantial energy reduction with higher coordination. At the H site, the calculated adsorption energy of In-SLG (− 0.75 eV at an equilibrium distance of 2.5 Å) was much smaller than that of As-SLG (− 0.13 eV at an equilibrium distance of 3.0 Å), suggesting that the initial substratum of In atoms placed close to the SLG was energetically stable.

Based on STEM analysis, they assumed that the InAs nanostructures grew along the ZB[111]B direction perpendicular to the graphene surface. The total energy values of the reconfigured interfaces (and surfaces) with In vacancies were calculated to be lower than the unreconstructed energies not only for the InAs(111)A free surface, but also for the vdW epitaxial InAs(111)/SLG heterostructure. The presence of In vacancies at each (2 × 2)InAs (111)A interface was desirable from energetic considerations: the total energy of the (2 × 2)-InAs (111)A surface with In vacancies ($= E^{v}_{InAs}$) was 1.151 eV lower than that of the blank InAs (111)A ($= E^{c}_{InAs}$); heterogeneous interfaces reconstructed for the the total energy of InAs(111)/SLG ($= E^{c}_{interface}$) was 0.931 eV lower than that of the unreconstructed heterostructure ($= E^{c}{}_{interface}$) (total energy values were shown in Table 2.2). Using thermodynamic calculations [59], the surface Gibbs free energy of the reconstructed InAs(111)/SLG heterointerface with In vacancies ($G^{v} = E^{c}_{interface} - (N_{In}\text{-}1)\mu_{In} - N_{As}\mu_{As} - N_{C}\mu_{C}$) was calculated to be lower than that of the non-reconstructed hetetointerface ($G^{c} = E^{c}{}_{interface} - N_{In}\mu_{In} - N_{As}\mu_{As} - N_{C}\mu_{C}$), suggesting that InAs(111)A junction with SLG tended to form In vacancies to minimize the energy. The theoretically simulated atomic configurations of the vdW heterointerfaces were shown in Fig. 2.14(a, b), presenting vdW epitaxial InAs/SLG heterointerfaces with In vacancies. The interfacial layers clearly showed that the As atoms near the In vacancies were strongly deformed along the -z direction, which induced structural relaxation, the so-called III-V vacancy buckling [60] verified by reflection high-energy electron diffraction and scanning tunneling electron microscopy. The InAs(111)A surface was cleaned under free surface vacuum conditions [61, 62]. Since the planar reconstructed InAs interfacial layer with In vacancies was much less polar along the z-direction than the bulk InAs with a zigzag (111) In-As configuration, the reconstructed initial (111)A substrate was hybridized with the nonpolar graphene surface as far as its electron orbitals are concerned. Importantly, since not only the In atoms but also the As atoms generated a stronger vdW gravitational force in the final reconstructed (111)A interface, the calculated reconstructed binding energy (64 meV) was about 1.5 times higher than that of the unreconstructed binding energy (46 meV). Comparison of the total energy values of the heterogeneous interfaces suggested that the flatly reconstructed (111)A heterogeneous interface was formed preferentially. At the top of the initial (111)A interfacial layer, the InAs (111) layer stacked along the sp^3 bond along the [111] B direction during MOVPE. This further confirmed the formation of InAs nanostructures along the ZB [111] B direction perpendicular to the graphene surface.

The energy values are in eV unit. The reconfigured InAs(111)A surface (or interface) had almost flatten In-As configuration with In vacancy in every (2 × 2) supercell; The bottommost layer of the heterostructure was InAs(111)A.

Table 2.2 Total energy values of vdW epitaxial InAs/SLG heterostructures [52]

	Clean InAs (111) A surface		InAs(111)/SLG heterogeneous interface	
	Reconstruction with In vacancies in each (2 × 2) InAs (111)A	No vacancies (no reconfiguration)	Reconstruction with In vacancies in each (2 × 2) InAs (111)A	No vacancies (no reconfiguration)
Total energy	– 41.726 (= E^{V}_{InAs})	– 40.572 (= E^{C}_{InAs})	– 254.272 (= $E^{V}_{Interface}$)	– 253.341 (= $E^{C}_{interface}$)
vdW binding potential	–	–	-0.064 (= EvdW)	–

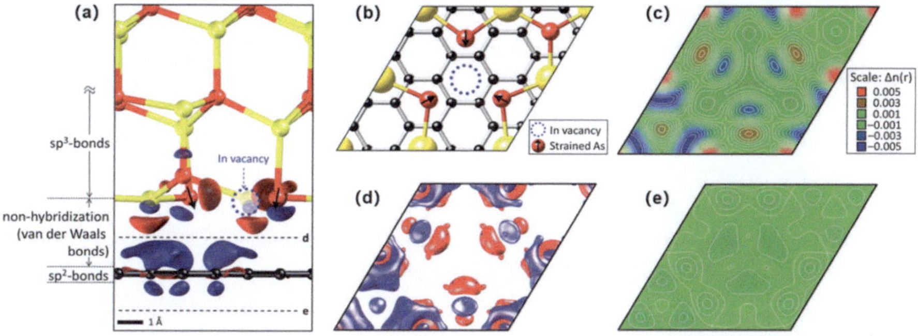

Fig. 2.14 Electronic structure of the vdW epitaxial InAs/SLG heterointerface in equilibrium. **a** Molecular structure and CDD isosurfaces at the periphery of the vdW epitaxial InAs/SLG heterointerface. The contours of the red and blue CDD isosurfaces were ± 0.005eÅ^{-3}, indicating electron accumulation and depletion, respectively. **b** Plan view of the atomic configuration of the epitaxial InAs(111)A/graphene heterointerface. **c** Plot of the CDD isosurfaces with an equivalent value of ± 0.005 eÅ^{-3}, and **d**, **e** contours of the CDD calculated at ± 1.55 Å from the SLG surface, as shown in (**a**) [47]

Hongseok Oh et al. [63] treated the surface of h-BN using oxygen or argon plasma to allow the formation of hydrogen dangling bonds for subsequent ZnO growth. Thereafter, the ZnO growth results of h-BN substrates with or without plasma treatment were significantly different. They provided selective treatment growth to control the shape and position of the ZnO nanowalls by special treatment of different regions of the substrate. They simulated a ball-and-rod model of the atomic configuration at the vdW ZnO/h-BN heterojunction interface based on TEM observations, assuming that the ZnO/h-BN interfacial layer was fully relaxed due to the weak heterogeneous external bonding of the vdW. The ball-and-stick images at low magnification clearly showed regular triangular moiré patterns with a periodicity of ~ 1.0 nm in Fig. 2.15a, c. The corresponding simulated diffractograms were obtained by Fourier transform from the ball-and-stick images of Fig. 2.15a, b. Within the unitary Moore pattern (marked with triangles in Fig. 2.15c), two different heterogeneous epitaxial atomic configurations were shown: ball-to-ball and ball-to-hollow epitaxial atomic configurations in Fig. 2.15d. According to the ball-and-stick model, there were at least some reasonable epitaxial arrangements due to inappropriate epitaxial relationships, as shown in Fig. 2.15e.

They explored the vdW interfacial binding properties of epitaxial ZnO/hBN heterostructures by DFT total energy and electronic structure calculations. Two representative supercells were shown in Figs. 2.16a, b: a supercell in a spherical top-matched configuration, where two Zn atoms were on top of the B and/or N atoms (type I of Fig. 2.16a); and a supercell in a ZnO flat plate (3 × 3) on a BN flat plate (4 × 4) in a hollow-matched supercell (3 × 3) consisting of two cavities in the ZnO hexagonal lattice on top of the two B or N atoms. DFT calculations of these two supercells showed some important interfacial features. First, the interaction energy ($\Delta E = E_{total}\text{-}E_{ZnO}\text{-}E_{h\text{-}BN}$) values for the type I and type II supercells were calculated to be 2.170 and 2.165 $eVnm^{-2}$, respectively. These values were one or two orders of magnitude smaller than those for typical covalent binding. Importantly, the difference in interfacial interaction energies between type I and type II supercells was very small (5 $meVnm^{-2}$). Second, the equilibrium vdW spacing of ZnO-hBN was 3.4 Å in Fig. 2.16c, which was larger than the typical distances for chemical first-order bonds (e.g., covalent or ionic bonds). Third, the heterogeneous interfaces of both type I and type II supercells showed no bonding, and valence electrons were shared or provided in Fig. 2.16d. The contours of the electron density difference of ± 0.0002 $e\text{Å}^{-3}$ and Bader's analysis in Fig. 2.16d showed that there was almost no electron orbital hybridization (charge transfer) through the vdW heterointerfaces. At equilibrium, little buckling (or deformation) was observed at the heterogeneous interface. It was noteworthy that the calculated values for type I and type II supe-runit heterointerfaces had negligible differences in binding energies, equilibrium vdW distances, and electronic structures. This result suggested that there was no major epitaxial relationship between the type I and type II superunit heterointerfaces. Therefore, all the calculated heterointerfacial results indicated that the binding interactions were almost homogeneous throughout the ZnO/hBN heterointerface.

2.3 Summary

This chapter is dedicated to the theoretical investigation of the qvdWE interface in nitrides, with an in-depth exploration of the adsorption and migration characteristics of nitride metal atoms on 2D materials. Specifically, the study examines the adsorption and migration behaviors of single metal atoms on graphene substrates, as well as the mechanisms governing the adsorption and migration of nitride semiconductor metal atoms on graphene. A detailed analysis of the critical factors influencing these processes is presented. Furthermore, the research extends to the atomic configuration analysis of nitride interfaces on 2D materials, encompassing not only graphene but also other 2D materials, along with the atomic configurations of other III-V

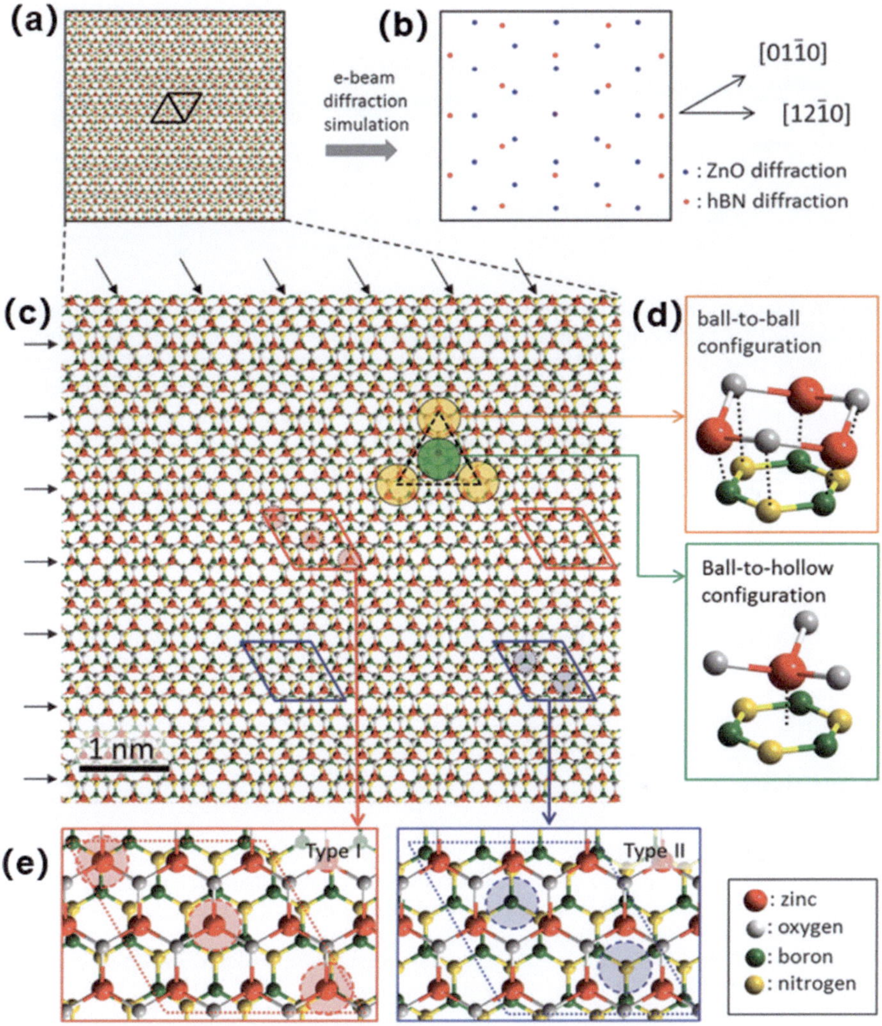

Fig. 2.15 Ball-and-stick model of Quasi van der Waals epitaxial ZnO/h-BN heterostructure. **a** Low-magnification model of the (000 $\bar{1}$)[10 $\bar{1}$ 0] ZnO ∥ (0001)[10 $\bar{1}$ 0] h-BN heterostructure observed along [0001]. The image clearly shows periodic triangular moiré fringes with a period of ~ 1.0 nm. **b** Diffraction pattern obtained by Fourier transform-based simulation. **c** High-magnification ball-and-stick model. **d** Atomic epitaxial relationships of two heterogeneous interfaces appearing in the Moore pattern for ball-to-ball (top panel) and ball-to-hollow (bottom panel) configurations. **e** Ball-top-matched (type I) and hollow-matched (type II) heterojunctions of two supercellular configurations of (3 × 3)-ZnO/(4 × 4)-BN, used for the unequal vdW ZnO/h-BN heterostructure [63]

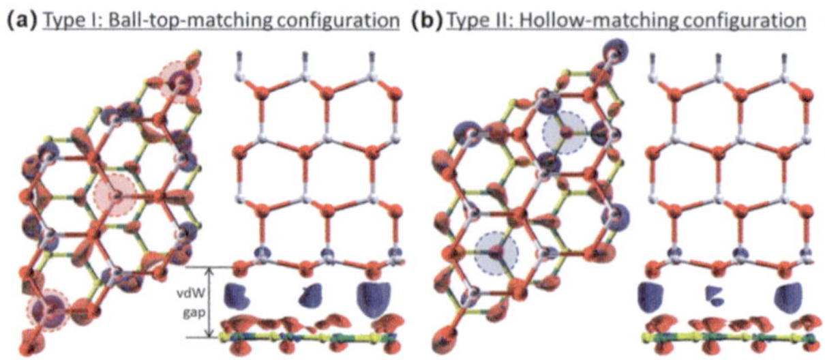

(c)

	ΔE [eV]	ΔE/unit area [eV nm^{-2}]	vdW gap [Å]
Type I	−1.894	−2.170	3.4
Type II	−1.890	−2.165	3.4

(d)

	no binding [electrons]	vdW epitaxy [electrons]
ZnO	657	656.79 (type I) 656.85 (type II)
hBN	128	128.21 (type I) 128.15 (type II)

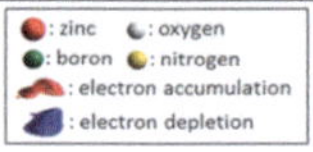

Fig. 2.16 Electronic structure and adhesion energy of vdW epitaxial ZnO/h-BN heterointerfaces calculated by DFT method. **a** Atomic model structures and electron density difference isosurfaces at the periphery of the heterogeneous interface of spherical matches of ZnO/h-BN. The electron density isosurfaces were ± 0.0002 eÅ^{-3}. **b** Atomic model structures and electron density difference isosurfaces at the periphery of the heterogeneous interface of spherical matches of hollow-match configurations. The electron density isosurfaces were ± 0.0002 eÅ^{-3}. **c** Values of vdW heterointerfacial adhesion energies and equilibrium vdW distances for type I and type II supercells. **d** Electronic populations in the ZnO and h-BN portions of the supercells before and after vdW epitaxy calculated by Bader analysis [63]

compound interfaces on 2D materials. These investigations provide robust theoretical support for the epitaxial growth of nitrides on 2D materials, with the aim of optimizing the epitaxial process and enhancing crystal quality.

References

1. Geim AK, Novoselov KS (2007) The rise of graphene. Nat Mater 6:183
2. Geim AK, MacDonald AH (2007) Graphene: exploring carbon flatland. Phys Today 60:35
3. Koma A (1999) van der Waals epitaxy for highly lattice-mismatched systems. J Cryst Growth 201:236
4. Munshi AM, Weman H (2013) Advances in semiconductor nanowire growth on graphene. Phys Status Solid-R 7:713
5. Widjaja H, Altarawneh M, Jiang Z-T (2016) Trends of elemental adsorption on graphene. Can J Phys 94:437
6. Nakada K, Ishii A (2011) Migration of adatom adsorption on graphene using DFT calculation. Solid State Commun 151:13
7. Kresse G, Furthmuller J (1996) Efficiency of Ab-Initio total energy calculations for metals and semiconductors using a plane-wave basis set. Comp Mater Sci 6:15
8. Kresse G, Furthmuller J (1996) Efficient iterative schemes for Ab initio total-energy calculations using a plane-wave basis set. Phys Rev B 54:11169
9. Kresse G, Hafner J (1993) Abinitio molecular-dynamics for liquid-metals. Phys Rev B 47:558
10. Ito T, Shiraishi K (1996) A theoretical investigation of migration potentials of ga adatoms near kink and step edges on Gaas(001)-(2x4) surface. Jpn J Appl Phys 35:L949
11. Shiraishi K, Ito T (1998) Ga-Adatom-Induced as rearrangement during gaas epitaxial growth: self-surfactant effect. Phys Rev B 57:6301
12. Chan KT, Neaton JB, Cohen ML (2008) First-Principles study of metal adatom adsorption on graphene. Phys Rev B 77

13. Perdew JP, Burke K, Ernzerhof M (1996) Generalized gradient approximation made simple. Phys Rev Lett 77:3865
14. Blochl PE (1994) Projector augmented-wave method. Phys Rev B 50:17953
15. Kresse G, Joubert D (1999) From ultrasoft pseudopotentials to the projector augmented-wave method. Phys Rev B 59:1758
16. Slater JC (1964) Atomic radii in crystals. J Chem Phys 41:3199
17. Makov G, Payne MC (1995) Periodic boundary-conditions in Ab-initio calculations. Phys Rev B 51:4014
18. Neugebauer J, Scheffler M (1992) Adsorbate-substrate and adsorbate-adsorbate interactions of Na and K Adlayers on Al(111). Phys Rev B 46:16067
19. Meng S, Gao S (2006) Formation and interaction of hydrated alkali metal Ions at the graphite-water interface. J Chem Phys 125
20. Rytkonen K, Akola J, Manninen M (2007) Density functional study of alkali-metal atoms and monolayers on graphite (0001). Phys Rev B 75
21. Valencia F, Romero AH, Ancilotto F, Silvestrelli PL (2006) Lithium adsorption on graphite from density functional theory calculations. J Phys Chem B 110:14832
22. Lugo-Solis A, Vasiliev I (2007) Ab Initio study of K adsorption on graphene and carbon nanotubes: role of long-range ionic forces. Phys Rev B 76
23. Ribeiro FJ, Neaton JB, Louie SG, Cohen ML (2005) Mechanism for bias-assisted indium mass transport on carbon nanotube surfaces. Phys Rev B 72
24. Maurice V, Marcus P (1992) Stm study of sputter-deposited Al clusters in chemical interaction with graphite (0001) surfaces. Surf Sci 275:65
25. Srivastava S, Almlof J (1992) Chemisorption of aluminum atoms on a graphite surface—cluster convergence and effects of surface reconstruction. Surf Sci 274:113
26. Moullet I (1995) Ab-Initio molecular-dynamics study of the interaction of aluminum clusters on a graphite surface. Surf Sci 331:697
27. Ma Q, Rosenberg RA (1997) Interaction of Al clusters with the (0001) surface of highly oriented pyrolytic graphite. Surf Sci 391:L1224
28. Ganz E, Sattler K, Clarke J (1989) Scanning tunneling microscopy of Cu, Ag, Au and Al adatoms, small clusters, and islands on graphite. Surf Sci 219:33
29. Zhang Y, Franklin NW, Chen RJ, Dai HJ (2000) Metal coating on suspended carbon nanotubes and its implication to metal-tube interaction. Chem Phys Lett 331:35
30. Caragiu M, Finberg S (2005) Alkali metal adsorption on graphite: a review. J Phys-condens Mat 17:R995
31. Zhu ZH, Lu GQ, Wang FY (2005) Why H Atom prefers the on-top site and alkali metals favor the middle hollow site on the basal plane of graphite. J Phys Chem B 109:7923
32. Lindan, P, Duplock, E, Zhang, CJ, Thomas, M, Chatten, R, Chadwick, A (2004) The interdependence of defects, electronic structure and surface chemistry. Dalton T. 3076
33. Yagi Y, Briere TM, Sluiter MHF, Kumar V, Farajian AA, Kawazoe Y (2004) Stable geometries and magnetic properties of single-walled carbon nanotubes doped with 3d transition metals: a first-principles study. Phys Rev B 69:075414
34. Rojas MI, Leiva EPM (2007) Density functional theory study of a graphene sheet modified with titanium in contact with different adsorbates. Phys Rev B 76:155415
35. Hiruma K, Katsuyama T, Ogawa K, Koguchi M, Kakibayashi H, Morgan GP (1991) Quantum size microcrystals grown using organometallic vapor-phase epitaxy. Appl Phys Lett 59:431
36. Yazawa M, Koguchi M, Muto A, Ozawa M, Hiruma K (1992) Effect of one monolayer of surface gold atoms on the epitaxial-growth of inas nanowhiskers. Appl Phys Lett 61:2051
37. Munshi AM, Dheeraj DL, Fauske VT, Kim D-C, van Helvoort ATJ, Fimland B-O, Weman H (2012) Vertically aligned gaas nanowires on graphite and few-layer graphene: generic model and epitaxial growth. Nano Lett 12:4570
38. Chang H, Chen Z, Li W, Yan J, Hou R, Yang S, Liu Z, Yuan G, Wang J, Li J, Gao P, Wei T (2019) Graphene-Assisted Quasi van der Waals epitaxy of aln film for ultraviolet light emitting diodes on nano-patterned sapphire substrate. Appl Phys Lett 114:091107

39. Zhu WG, de Mongeot FB, Valbusa U, Wang EG, Zhang ZY (2004) Adatom ascending at step edges and faceting on Fcc metal (110) surfaces. Phys Rev Lett 92:106102
40. Klimes J, Bowler DR, Michaelides A (2011) van der Waals density functionals applied to solids. Phys Rev B 83:195131
41. Mills G, Jonsson H, Schenter GK (1995) Reversible work transition-state theory—application to dissociative adsorption of hydrogen. Surf Sci 324:305
42. Monkhorst HJ, Pack JD (1976) Special points for brillouin-zone integrations. Phys Rev B 13:5188
43. Wang XG, Chaka A, Scheffler M (2000) Effect of the environment on Alpha-Al_2O_3 (0001) surface structures. Phys Rev Lett 84:3650
44. Alaskar Y, Arafin S, Wickramaratne D, Zurbuchen MA, He L, McKay J, Lin Q, Goorsky MS, Lake RK, Wang KL (2014) Towards van der Waals epitaxial growth of gaas on Si using a graphene buffer layer. Adv Funct Mater 24:6629
45. Al Balushi ZY, Miyagi T, Lin Y-C, Wang K, Calderin L, Bhimanapati G, Redwing JM, Robinson JA (2015) The impact of graphene properties on gan and Aln nucleation. Surf Sci 634:81
46. Liang D, Wei T, Wang J, Li J (2020) Quasi van der Waals epitaxy nitride materials and devices on two dimension materials. Nano Energy 69:104463
47. Chen Z, Liu Z, Wei T, Yang S, Dou Z, Wang Y, Ci H, Chang H, Qi Y, Yan J, Wang J, Zhang Y, Gao P, Li J, Liu Z (2019) Improved epitaxy of aln film for deep- ultraviolet light-emitting diodes enabled by graphene. Adv Mater 31:1807345
48. Reddy ALM, Srivastava A, Gowda SR, Gullapalli H, Dubey M, Ajayan PM (2010) Synthesis of nitrogen-doped graphene films for lithium battery application. ACS Nano 4:6337
49. Lin Z, Waller GH, Liu Y, Liu M, Wong C-P (2013) 3d nitrogen-doped graphene prepared by pyrolysis of graphene oxide with polypyrrole for electrocatalysis of oxygen reduction reaction. Nano Energy 2:241
50. Qi Y, Wang Y, Pang Z, Dou Z, Wei T, Gao P, Zhang S, Xu X, Chang Z, Deng B, Chen S, Chen Z, Ci H, Wang R, Zhao F, Yan J, Yi X, Liu K, Peng H, Liu Z, Tong L, Zhang J, Wei Y, Li J, Liu Z (2018) Fast growth of strain-free Ain on graphene-buffered sapphire. J Am Chem Soc 140:11935
51. Wu Q, Yan J, Zhang L, Chen X, Wei T, Li Y, Liu Z, Wei X, Zhang Y, Wang J, Li J (2017) Growth mechanism of Aln on hexagonal Bn/sapphire substrate by metal- organic chemical vapor deposition. CrystEngComm 19:5849
52. Hong YJ, Yang JW, Lee WH, Ruoff RS, Kim KS, Fukui T (2013) van der Waals epitaxial double heterostructure: InAs/single-layer graphene/InAs. Adv Mater 25:6847–6853
53. Dayeh SA, Susac DA, Kavanagh KL, Yu ET, Wang D (2009) Structural and room-temperature transport properties of zinc blende and wurtzite Inas nanowires. Adv Funct Mater 19:2102
54. Tomioka K, Motohisa J, Hara S, Fukui T (2007) Crystallographic structure of InAs nanowires studied by transmission electron microscopy. Jpn J Appl Phys 46:L1102
55. Akiyama T, Sano K, Nakamura K, Ito T (2006) An empirical potential approach to wurtzite-zinc-blende polytypism in group Iii-V semiconductor nanowires. Jpn J Appl Phys 45:L275
56. Algra RE, Verheijen MA, Borgstrom MT, Feiner L-F, Immink G, van Enckevort WJP, Vlieg E, Bakkers EPAM (2008) Twinning superlattices in indium phosphide nanowires. Nature 456:369
57. Johansson J, Dick KA, Caroff P, Messing ME, Bolinsson J, Deppert K, Samuelson L (2010) Diameter dependence of the wurtzite-zinc blende transition in inas nanowires. J Phys Chem C 114:3837
58. Utama MIB, de la Mata M, Magen C, Arbiol J, Xiong Q (2013) Twinning-, polytypism-, and polarity-induced morphological modulation in nonplanar nanostructures with van der Waals epitaxy. Adv Funct Mater 23:1636
59. Reuter K, Scheffler M (2002) Composition, structure, and stability of Ruo2 (110) as a function of oxygen pressure. Phys Rev B 65:035406
60. Tong SY, Xu G, Hu WY, Puga MW (1985) Vacancy buckling model for the (111) surface of Iii-V-compound semiconductors. J Vac Sci Technol B 3:1076

61. Ohtake A, Ozeki M, Nakamura J (2000) Strain relaxation in inas/gaas (111) a heteroepitaxy. Phys Rev Lett 84:4665
62. Taguchi A, Kanisawa K (2006) Stable reconstruction and adsorbates of Inas (111) a surface. Appl Surf Sci 252:5263
63. Oh H, Hong YJ, Kim K-S, Yoon S, Baek H, Kang S-H, Kwon Y-K, Kim M, Yi G-C (2014) Architectured van der Waals epitaxy of Zno nanostructures on Hexagonal Bn. Npg Asia Mater 6:e145

Chapter 3
Nitride Quasi van der Waals Epitaxy Bonding and Nucleation on 2D Materials

Low nucleation density makes the direct growth of thin-film compound semiconductors on 2D substrates difficult to achieve. The previous section has introduced the interface structure and theory of qvdWE growth on 2D materials. For the III-V compound and graphene system, the ratio of the adsorption energy of the compound on the graphene surface to the aggregation energy of the bulk material (i.e., the energy released when atoms combine into bulk material) should be considered. The larger this ratio, the easier it is for this III-V compound to adsorb and nucleate on graphene. It also exhibits a 2D planar growth mode, that is, it is easier to obtain a smooth, flat film of this material through qvdWE growth. In addition, the migration barrier of atoms on the graphene surface should also be considered.

Compared to 3D materials, graphene has a lower surface free energy and a lower migration barrier for the atoms, so the migration rate of atoms on the graphene surface is much larger than that on bulk materials. For a certain compound, it is necessary to analyze its adsorption energy and migration barrier on graphene to determine the difficulty of the initial nucleation stage of epitaxy and the dominant atoms. The adsorption energy should not be too small, otherwise it is not conducive to the nucleation of atoms on the graphene surface. The migration barrier should not be too high, otherwise it is easy to induce 3D island-like nucleation growth, which is not conducive to the generation of continuous films. Therefore, for the nucleation stage, by analyzing the adsorption energy and migration barrier of different compounds and different atoms, the epitaxial growth condition parameters can be more reasonably selected. The evolution of their morphology during growth is influenced by kinetic factors, which determine whether newly deposited adatoms are likely to form a new nucleus or grow on an existing nucleus [1]. When the diffusion barriers of adatoms are low, adatoms diffuse rapidly on the surface to form large crystalline islands. Conversely, when the diffusion barriers of adatoms are high, the nucleus forms finely dispersed islands with a small volume.

This chapter mainly introduces the nucleation mechanism of nitride qvdWE, describes the impact of the surface morphology of 2D materials and the number of

T. Wei et al., *Quasi van der Waals Epitaxy of Nitride Semiconductor and Its Applications*, Wide Bandgap Semiconductors,
https://doi.org/10.1007/978-981-95-3469-2_3

layers of 2D materials on the nucleation of qvdWE, as well as the interface treatment methods of nitride qvdWE nucleation. A systematic explanation of the nucleation mechanism of nitride qvdWE on 2D materials is also provided.

3.1 Nucleation Mechanism of Nitride Quasi van der Waals Epitaxy

For the epitaxial growth of nitrides, due to the lack of intrinsic substrates, nitride semiconductors mainly grow on hetero-substrates, including sapphire substrate, silicon substrate, silicon carbon substrate, etc. [2–6] However, large lattice mismatches and thermal expansion mismatches become the key issues in the process of hetero-epitaxial growth [7, 8]. During heteroepitaxy, stresses accumulate in the nitride epilayer, making the epilayer vulnerable to cracking or dislocations due to the tendency for stresses to be released when the critical thickness is exceeded, negatively affecting device performance. For the large lattice mismatch ones, misfit dislocations would immediately occur at the heterointerface to accommodate the large mismatch induced strain. As a result, in principle, the crystalline quality of the hetero-epilayer is not as satisfied as that of the homoepitaxial ones. Usually, a large density of dislocations (including point dislocations, screw dislocations, and planar dislocations) and residual stress are generated in the grown thin film, which are the fundamental reasons affecting the performance of nitride devices. As a type of layered material, 2D materials usually exhibit a thickness of one or more atomic layers, with atoms covalently bonded, and each layer is held together by weak vdW forces [9–11]. 2D materials can be used as buffer layers for the growth of nitride qvdWE to reduce the stress of the epitaxial thin film and improve the crystal quality.

In 1984, Koma et al. first realized the hetero-crystal epitaxy of $NbSe_2/MoS_2$ through vdWE on a dangling-bonds-free surface [12], which has attracted increasing attention. In this scenario, the epilayer is not constrained to the substrate by covalent interaction, thus the epilayer tends to preserve its own lattice structure and not establish the epitaxial relationship with the substrate as that happens in the traditional covalent epitaxy. This growth paradigm offers hope for fully breaking the substrate restriction and is favorable for implementing dissimilar material integration. In order to expand the application range of nitride-based devices, 2D materials have also been introduced into nitride growth [13], and the presence of 2D materials can be seen at the interface between the substrate and the nitride. Specifically, chemical bonds are formed at the nucleation positions at the interface between the 2D material and the nitride, while vdW forces are maintained at other non-nucleation positions. However, the nucleation of adsorbed atoms on the surface of sp^2 bonded 2D materials is suppressed, and the growth of 3D single-crystal thin films on 2D materials is still a huge challenge. To better understand the growth mechanism, it is necessary to clearly study the physical and chemical changes at the interface. Therefore, the nucleation growth of nitrides on 2D materials is particularly important. Below, we will introduce from several aspects such as the nucleation mechanism of nitrides on the surface of 2D materials, the impact of the surface morphology of 2D materials (defect situation and the number of layers of 2D materials) on nitride nucleation.

For the conventional MOCVD epitaxial growth of nitrides, uniform nucleation of GaN is an essential process for growing high-quality single crystalline GaN. The growth of GaN thin films needs to be carried out at high temperatures (usually > 1040 °C). Before this process, a low-temperature (< 600 °C) nucleation layer (Nucleation Layer, NL) usually needs to be grown on the substrate surface first. A schematic diagram of GaN nucleation on a specific interlayer is shown in Fig. 3.1 [14]. Low-temperature (LT) GaN or AlN are usually used as the nucleation layer for high-temperature (HT) GaN growth, with a thickness in the range of 30–100 nm. Therefore, whether the intermediate layer (including 2D materials) on the substrate surface forms an effective nucleation layer for HT GaN growth is crucial. The metal titanium (Ti) intermediate layer has been proved to be a poor NL. For example, in the radio frequency plasma-assisted MBE system, the Ti layer is used as a mask for GaN growth, but it is difficult for HT GaN to nucleate on Ti [15, 16]. On the contrary, LT GaN (56 nm nucleation layer) can be formed on the Ti interlayer at a temperature of 560 °C, serving as a buffer layer for subsequent HT GaN. Therefore, only at sufficiently low temperatures, the surface diffusion of the Ga source and the desorption of Ga can be minimized, in order to react with the nitrogen source (such as NH_3) to form nucleation sites on the growth surface. However, the low growth temperature significantly deteriorates the crystal quality of the epitaxial layer.

For 2D layered materials (such as graphene and h-BN, etc.) as the interlayer (Interlayer, IL) for the growth of GaN thin film materials, uniform nucleation of GaN is much more difficult. This is because there are no vertical dangling bonds on the surface of these 2D materials, resulting in extremely low surface energy (γ) of the growth material in vacuum or air. Several common substrate materials and nitride materials' γ are listed in Table 3.1. On 2D materials, the adsorption or nucleation of foreign atoms is thermodynamically unfavorable because it increases the total free energy of the system, which is expressed as follows [14]:

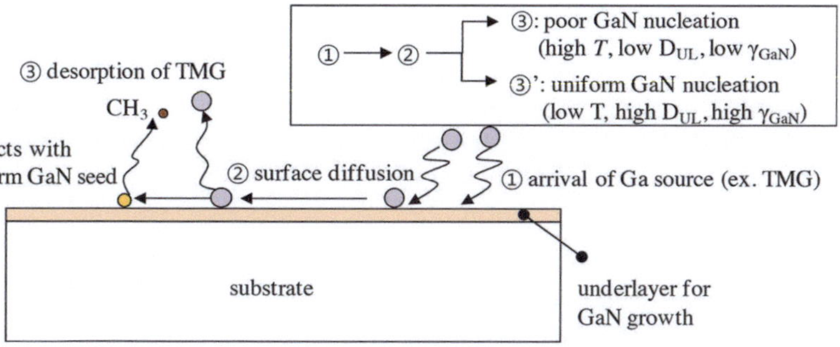

Fig. 3.1 Schematic diagram of GaN nucleation process on the specific underlayer surface [14]

$$\Delta G = G_f - G_i = \gamma(NL) + \gamma_{IL-NL} - \gamma(2DIL) \gg 0 \quad (3.1)$$

where G_i, G_f are the total Gibbs free energies before and after the formation of a uniform NL, respectively, $\gamma_{(NL)}$, $\gamma_{(2D\text{-}IL)}$ and $\gamma_{(IL\text{-}NL)}$ are surface energies of the NL and IL in vacuum and the interfacial energy between NL and IL, respectively. Therefore, the nucleation of GaN on such a 2D material is extremely challenging. Indeed, it has been found that direct LT-GaN growth (530 °C) on highly oriented pyrolytic graphite leads to nucleation only at grain boundaries. One way to circumvent this problem is to generate defects in the 2D materials. Kunook Chung et al. used ZnO nanowalls as NL for GaN on O_2-plasma-treated graphene [17], which may enhance Ga adsorption and GaN nucleation. Following the defect generation process, γ of the graphene IL will increase significantly, such that ΔG will be close to zero. Hence the nucleation will be energetically favorable. In addition, Kobayashi et al. also reported that an $Al_xGa_{1-x}N$ ($x > 0.1$) or AlN layers can be uniformly formed on *h*-BN [18], resulting in an excellent NL for GaN and a significantly improved growth morphology. Note that the NLs discussed here, including ZnO, AlN, and $Al_xGa_{1-x}N$, have sufficient dangling bonds, and hence have comparable γ to that of GaN, enabling uniform nucleation of GaN on 2D materials.

In addition, the adhesion energies and diffusion barriers of adatoms are also important parameters in the nucleation process of nitride growth, especially for qvdWE [19]. The morphology of the nucleus is determined based on the competition between the adhesion energy (E_{ads}) of the adatom to substrate and the cohesive energy (E_{coh}) of adatom to adatom [1]. The evolution of their morphology during growth is influenced by kinetic factors. Adatoms may diffuse rapidly on the surface to form large crystalline islands if the diffusion barriers of adatoms are low. Conversely, when the diffusion barriers of adatoms are high, the nucleus forms finely dispersed islands with a small volume. In the thermodynamic limit, epitaxial growth can follow three major growth types according to E_{ads}/E_{coh} and the diffusion barriers as shown in Fig. 3.2a: (1) Frank-van der Merwe type, which consists of layer-by-layer growth of the epilayers for large E_{ads}/E_{coh} and low diffusion barriers; (2) Volmer-Weber type,

Table 3.1 Surface energies of different materials [14]

Materials	Surface free energy (mJ m^{-2})
Si (111)	1467
Sapphire (001)	4800 (before relaxation), 1850 (after relaxation)
w-ZnO (001)	2025–2040
w-GaN (001)	1970
w-AlN (001)	5840
h-BN (001)	37–57 (cubic BN)
graphene (001)	52 (multilayer)
α-Ti (001)	2048
glass	2000–4000

which describes the 3D clustering of adatoms on a bare substrate for small E_{ads}/E_{coh} and low diffusion barriers; and (3) Stranski–Krastanov type, which is an intermediary growth type characterized by both Frank-van der Merwe and Volmer-Weber types [1]. Transition from the layer-by-layer growth to island-based growth occurs at a critical layer thickness, which is related to the interface strain between the substrate and epilayer. In general, 2D materials provide low diffusion barriers and Eads to adatoms due to the dangling bond-free inert surface. Therefore, adatoms can quickly diffuse and add on to the fastest growth front at the growing epilayer to extend it. Meanwhile, the Frank-van der Merwe type growth mode during 3D/2D qvdWE is difficult to see because of the low Eads of adatoms in Fig. 3.2b. For example, in qvdWE of metals on 2D substrates, only Li, Na, and K are predicted to follow Frank-van der Merwe-type growth mode due to the large E_{ads}/E_{coh} and low diffusion barrier energy [19]. In 2D/2D vdWE, E_{coh} of adatoms is anisotropic for growing 2D materials, which results in the layer-by-layer growth of the 2D epilayers. The edge site of the 2D epilayer is more active than that of the plane [20]. Consequently, adatoms preferentially merge at the edge site of the growing 2D epilayer, leading to the extension of the 2D epilayer in ultrathin structures following the Frank-van der Merwe type growth mode. By contrast, the 3D substrates possess high surface roughness and dangling bonds on the surface, which result in high diffusion barriers that prevent the layer-by-layer growth of 2D epilayers and increase the thickness of the 2D epilayer and the number of 2D islands. For example, AlN nucleation growth follows the above features.

In 2018, Qi et al. [21] studied the nucleation mechanism of AlN material grown on sapphire substrates with graphene as a buffer layer. In the initial stage of growth, the graphene covering the surface showed a significant impact on the nucleation density and individual growth rate of AlN. The scanning electron microscope (SEM)

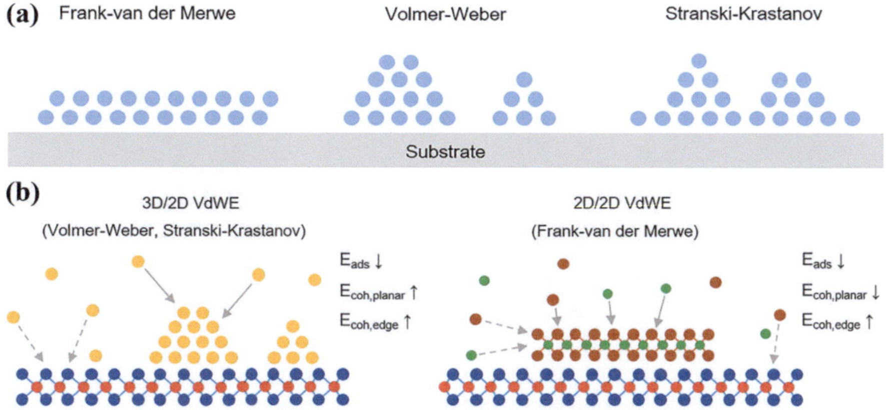

Fig. 3.2 **a** Schematic illustrations of the basic growth modes including Frank-Van der Merwe, Volmer-Weber, and Stranski–Krastanov growth. **b** Schematic illustrations of the 3D/2D qvdWE and 2D/2D vdWE. Cohesive energy difference between 3 and 2D epilayer induces different growth modes [1]

image in Fig. 3.3a showed AlN nucleation on bare sapphire (top right) and AlN nucleation on graphene-covered sapphire (bottom left). The two typical regions can be easily distinguished from the different nucleation density and domain size of AlN, as schematically shown in Fig. 3.3b. The enlarged SEM statistical data in Fig. 3.3c, d showed that the nucleation density of AlN on sapphire was 41/μm^2, while on graphene/sapphire it was 23/μm^2. The results suggested that graphene can inhibit the nucleation of AlN, which was likely due to the lack of dangling bonds on the surface of the high-quality transferred graphene. Moreover, on sapphire and graphene/sapphire, the densities of AlN nucleation islands with diagonal distances greater than 200 nm were 3/μm^2 and 11/μm^2, respectively (as shown in Fig. 3.3e), indicating that the lateral growth rate of AlN nucleation islands grown on graphene was much larger than that on sapphire. The reason for the different growth rates of AlN on the two types of substrates is that graphene can reduce the diffusion barrier of surface metal atoms, making the adsorbed atoms easier to diffuse with a larger diffusion length, thus accelerating the lateral growth of islands on the graphene-covered sapphire.

Furthermore, Hongliang Chang et al. also studied the nucleation mechanism of AlN on the graphene surface [22]. On nitrogen plasma-treated graphene, nitrides can effectively nucleate using pyrrole nitrogen. To explore the agglomeration behavior between AlN nucleation points after nucleation growth, they used first-principles to

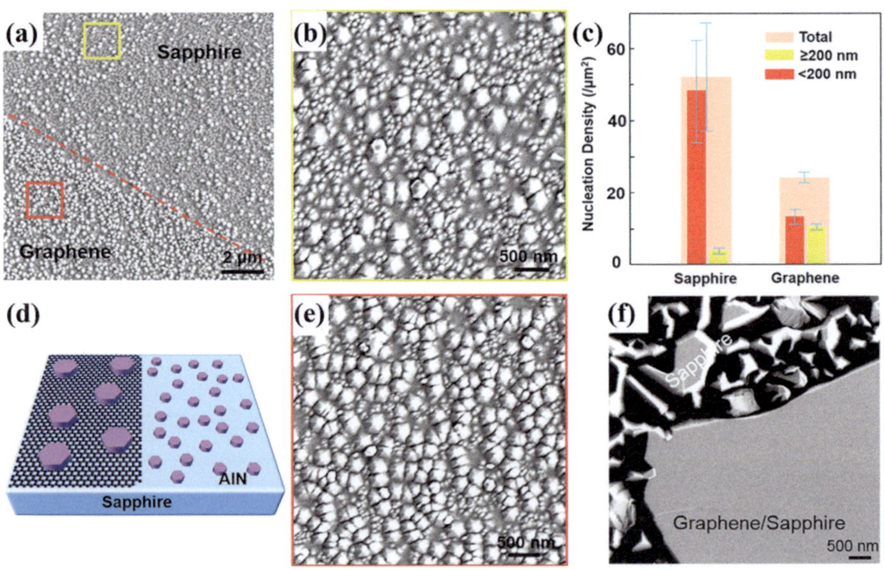

Fig. 3.3 Effects of the graphene interlayer on the AlN nucleation. **a** Large-scale SEM image showing the nucleation of AlN on bare sapphire and graphene-buffered sapphire (growth duration: ~ 6 min). **b** Corresponding schematic. **c, d** Magnified SEM images showing the distributions of the AlN nucleus on bare sapphire (**c**) and graphene-buffered sapphire (**d**). **e** Statistics of the density and size of the AlN nucleus on sapphire and graphene/sapphire. **f** SEM image for the noncoalesced and coalesced AlN domains on sapphire and graphene/sapphire, respectively [21]

study the adsorption and diffusion of Al atoms on small AlN clusters on graphene. Firstly, considering the growth of AlN on the Al_2O_3 (0001) surface, four Al atoms (three located at high Al positions, one at the vacancy-Al position) and three N atoms were added to the Al_2O_3 3×3 surface unit to form a small $Al_{10}N_3$ cluster (see the black triangle in Fig. 3.4a), and then another Al atom was introduced into the cluster. The most stable adsorption sites on the Al_2O_3 surface and near the edge of the AlN cluster were the nearby high Al positions, as shown by the solid red circle in Fig. 3.4a. However, the position above the N atom was more stable than this position by 0.085 eV. Although the energy difference was small, the diffusion barrier from the top position to the high Al position was calculated to be 1.107 eV (Fig. 3.4b). Therefore, the incoming Al atoms were energetically favored to adsorb on the top of the existing AlN cluster and were difficult to jump down to the Al_2O_3 surface. In other words, the AlN nucleation pointed on the Al_2O_3 surface tended to expand in a 3D mode, but showed a different growth mode on the graphene surface. Consider the aluminum atom on a larger Al_9N_4 adsorption on clusters, as shown in t1 and t2 in Fig. 3.4c, the adsorption energy of two top positions stabilized around 1.8–1.9 eV, and the top positions of the other two N atoms were also unstable. On graphene, there were two types of adsorption sites near the edge of the Al_9N_4 cluster, as shown in Fig. 3.4c, one type was the a_1 and a_2 positions next to the N atoms in the cluster, and the other type was the b_1 and b_2 positions where there were no N atoms nearby. The adsorption of Al on b_1 and b_2 was slightly better than the top position, with an adsorption energy of 1.9–2.1 eV. However, the most stable adsorption sites for Al atoms were a_1 and a_2, with a larger adsorption energy of 2.7–2.9 eV. In addition, the potential barrier from the top t1 point to the side a1 point was calculated to be only 0.005 eV (Fig. 3.4d). Considering such a small barrier, the Al atom at the t_1 position can easily jump to the graphene surface. Therefore, it was easy to draw a conclusion that AlN on graphene tended to grow laterally in quasi-2D mode in terms of energy and kinetics at the initial growth stage.

3.2 The Influence of the Surface Morphology of Two-Dimensional Materials on the Nucleation of Quasi van der Waals Epitaxy

The bonds of nitrides are the results of ionic and covalent interactions, while the bonds of 2D materials originate from vdW forces. Due to the lack of chemical activity on the surface of 2D materials, it is very difficult to grow nitrides directly on 2D materials. Therefore, this means that to achieve high-quality growth of nitrides, the surface chemical activity of 2D materials must be greatly improved [23–29]. Typically, the surface of 2D materials has certain steps and boundaries, with the size of the crystal domains ranging from tens of nanometers to several micrometers. Therefore, the nucleation of nitrides usually occurs at places where the surface state changes, such as at steps and wrinkles. However, since epitaxial growth only occurs

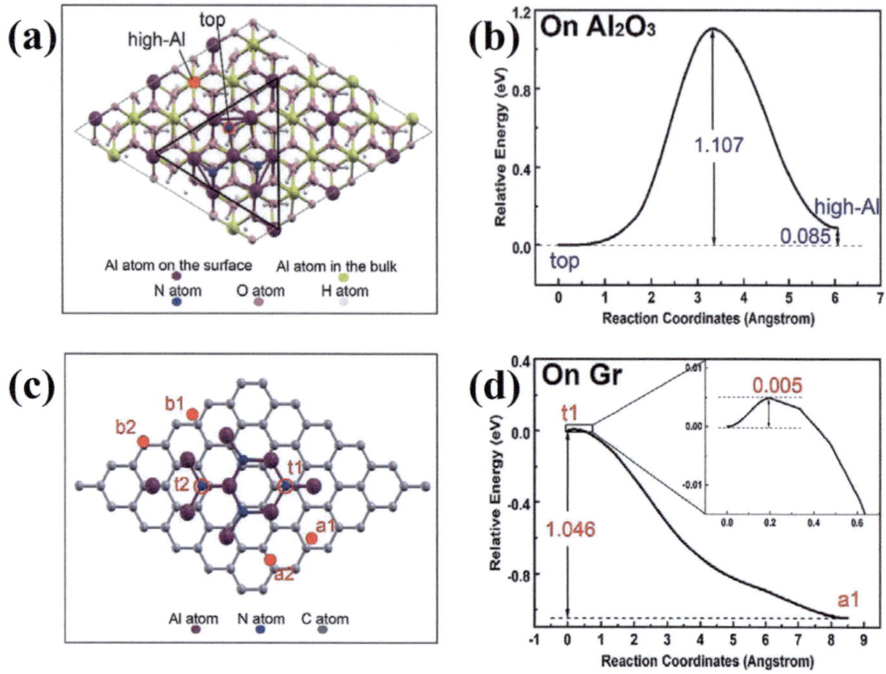

Fig. 3.4 First-principles calculation and schematic diagram of the growth of AlN on Al_2O_3 and graphene. **a** Adsorption sites of Al atom to the $Al_{10}N_3$ cluster on Al_2O_3 (0001) surface. The $Al_{10}N_3$ cluster is denoted by a triangle. The solid red circle represents the high-Al adsorption site that is on the Al_2O_3 surface and next to the $Al_{10}N_3$ cluster, while the empty red circle represents the top site on top of an N atom in the $Al_{10}N_3$ cluster. **b** The energy barrier of Al atom diffusing from the top site to the high-Al site. **c** Adsorption sites of Al atom to the Al_9N_4 cluster on graphene surface. The solid red circles (a and b sites) represent the Al adsorption sites that are on the graphene surface and next to the Al_9N_4 cluster, while the empty red circle represents the top site on top of N atoms in the Al_9N_4 cluster. **d** The energy barrier of Al atom diffusing from the t1 site to the a1 site. The small energy barrier is enlarged in the insect [22]

at steps and boundaries, it is difficult to achieve high-quality nitride film on 2D materials. Therefore, for the growth of nitrides on 2D materials, it is crucial to enhance their adsorption ability for Al, Ga, N, and other atoms by introducing other defects and forming dangling bonds. In this case, the surface of the 2D material is usually pretreated or in-situ pretreated in the growth chamber.

In 2012, Jae-Kyung Choi et al. found that GaN materials can easily nucleate on natural graphene ridges [30], and the sapphire template covered with a graphene layer was prepared by the Diffusion-assisted Synthesis (DAS) process at low temperatures. Notably, the graphene layer formed the ridges at the corresponding positions of the grain boundaries of the polycrystalline Ni film, which can serve as nucleation sites for nitride growth. The graphene coating improved the wetting between GaN and sapphire, and, with as little as 0.6 nm of graphene coating, the overgrown GaN layer on sapphire became continuous and flat. Here, the use of graphene as a coating layer allowed the one-step growth of heteroepitaxial GaN films on sapphire in a MOCVD reactor, simplifying the GaN growth process, as shown in Fig. 3.5a–d. In addition, Yuqing Song et al. presented that the as-fabricated graphene wrinkle arrays would function as the buffer layers for the epitaxial growth of AlN, in which the presence of wrinkles would promote the nucleation and uniform growth of AlN with low strain level and improved lateral growth rates. The temporal evolution of the as-grown AlN epitaxial layer was presented in Fig. 3.5e, f, where the different growth behaviors of AlN on the region in presence of graphene wrinkles, flat graphene region, and bare sapphire region were clearly observable. This difference of growth rates should be caused by the enhanced adsorption of coming AlN particles owing to the presence of strain-rich wrinkles. After 20 min growth, a flat AlN film had been already formed on the region with graphene wrinkles, while in the flat graphene region, the coalescence of individual AlN domains into a flat film is still not observable. Furthermore, the formation of graphene wrinkle arrays would also reduce AlN strain caused by the lattice and thermal mismatch between substrate and epitaxial layer in Fig. 3.5g–h.

Unlike graphene grown on metals with boundaries, graphene produced on SiC substrates usually accompanies the formation of steps. Kim et al. [32] grew high-quality GaN films on SiC substrate covered with graphene. Figure 3.6a showed the Atomic Force Microscope (AFM) image of the step-like morphology of the graphene-converted SiC surface, where the step height and platform width of the SiC surface covered with graphene are 5–10 nm and 5–10 μm, respectively. It was speculated that GaN nucleated at the bar-shaped steps and then grew laterally to form a thin film. Figure 3.6b–d showed the SEM images of GaN films epitaxially grown on graphene under different growth conditions by MOCVD. During the MOCVD

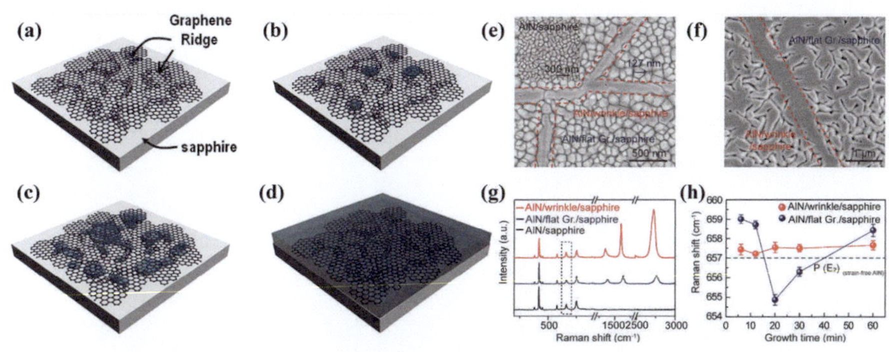

Fig. 3.5 Scheme of one-step growth of epitaxial GaN films using ultrathin graphene coating layers: **a** graphene coating on a sapphire by the DAS process; **b** GaN nucleation along naturally-formed multilayer graphene ridges; **c** lateral growth of GaN on a graphene layer; **d** full coalescence of GaN on a graphene/sapphire. Effects of the graphene wrinkle arrays on the growth speed and strains in AlN [30]. **e** SEM image of the as-grown AlN on graphene wrinkle arrays after 10 min epitaxial growth of AlN. **f** SEM image of the as-grown AlN on graphene wrinkle arrays after 20 min epitaxial growth of AlN. **g** Raman spectra of flat region (blue), graphene wrinkles region (red), and bare sapphire region (black) after the 6 min epitaxial growth of AlN. **h** Raman peak position of E_2 (high) of AlN on flat region (blue), graphene wrinkles region (red) at different growth time [31]

growth process, the influence of temperature was very important, and the temperature of MOCVD growth was the highest among all methods (other growth methods include CVD, MBE, HVPE, etc.). The experimental conditions of this study were as follows: conventional two-step growth (nucleation at 580 °C, growth at 1150 °C), one-step growth under 1100 °C conditions, and improved two-step growth (nucleation at 1100 °C, growth at 1250 °C). These results indicated that due to the limited migration rate of Ga atoms on the surface of 2D materials at low temperatures, a temperature of 580 °C was not conducive to the nucleation and growth of GaN on graphene. The line dislocation density of the GaN thin film grown by the improved two-step method was 1×10^9 cm^{-2}, which was comparable to the value obtained when AlN was used as a buffer layer on sapphire or SiC substrates (5×10^8 cm^{-2}–8×10^9 cm^{-2}). In addition, as shown in Fig. 3.6e–f, transmission electron microscopy (TEM) analysis of the GaN/graphene/SiC interface showed that after growth in the chamber under high temperature and complex gas conditions, graphene still existed at the interface position.

In addition, the surface morphology of 2D materials can be artificially altered, such as introducing steps and boundaries to control the growth of nitrides on 2D materials. Al Balushi et al. [33] introduced hydrogen gas to etch the surface of

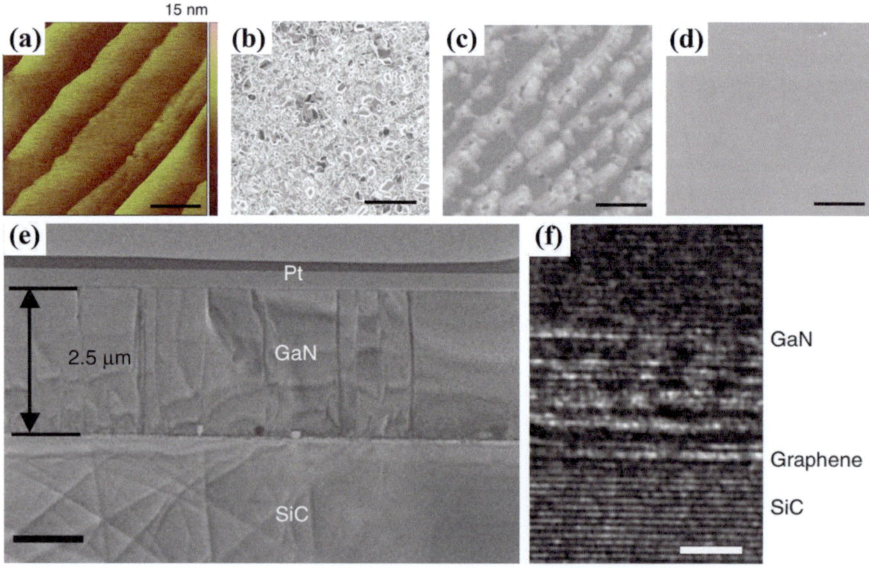

Fig. 3.6 Surface morphology of graphene/SiC substrate and GaN grown on graphene. **a** AFM image of the surface of a graphitized SiC substrate (scale bar, 10 mm). **b–d** Plan-view SEM images of GaN films grown on graphene (scale bars, 10 mm) by (**b**) the conventional two-step growth (nucleation at 580 °C and growth at 1150 °C), **c** one-step growth at 1100 °C, and **d** modified two-step growth (nucleation at 1100 °C and growth at 1250 °C). **e** Low-magnification cross-sectional TEM image of 2.5-mm-thick GaN grown on a graphene/SiC substrate (scale bar, 1 mm). **f** HRTEM taken at the interface of GaN/graphene/SiC (scale bar, 2 nm) [32]

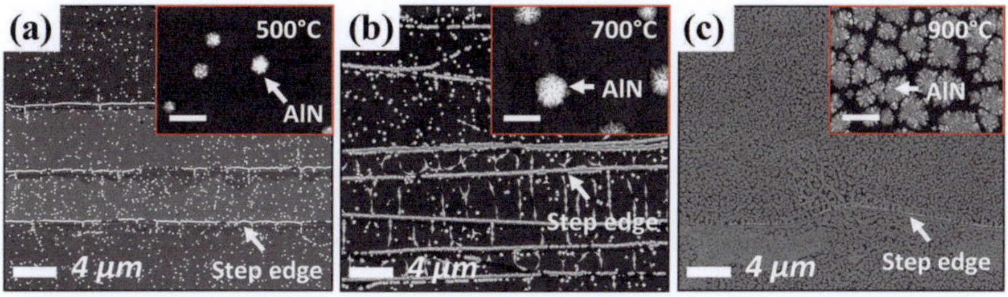

Fig. 3.7 SEM images following the nucleation of AlN at **a** 500 °C, **b** 700 °C and **c** 900 °C. Preferential nucleation of AlN on (110n) EG coated step edges is indicated in the image, where the insets illustrate the size and density of the AlN nuclei on the (0001) terrace region. The scale bar of the insets is 400 nm [33]

graphene, preparing a graphene template with high-density steps and wrinkles. As shown in Fig. 3.7a–c, the growth results showed that the nucleation of AlN mainly occured at the step edges of graphene, and the size of AlN nuclei increased with the increase of temperature, which was attributed to the increase of the migration rate of adsorbed atoms with the increase of temperature. In addition, the density of AlN nuclei on the graphene steps increased with the increase of nucleation temperature (the growth temperature increased from 500 to 900 °C). According to Lin et al.'s explanation, this phenomenon was related to the reduction of the surface energy of the materials [34], which affected the chemical activity of the surface of the 2D materials. Through surface potential force microscopy (SPFM) measurements, these locations mainly occured at the step edges of the SiC substrate covered by graphene.

When nitrides are directly deposited on graphene without any surface treatment, rough multi-faceted nitride clusters are formed on the surface. However, by pre-treating the intrinsic graphene with ammonia, a significant increase in the number of nitride nucleations can be observed. Yuxia Feng et al. had revealed that the nucleation of III-nitrides on graphene was triggered by the formation of sp^3 hybrid C-N bonds on graphene surface [35]. They achieved the epitaxy of continuous single crystalline GaN film on Si(100) by using single crystalline graphene and SiO_2 as interlayers. To reveal the nucleation mechanism of nitrides on sp^2-bonded graphene, they used XPS to identify the potential bonds between the nitrides and graphene. As shown in Fig. 3.8a, the N 1 s of AlN on graphene showed a broad peak in the binding energy (BE) range of 394–402 eV, which was deconvoluted into three individual peaks, located at 397.6, 398.7, and 399.6 eV, respectively. The peak at 397.6 eV is also observed for the AlN on conventional Si(111), as shown in Fig. 3.8b, which could originate from the oxidation of AlN. The peaks at 398.7 and 399.6 eV came from the nitrogen-doped graphene and did not correlate with the underneath layer. The peak at 398.7 eV may be assigned to pyridine-like N, corresponding to the sp^2 C-N bonding configuration, while the peak at 399.6 eV indicated the presence of pyrroline-like N corresponding to sp^3 C-N. After the NH^3 pretreatment of graphene, both sp^2 C-N and sp^3 C-N were formed, as shown schematically in Fig. 3.8c. Note that the pyrroline-like N in this work located at a little lower BE than that reported in previous work

(399.8–401.2 eV) [36]. That was mainly attributed to the growth of AlN on sp^3 C-N because the sp^3 C-N was more reactive than the sp^2 C-N. A three-component system C-N-Al leaded to a different bonding environment of N during the epitaxy process. Therefore, they believed that it was the NH^3 pretreatment of graphene that generated the sp^3 C-N and hence triggered the nucleation of the nitrides by the formation of C-N-Al.

Here, the surface morphology of the transferred graphene exhibited high-density of wrinkles in Fig. 3.8d due to the thermal mismatch between the CVD process and transfer process. After depositing AlN, the nucleation occured mainly along the wrinkles on the graphene, and there were only a few islands on the flat regions

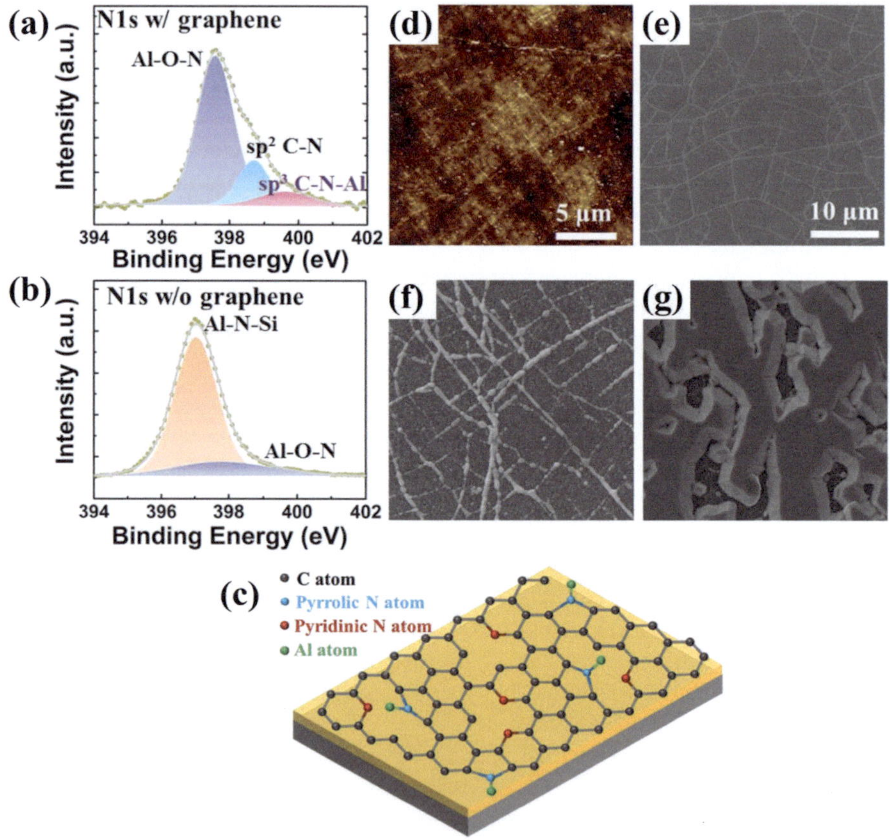

Fig. 3.8 Nucleation mechanism and epitaxy process of GaN film on Si(100) with single-crystalline-graphene/SiO_2 interlayers. XPS core level spectra with deconvolution of N 1 s of **a** AlN/graphene/SiO_2/Si(100) and **b** AlN/Si (111). **c** Schematic diagram of the transferred nitrided graphene, the gray, blue, red, and green spheres represent the C, pyrrolic N, pyridinic N, and Al atoms, respectively. **d**–**g** Surface morphology of each growth step, **d** AFM image of transferred graphene, **e** AlN nucleated on graphene, **f** initial growth of GaN on AlN/graphene, and **g** a 2.5 μm thick GaN grown with normal V/III

in Fig. 3.8e. Sequentially, at the initial growth stage of GaN (Fig. 3.8f), it also deposited on the AlN mainly along the wrinkles and formed GaN stripes. The GaN grown at conventional GaN growth conditions (V/III = 1500) showed an uncoalesced surface morphology with discrete stripes. In contrast, GaN grown at optimized growth conditions (V/III = 3000) exhibited continuous and smooth surface morphology, as shown in Fig. 3.8g. This could be attributed to the enhanced lateral growth rate at an increased V/III. Recently, Dongdong Liang et al. had also realized continuous and smooth GaN film on the amorphous SiO_2/Si(100) using transfer-free graphene as the interlayer and high temperature AlN as the seeding layer [37]. They found the out-of-plane orientation of GaN is guided by graphene and multi-step growth, while the in-plane orientation is aligned through the slip and rotation process of AlN nucleation islands due to the weakened qvdW interface interaction.

3.3 The Influence of Layer Number of Two-Dimensional Materials on the Nucleation of Quasi van der Waals Epitaxy

The surface characteristics of 2D materials largely depend on the number of layers, which will change the work function of the 2D material surface, surface potential, and oxidation characteristics of the 2D material surface, etc. It may ultimately affect the surface chemical activity of the 2D material, thereby changing the nucleation position of the nitride epilayer. On the other hand, 2D materials are usually used as an interlayer between nitride materials and substrates such as sapphire, silicon, silicon carbide, and glass, and their thickness usually does not exceed tens of nanometers. Therefore, when the 2D material reaches a certain thickness (usually less than four layers), it can shield the influence of the underlying substrate on the growth of the nitride.

In 2012, Choi et al. [30] first demonstrated the thickness of graphene could affect the quality of nitride epitaxial layers. In the case of direct growth of GaN on sapphire (Fig. 3.9a), the morphologies were rough, with the films not always being continuous on the whole sapphire surface. This can be ascribed to a 'wetting' problem between GaN and sapphire, due to both the lattice mismatch and the surface chemical differences. This problem can be drastically reduced by the use of an ultrathin graphene coating layer, which improved the wetting between the sapphire substrate and the overgrown GaN. With a nanocrystalline graphene coating layer as thin as ~ 0.6 nm, the overgrown GaN layer became continuous and flat in Fig. 3.9b. It was interesting to note that crystalline quality may further improved by increasing the thickness of the graphene coating layer. By insertion of a ~ 7.9 nm-thick layer of nanocrystalline graphene, a nearly surface pit-free, mirror-like GaN layer was achieved under the same growth conditions (Fig. 3.9c). The characteristic XRD θ–2θ diffraction pattern of the GaN/graphene/sapphire heterosystem was demonstrated, as shown in Fig. 3.9d, representing the single crystalline hexagonal structure of GaN on the graphene coating layer. In contrast, Zakaria Y. Al Balushi et al. studied the epitaxial growth of nitrides on multilayer epitaxial graphene prepared on the Si-polar face of 6H-SiC [33], and found that nitrides selectively nucleated on a few layers (≤ 3 layers) of epitaxial graphene, while they rarely nucleated on thick graphene layers (> 3 layers). The large difference in these experimental results is related to the adjustment of the pretreatment of the graphene surface before the epitaxial growth of the nitride thin film.

Besides MOCVD method, Sergio Fernández-Garrido et al. investigated the single-crystalline GaN nanowires on graphene by plasma-assisted MBE (PAMBE) [38]. Several types of epitaxial graphene layer structures were synthesized on SiC. PA-MBE enabled the growth of GaN nanowires (NWs) with typically smaller diameters than those attainable by MOCVD (down to 15–20 nm). Figure 3.10a–c, d–f showed schematic illustrations and atomic force micrographs, respectively, of the three types of graphene layer structures used in this work for the growth of GaN

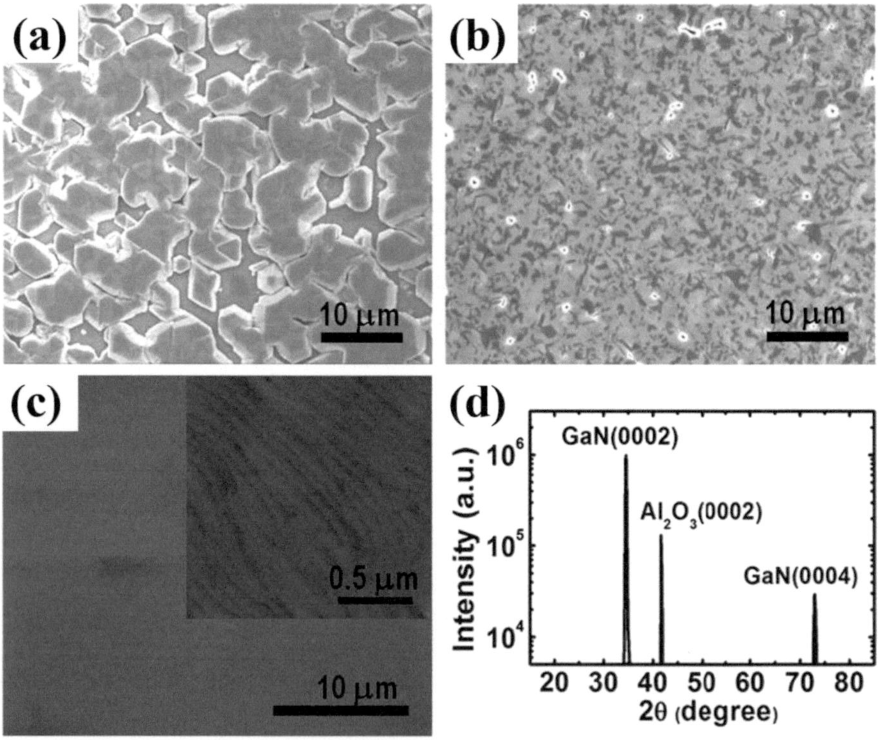

Fig. 3.9 Typical overgrown GaN layer morphologies on sapphire grown **a** without graphene coating layer, **b** with a 0.6 nm-thick and **c** with a 7.9 nm-thick graphene coating layers under the same growth conditions. The growth temperature and growth time of GaN layers were kept at 1040 _C and 1.3 h, respectively. The inset in **c** shows an AFM surface morphology of the GaN film grown on a 7.9 nm-thick graphene coating layer. **d** XRD data of GaN/graphene/sapphire heterosystem with a graphene coating thickness of 7.9 nm, identifying the grown material as the single crystalline GaN [30]

NWs. Graphene lied in this case on the top of the so-called buffer layer, which was a C layer isomorphic to graphene. Figure 3.10g, j presented bird's eye view SEMs with different magnifications of the GaN NWs grown on GLS I structure. For the growth conditions employed in this work, NWs only formed along the step edges, where the pristine substrate was covered by more than 1 layer of graphene in Fig. 3.10a. Hence, this type of substrate enabled the formation of well-separated rows of GaN NWs. The NWs were about 1 μm in length, exhibited a diameter of approximately 100 nm, and nucleated very close to each other leading to the formation of coalesced aggregates. For lower substrate temperatures than the one used in this work, they also observed GaN nucleation on the terraces but in the form of a compact and faceted layer, as those commonly reported when similar growth conditions were used on AlN-buffered SiC(0001) and Si(111) substrates. Interestingly, in the present case, the NWs formed at the step edges were highly tilted along a well defined direction in Fig. 3.10j. Since GaN NWs typically grew perpendicularly to the substrate normal, this result suggested that here NWs nucleated exclusively on the few-layer thick graphene regions created on the semipolar facets of the SiC substrate in Fig. 3.10a. It was worth noting that when GaN grew on SLG and BLG, the signal peak of graphene cannot be detected by Raman testing. When GaN was grown under the same conditions for NWs growth, SLG was etched away by combining with Ga or active N_2 in the early stage during PA-MBE growth. Correspondingly, MLG was etched during PA-MBE growth, forming new defect positions, which became the nucleation sites for GaN nanowire growth. Here, the synthesis of GaN NWs on epitaxial graphene does not only enable the fabrication of hybrid GaN/graphene devices on large area wafers without the technical inconveniences inherent to the graphene transfer process, but also improves the morphological and structural properties of the final NW ensembles.

In addition, 2D materials with different layers can also be used to shield the influence of the substrate on the growth of nitrides. Conventional nitride epitaxy usually does not insert a 2D material layer, but the growth of nitrides is affected by various factors such as lattice matching, thermal mismatch, and chemical properties of the substrate surface. After introducing the 2D material interlayer, the growth of nitrides is directly realized on the surface of the2D material, so how the initial substrate affects the growth of nitrides becomes a key issue. Wei Kong et al. reported that remote atomic interaction through 2D materials was governed by the binding nature [39], that was, the polarity of atomic bonds, both in the underlying substrates and in 2D material interlayers. They proposed the concept of RE. They compared the remote interaction penetration of Si, GaAs, GaN and LiF, increasing ionic bonding characters of 0%, 31%, 50% and 90%, respectively. With Density functional theory (DFT), they calculated the electrostatic potential fluctuation transmitted from Si, GaAs, GaN and LiF substrates through 1 ML-3 ML graphene based on the atomic models in Fig. 3.11a. The electrostatic potential fluctuation of the Si and GaAs substrates through 2 ML graphene was negligible in Fig. 3.11b, c. The potential fluctuation of the GaN substrate can effectively penetrate through 2 ML graphene, but it is significantly dampened by 3 ML. Therefore, single-crystalline GaN was obtained with a 2 ML graphene interlayer but not with 3 ML graphene and single-crystalline LiF was obtained even with a 3 ML graphene interlayer. The surface morphologies

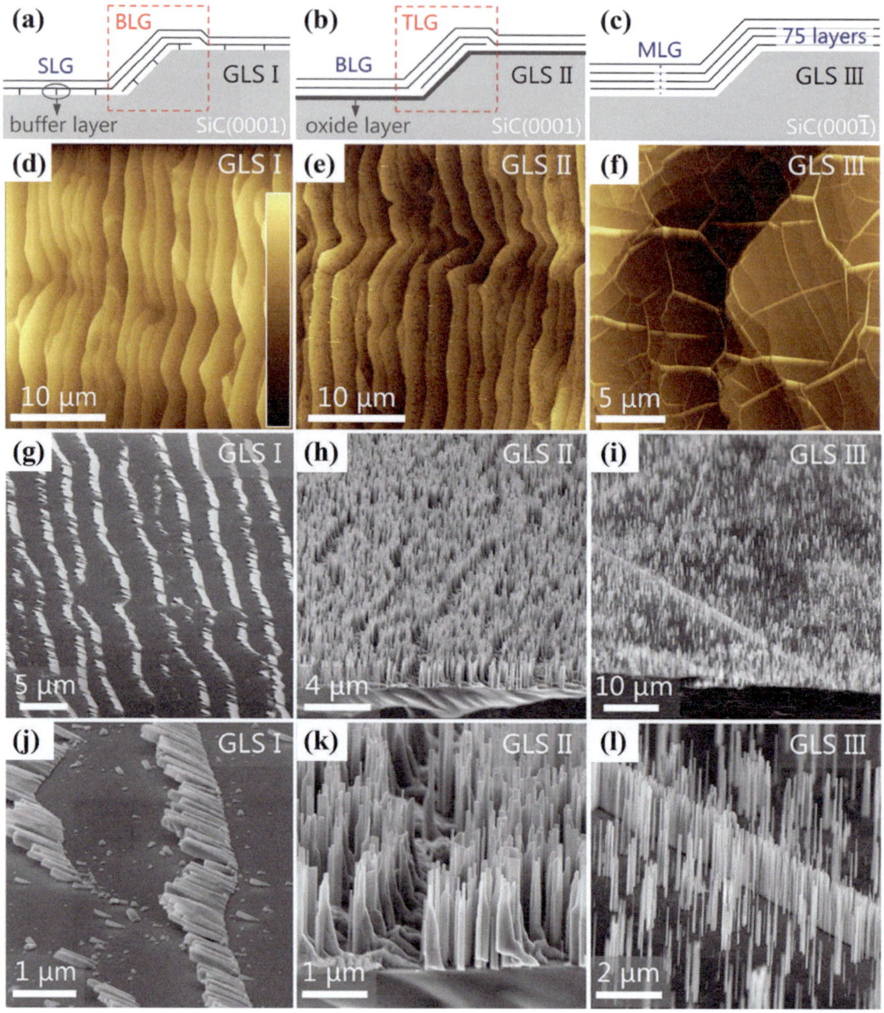

Fig. 3.10 Sketches of the **a** Type I, **b** II, and **c** III epitaxial graphene layer structures (GLS) prepared on 6H-SiC substrates. SLG, BLG, TLG, and MLG stand for single-layer, bilayer, trilayer, and multilayer graphene, respectively. **d–f** AFMs of the corresponding pristine graphene layer structures. The linear color-code scale bar included in **d** represents 7, 15, and 60 nm in figures **d**, **e** and **f**, respectively. (g – l) Bird's eye view SEMs of the GaN NW ensembles prepared on the three different types of graphene layer structures, as indicated in the images [38]

of Si, GaAs, GaN and LiF grown on 1 ML-3 ML graphene were shown in Fig. 3.11d. Figure 3.11e, f quantitatively summarized the potential fluctuation penetration of the Si, GaAs, GaN and LiF substrates as a function of distance (number of graphene layers), and clearly showed the enhanced penetration of the electrostatic potential from the substrates through graphene by increasing ionic bond characters.

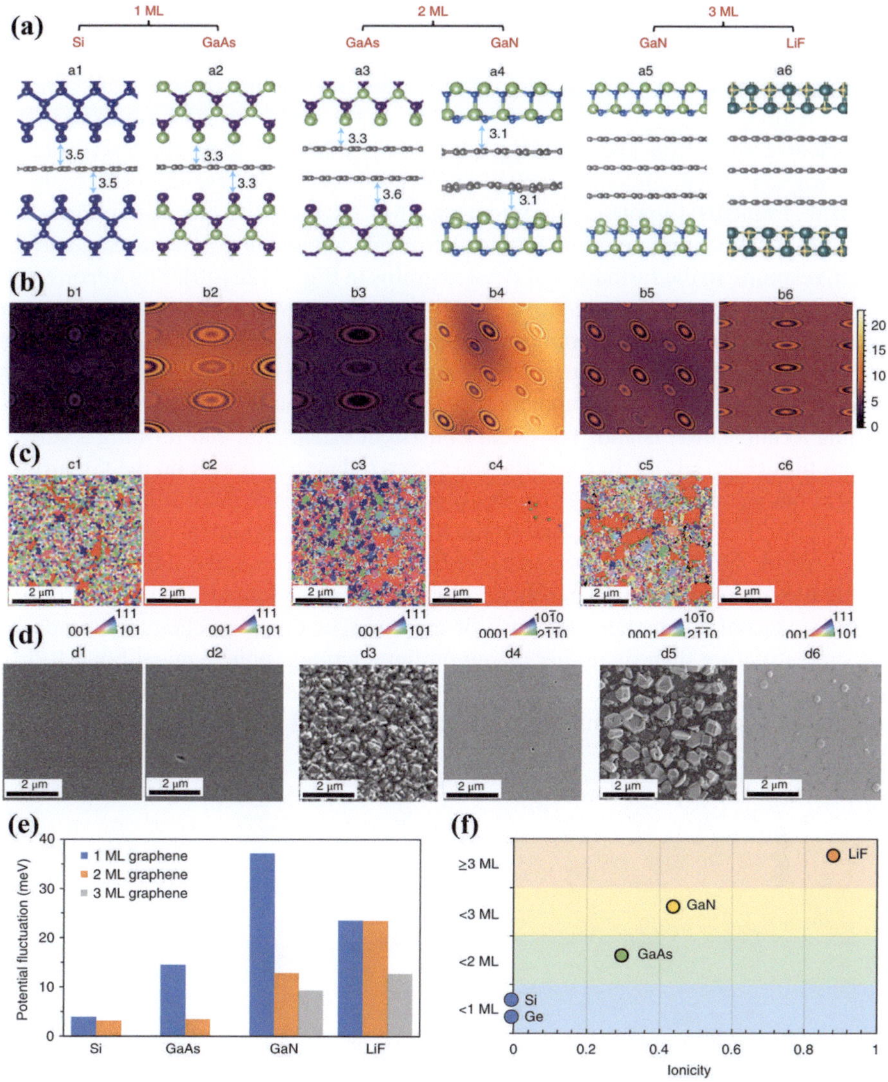

Fig. 3.11 Penetration distance of the potential fluctuations from the Si, GaAs and GaN substrates. **a** DFT-modelled atomic structures of Si, GaAs, GaN and LiF on 1-ML-, 2-ML (bilayer)- or 3-ML (trilayer)-graphene-coated Si, GaAs, GaN and LiF, respectively. **b** Maps of potential fluctuation on the same scale (0–25 meV) for cross comparison. **c** Electron backscatter diffraction (EBSD) of released surfaces. **d** SEM morphology of as-grown surfaces. Within the same column of (**a**–**d**), the atomic structure remains identical, being Si/1-ML-graphene/Si for (a1), (b1), (c1) and (d1), GaAs/1-ML-graphene/GaAs for (a2), (b2), (c2) and (d2), GaAs/2-ML-graphene/GaAs for (a3), (b3), (c3) and (d3), GaN/2-ML-graphene/GaN for (a4), (b4), (c4) and (d4), GaN/3-ML-graphene/GaN for (a5), (b5), (c5) and (d5) and LiF/3-ML-graphene/LiF for (a6), (b6), (c6) and (d6). **e** Comparison of the effective potential energy fluctuation on 1-ML-, 2-ML- and 3-ML-graphene-coated substrates. **f** The schematic of the remote interaction penetration depth depending on ionicity across groups IV, III-V and I-VII materials showed that graphene transparency increased with material ionicity [39]

It is worth noting that 2D materials with polar bonds have better ability to shield the substrate electric field than graphene. However, the binding energy fluctuation of GaN on h-BN increased by fivefold in comparison to that on graphene. Such a strong binding energy fluctuation for GaN nuclei allowed 3 nm thick epitaxial h-BN (sufficiently away from the substrate field) to seed the growth of single-crystalline GaN, whereas thick epitaxial graphene failed to host single-crystalline GaN. Fang Liu et al., found the remote interaction from 3 ML graphene/III-nitride templates can constrain the out-of-plane orientation of III-nitride nuclei but not the in-plane orientation, resulting in the formation of twisted grains in Fig. 3.12a [40]. The misoriented grains expanded and formed a polycrystalline film, which was confirmed by planar scanning transmission electron microscopy (STEM) measurement in Fig. 3.12c. On the other hand, qvdWE III-nitride films can establish an epitaxial relationship with irradiated 3 ML graphene through interfacial covalent bonds as shown in Fig. 3.12b, leading to nucleation islands with near-consistent orientation and forming a single-crystal film (Fig. 3.12d). In this case, unsaturated dangling bonds on the Gr surface acted as nucleation sites, triggering highly oriented III-nitride nucleation. The importance of interfacial chemical bonds in the formation of the epitaxial registry between the III-nitride film and graphene was emphasized. In this case, the single-crystal III-nitride film grew on wet-transferred graphene and was no longer strictly constrained by the underlying substrate. Because of the stronger effect of chemical bonds on the superstructure than the effect of remote interactions (nonchemical bond form), qvdWE is more inclusive in the choice of graphene and substrate materials than RE, making it easier to achieve high-quality transferrable III-nitride films.

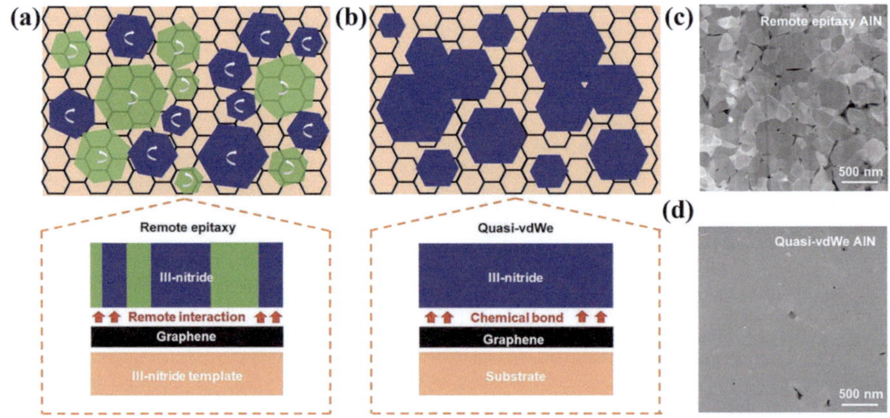

Fig. 3.12 Schematic illustration of III-nitride films grown on 3 ML wet-transferred graphene. **a** RE of III-nitrides on graphene/III-nitride templates. **b** qvdWE of III-nitrides on single-crystal graphene. **c** Planar HAADF-STEM image of AlN grown on a wet-transferred one-ML graphene/AlN(0001) template by RE. **d** Planar HAADF-STEM image of AlN film grown on a wet-transferred 3-ML graphene/AlN(0001) template by QvdWE [40]

3.4 Nucleation Interface Treatment Method of Quasi van der Waals Epitaxy Nitride

The regulation of the in-plane orientation and crystal quality of nitride growth on 2D materials becomes a key issue that needs to be resolved for nitride qvdWE growth. In addition, the chemical inertness and low migration rate of metal atoms on 2D materials are also the main obstacles. The previous section systematically presented the influence of the surface morphology of 2D materials (including surface defects, the number of layers of 2D materials) on qvdWE. To solve these problems, different buffer layers have been proposed on 2D materials in recent years, such as ZnO nanowalls, AlN buffer layers, MOCVD in-situ NH_3 treatment, O_2 plasma and N_2 plasma treatment of graphene surfaces, etc. By improving the chemical reactivity of the surface of 2D materials, the nucleation sites are significantly increased, therefore improving the nitride crystal quality. This section will elaborate on several commonly used growth treatment methods for nitride qvdWE.

3.4.1 Oxygen Plasma Treatment

The biggest challenge for nitride qvdWE is that surface characteristics of 2D materials without dangling bond inhibit the nucleation of nitrides. In order to promote better epitaxial growth of nitrides on 2D materials, plasma treatment is usually used to enhance the surface activity. Here, the plasma source includes oxygen and nitrogen, etc. In early 2010, Chung et al. [17] first studied the growth of nitrides on graphene and found that GaN can grow directly on untreated graphene, but only polycrystalline GaN was obtained. As a solution, the substrate can be treated with oxygen plasma, and then high-density ZnO nanowires can be grown on the graphene layer treated with oxygen plasma. Although oxygen plasma treatment can enhance the surface activity of graphene, it is still difficult to directly grow high-quality GaN crystal on 2D materials [41]. Therefore, ZnO nanoparticles with high nucleation density are introduced as the intermediate layer for subsequent GaN growth.

To further study the effect of oxygen plasma treatment on graphene, Al Balushi et al. [33] used Raman spectroscopy to analyze the change in graphene. They chose epitaxially grown graphene on (0001) 6H-SiC substrates, which were treated with O_2 plasma for 90 s under 100 W and 500 mTorr conditions to increase the point defect density in graphene. The point defect density in the graphene layer can be determined by the ratio of the Raman intensities of the D peak and G peak of graphene, i.e., $I_{(D}/I_{(G)}$. The increase in graphene defects after plasma treatment would lead to an increase in the $I_{(D)}/I_{(G)}$ ratio. Here, the G peak in graphene corresponded to the high-frequency E_{2g} phonon vibration of sp^2 hybridized carbon atoms, while the D peak corresponded to the vibration of the disordered sp^2 hybridized graphene lattice caused by defects. After introducing defects on the graphene surface, the nucleation sites of AlN were evenly distributed on the graphene surface. Here, a ~ 200 × increase in graphene defect density correlateed strongly with the significant increase in AlN nucleation density across graphene.

In addition, to improve the crystal quality of the nitride epitaxial layer, Li et al. [42] introduced highly ordered pyrolytic graphene (HOPG), a new type of graphene material prepared by pyrolytic graphite under high temperature and high pressure, with the performance similar to single crystals. After treating graphene with oxygen plasma, AFM analysis showed that the roughness of the treated graphene was significantly higher than that of the untreated graphene, and the root-mean-square roughness (RMS) increased from 0.28 nm before treatment to 0.39 nm after treatment, indicating an increase in defects on the graphene surface. GaN nuclei and islands were grown on the graphene surface and then annealed at high temperatures. Notably, the GaN nuclei were only decomposed and not recrystallized, as most of the annealed GaN nuclei and islands cannot reach the critical size for Ostwald ripening. Generally, high temperatures have a trade-off effect on the nucleation process. While high temperatures can promote the migration of atoms, it is known that the nucleation rate (dN/dt) decreases at high temperatures according to the nucleation rate formula:

$$\frac{dN}{dt} \infty \exp\left[\frac{E_{\rm d} - E_{\rm s} - \Delta G^*}{kT}\right] \tag{3.2}$$

where N is the number density of growing centers, E_d is adsorption energy, E_S is activation energy for migration, ΔG^* is the nucleation barrier, T is the absolute temperature, and k is the Boltzmann constant [43]. Here, high-temperature conditions reduced the sticking coefficient of graphite. The low nucleation rate and sticking coefficient at high temperatures played a determinate role in the nucleation stage, preventing the GaN nuclei from forming on the graphite surface. According to the nucleation rate formula, the nucleation density can be increased at high temperatures by increasing the adsorption energy and reducing the migration barrier between the nucleation layer and graphene. Therefore, due to the high adsorption energy of Al atoms and the low migration barrier of graphene, AlGaN and AlN may be thought more likely to nucleate on the graphene surface.

In addition to using graphene, h-BN is also a commonly used 2D material for nitride QvdWE. Oh et al. [44] treated the h-BN surface with oxygen or argon plasma to form hydrogen dangling bonds on its surface, thereby promoting the subsequent nucleation of ZnO. The results showed that plasma treatment of the h-BN substrate had a significant effect on the growth of ZnO, and selective growth can be achieved by special treatment of different areas of the substrate, to control the shape and position of ZnO nanotubes. In addition, DFT calculations showed that the interaction energy ($\Delta E = E_{total} - E_{ZnO} - E_{h\text{-}BN}$) had a value (2.170 $eVnm^{-2}$ or 2.165 $eVnm^{-2}$) that was 1–2 orders of magnitude smaller than the typical covalent bond binding energy. The equilibrium distance between ZnO and h-BN was 3.4 Å, which was greater than the typical covalent or ionic bond distance. Further theoretical analysis did not find that the ZnO/h-BN epitaxial heterostructure demonstrated the main chemical bonding characteristics of large interfacial spacing and small binding energy.

Wu et al. [45] also studied the growth of AlN thin films on h-BN buffer. A single layer of h-BN was grown on copper foil by LPCVD, and transferred to a sapphire substrate. Then the h-BN on the sapphire was treated with oxygen plasma under 100 W radio frequency (RF) power for 3 min, and subsequent AlN epilayer was grown on the treated substrate. As shown in Fig. 3.13a–c, the nucleation of AlN on h-BN/sapphire, O_2 plasma-treated h-BN/sapphire, and sapphire substrate were compared. The nucleation morphology of AlN on O_2 plasma-treated h-BN/sapphire was similar to that on the sapphire substrate, but there were basically no nucleation points on h-BN/sapphire. This confirmed that AlN preferred nucleating on h-BN-O_2 rather than the h-BN layer, and the nucleation energy of AlN on h-BN-O_2 was lower than that on h-BN at low temperature. Al and N atoms preferred moving to the sites with the dangling bonds rather than those without the dangling bonds. Figure 3.13d–i demonstrated the process of AlN film growth on sapphire substrate and O_2 plasma-treated h-BN/sapphire substrate. It can be seen that after oxygen plasma treatment, the defects on the h-BN surface were conducive to the uniform nucleation of AlN on its surface. Initial nucleation islands grew longitudinally from the nucleus, merged horizontally, and finally expanded as a smooth and continuous thin film. The crystal quality of AlN on O_2 plasma-treated h-BN/sapphire was characterized by XRD. Based the oxygen plasma treatment, the full width at half maximum (FWHM) of (0002) and (10–12) of AlN grown on h-BN-O_2 were 409 arcsec and 963 arcsec, respectively, which were comparable to the values of AlN thin films directly grown on sapphire.

Based on the above experimental results, the rapid coalescence of the AlN film with the h-BN interlayer indicated a fast migration of the coming adatoms on h-BN. To investigate the effects of the h-BN interlayer on the initial nucleation of AlN, Lulu Wang et al. conducted first-principles calculations to study the adsorption and diffusion of Al and/or Natoms on h-BN monolayer [46]. The Al atom can adsorb above the hexagonal ring (H site), on top of the B atom (B_t site), and on top of the N atom (N_t site). As shown in Fig. 3.14a–c, the Al adsorption energy on B_n was − 0.45 eV, obviously stronger than those on the nearby B_t and H sites. Two more stable adsorption sites for the Al atom were the N_n1 and N_n2 sites near the top of the N atom with adsorption energies around − 0.48 eV. In contrast, the adsorption energies of the N atom on different sites were quite small (around − 0.15 eV). O_2 treatment can introduce O-related defects on h-BN and thus enhanced the adsorption of Al atoms, increasing the nucleation density of AlN. As shown in Fig. 3.14d, e, the largest energy barrier corresponded to the diffusion from the N_n2 to B_t site, which was calculated to be 0.14 eV. The Al diffusion barriers on h-BN were slightly larger than those on the graphene monolayer ($<$ 0.1 eV), but was much smaller than that on the bare sapphire substrate ($>$ 1 eV). The h-BN provided fast diffusion pathways and allowed for rapid migration of Al atoms. These fast diffusive Al atoms can quickly join in existing AlN nucleation centers on the h-BN surface, facilitating the fast lateral growth of the AlN film.

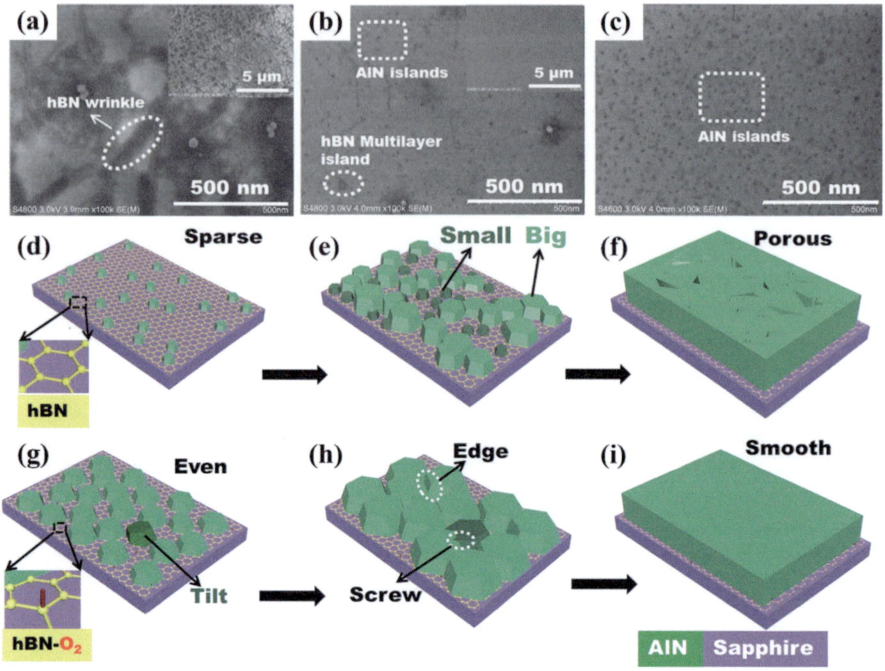

Fig. 3.13 SEM images of 2 min LT-AlN on **a** h-BN/sapphire, **b** h-BN-O_2/sapphire and **c** sapphire. Schematic illustration of the growth mechanism of AlN materials on h-BN/sapphire and h-BN-O_2/sapphire: **d** nucleation of LT-AlN, **e** nucleation of HT-AlN and **f** formation of the AlN films on h-BN/sapphire; **g** nucleation of LT-AlN, **h** nucleation of HT-AlN and **i** formation of the AlN films on h-BN-O_2/sapphire [45]

3.4.2 Nitrogen Plasma Treatment

Since nitrogen itself is a component of nitride materials, nitrogen plasma treatment has more advantages in growing high-quality nitride materials on 2D materials. In nitrogen plasma treatment, the impact of nitrogen atoms on the surface of 2D materials is crucial. Doping of pristine graphene can change its chemical and electrical properties. For example, Maxim Rybin et al. [47] reported the synthesis of transferred graphene on copper foil, and then transferred the graphene onto SiO_2/Si substrate for oxygen plasma treatment. The graphene samples treated in the RF discharge with an effective RF power of 10 W were carefully studied by SEM and Raman spectroscopy. After plasma treatment, the number of defects on the graphene surface significantly increased. Figure 3.15a showed the comparison of Raman spectra before and after plasma treatment of graphene films at different times (from 2 to 20 min). It can be seen that the D peak around 1350 cm^{-1} increased several times after plasma treatment, and it was not affected by the treatment time. Another peak was around 1617 cm^{-1}, called D′, which also corresponded to defects in the graphene lattice, and it increased with the increase of treatment time. Interesting results have been achieved after a

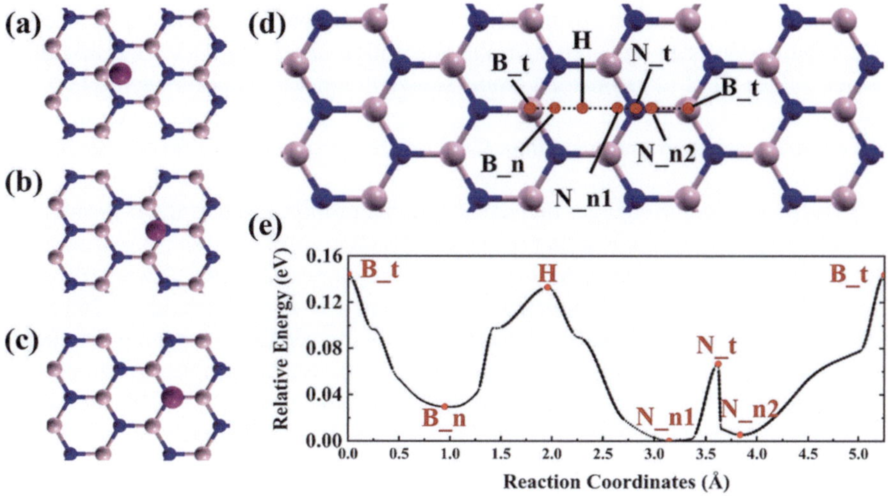

Fig. 3.14 Top views of the Al atom adsorbed on the **a** B_n site, **b** N_n1 site, and **c** N_n2 site on the pristine h-BN monolayer. **d** Adsorption sites and diffusion path for the Al atom on the h-BN monolayer. **e** Energy barriers of one Al atom along the diffusion path in panel **d** on the h-BN monolayer [46]

long-time treatment (more than 20 min) because in this case the amount of nitrogen atoms was fixed while the doping type was changed. In Fig. 3.15b, the XPS spectra of N1s peak were presented. The study of N1s line with binding energy around 400 eV in XPS spectra of samples treated with different times had revealed a redistribution of intensities of all components of the band during a long-time treatment. The peaks with energies 398.7 ± 0.2 eV and 402.0 ± 0.2 eV corresponding to pyridinic-N and quaternary nitrogen or graphitic-N accordingly decreased, while the other component with energies 400.1 ± 0.2 eV, corresponding to a pyrrolic-N increased. It was caused by a strong influence of ammonia radicals (such as NH_2 and NH), created in plasma on graphene film. The SEM images of the graphene film before and after plasma treatment also confirmed the damage of graphene as shown in Fig. 3.15c, d. Here, the dark area corresponded to graphene, and the bright area corresponded to the substrate. Some cracks were observed on the graphene surface before plasma treatment, and their number significantly increased after plasma treatment. Zhaolong Chen et al. found the similar phenomenon, which showed that after nitrogen plasma treatment, the content of pyrrole-N in graphene was 5.6%, while the XPS spectrum of untreated graphene did not show the corresponding peak [48]. The growth process of nitrides on graphene after nitrogen plasma treatment is roughly as follows: First, there are low-energy Al atoms and high-energy Al atoms on the sapphire surface, and the binding strength of low-energy Al atoms with the underlying O atoms is larger. According to DFT calculations [49], Al atoms had two adsorption positions on the sapphire substrate, high-energy Al and vacancies, with adsorption energies of 2.69 eV and 1.67 eV, respectively. When Al atoms moved from the low-energy

Al atom position to other positions, they must overcome an energy difference of 1.02 eV. In addition, compared with the adsorption energy (1.5 eV) and migration barrier (0.05 eV) of Ga atoms, Al atoms showed higher adsorption energy (1.7 eV) and lower migration barrier (0.03 eV) on the graphene surface, indicating that AlN was easier to grow on graphene than GaN, and they preferred to form large atomic nuclei.

To further study the changes in the nucleation interface, Chen et al. [48] conducted a detailed study of the nucleation process of AlN on graphene. After N_2 plasma treatment of graphene, the defect density in graphene increased from 2.13×10^{11} cm^{-2} to 3.23×10^{11} cm^{-2}. The XPS spectra showed a new N-sp^3 C peak (about 286.5 eV) and the formation of pyrrole-N. DFT calculations showed that the adsorption energy of Al atoms on the pyrrole-N formed on the plasma-treated graphene significantly increased, reaching 5.9–8.6 eV, while the adsorption energy at the wrinkles and steps of intrinsic graphene was only about 1.1 eV. Therefore, the reactivity of the

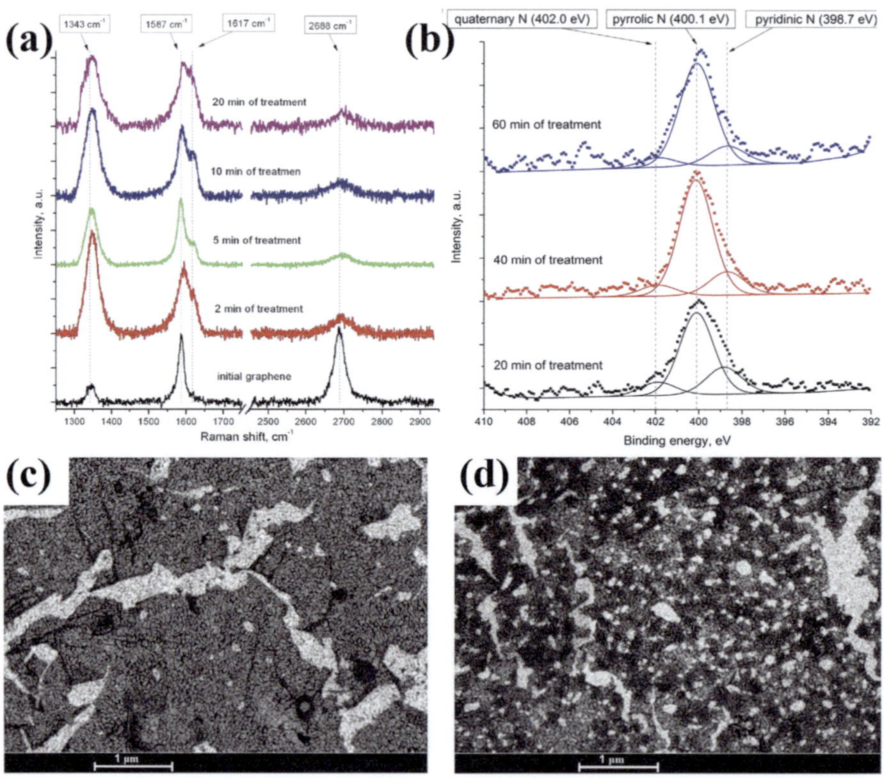

Fig. 3.15 **a** The Raman spectra of graphene film before and after the plasma treatment with different times. **b** The XPS measurements of N1s band in graphene films treated by ammonia plasma during different times. The UPS spectra of valence band of initial graphene and N-doped graphene after 20 min of plasma treatment. The SEM images of graphene before plasma treatment (**c**) and after 20 min of plasma treatment (**d**) [47]

graphene surface to AlN nucleation significantly increased. In addition, the adsorption of an Al atom partially restored the honeycomb structure near the N defect, but introduced significant structural distortion. To experimentally verify the superiority of AlN growth on N_2 plasma-treated graphene films, AlN nucleation experiments were conducted on sapphire, graphene/sapphire, and N_2 plasma-treated graphene/sapphire substrates under the same growth conditions. The results showed that, as shown in Fig. 3.16a–c, the AlN nucleation on the N_2 plasma-treated graphene was continuous. The nucleation density increased by nearly an order of magnitude, and the size of the nucleus islands significantly decreased. In contrast, on bare sapphire and untreated graphene/sapphire substrates, the AlN grains were larger and sparser. This was because AlN nucleation mainly occured at N-defect sites, which greatly promoted the 2D lateral growth of AlN islands by reducing the diffusion barrier of atoms on ideal graphene. Figure 3.16d presented schematic diagram of the nucleation and film growth of AlN on N_2 plasma-treated graphene/sapphire substrate. Thus, AlN islands can form a continuous film through 2D lateral growth due to the reduced diffusion barrier of adatoms on the ideal graphene. In summary, N_2 plasma treatment introduced more defect sites on the graphene surface, which can provide entry points for the lateral expansion of the metal layer. Therefore, in the process of GaN lateral growth, the nucleation layer showed perfect continuity, eventually growing into a continuous and high-quality nitride film.

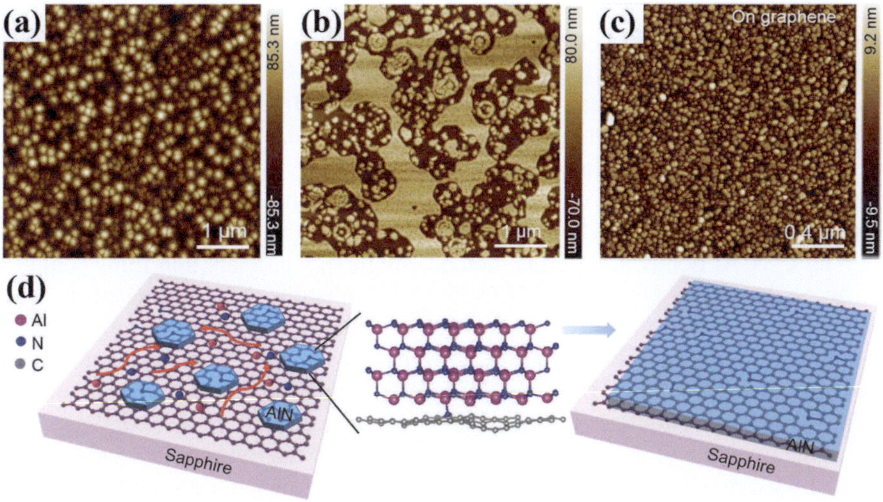

Fig. 3.16 AFM morphologies of AlN nucleation layers grown on **a** bare sapphire, **b** graphene/sapphire and **c** plasma treated-graphene/sapphire, respectively. **d** Schematic diagram of the nucleation and film growth of AlN on N_2 plasma-treated graphene/sapphire substrate [48]

3.4.3 NH_3 In-Situ Treatment

In addition to nitrogen and oxygen plasma treatment, NH_3 can also be used to treat the surface of 2D materials. The purpose of these treatments is usually to increase the surface defect density of 2D materials, that is, to enhance the surface chemical activity of 2D materials. One advantage of NH_3 treatment on the surface of 2D materials is that the gas in the MOCVD equipment chamber for nitride growth includes NH_3, so it can be pre-treated directly in the growth chamber to reduce contamination of the substrate. Al Balushi et al. [33] observed that the AlN nucleation density on the graphene surface grown at high temperature significantly increased, due to the introduction of defects into the graphene lattice when the 2D material was exposed to high-temperature NH_3 growth conditions. Usually, through XPS analysis, the observed sp^3 C-N bond indicated that a chemical reaction had occurred between graphene and AlN.

Subsequently, Heilmann et al. [50] grew single-crystal GaN nanorods on monolayer graphene using nano-sized AlGaN nucleation islands through MOCVD, and performed a nitridation step with NH_3 at high temperatures before the growth of the nanorods. Specifically, DFT calculations predicted chemical adsorption-like bonding for N atoms on graphene with a distance smaller than 2 Å, while for larger distances weaker physical adsorption-like bonding for adatoms was expected (e.g., for Al and Ga). In addition, as mentioned earlier, Al atoms showed higher adsorption energy and migration energy on graphene. On Si substrate covered with graphene, the density of AlGaN nuclei can be as high as about 1.2×10^{10} cm^{-2}, which was beneficial for the subsequent growth of GaN nanorods. Here, AlGaN nucleation islands on graphene enabled the vertically aligned, single crystalline growth of GaN NRs on different supporting substrates. Graphene as an atomically thin buffer layer can also prevent alloying between Ga and Si during the MOVPE growth of GaN NRs on Si. Furthermore, Feng et al. [35] also reported that graphene NH_3 pretreatment helped to form sp^3 hybridized C-N bond, thereby promoting the nucleation of nitrides by forming C-N-Al bonds. Using this method, single-crystal GaN thin films could be grown on single-crystal graphene/SiO_2/Si (100) substrate. The graphene thin layer at the interface can be characterized by Electron Energy Loss Spectroscopy (EELS). EELS showed that the O k-edge signal only appeared below C (graphene), and the N k-edge signal only appeared above C (graphene), implying that there was no mutual diffusion between graphene during the epitaxial process. The atomic bonding situation at the interface of nitride epitaxial growth on 2D materials can be characterized and analyzed by XPS, which helped to deeply analyze the nucleation mechanism of qvdWE on 2D materials and the theoretical mechanism of interface bonding.

3.5 Summary

Nitride materials, as an important third-generation semiconductor material, are primarily grown on hetero-substrate due to the lack of intrinsic substrate. This results in a poor crystal quality that is far inferior to mature Si substrate. The main reasons lie in the large lattice mismatch and thermal expansion coefficient mismatch faced during the heteroepitaxial growth of nitrides. Typically, a LT nitride buffer layer method is used for epitaxial growth, which further leads to a high density of dislocations and residual stress in the film during the epitaxial process, ultimately affecting the device performance of the nitrides. 2D materials usually have a thickness of one or more atoms layers, in which atoms within the layer connect by covalent bonds and layer-by-layer connects by weak vdW forces. Using qvdWE of nitrides, the strain of nitride epitaxial films can be reduced and the epitaxial quality can be improved by using 2D materials as buffer layers.

A lot of work has been done on the qvdWE of nitrides on different hetero-substrates with 2D materials as buffer layers, but great challenges still remain. The migration barrier of atoms on the surface of 2D materials is low, so their migration rate on the surface of 2D materials is high, which can easily lead to the growth of 2D smooth films during the growth process. The nucleation stage has a crucial impact on the subsequent film growth. Therefore, the nucleation process is regulated by studying the nucleation mechanism of qvdWE of nitrides on the surface of 2D materials. The effects of the surface morphology of 2D materials and the number of layers of 2D materials on the nucleation of qvdWE of nitrides are also studied. The issue of difficult nucleation is solved by using 2D material interface treatment methods. The qvdWE of high quality nitrides is ultimately achieved by optimizing the nucleation epitaxy process.

In conclusion, nitride materials have broad application prospects in the field of semiconductor devices. High-quality nitride film growth is expected to be achieved through qvdWE of nitrides on 2D material buffer layers, enabling more functionalized semiconductor electronic devices. This has a significant impact on the future development of the semiconductor industry and will also promote the progress of science and technology and economic development in our country.

References

1. Ryu H, Park H, Kim JH, Ren F, Kim J, Lee GH, Pearton SJ (2022) Two-dimensional material templates for van der Waals epitaxy, remote epitaxy, and intercalation growth. Appl Phys Rev 9:031305
2. Yamada A, Ho KP, Akaogi T, Maruyama T, Akimoto K (1999) Layered compound substrates for GaN growth. J Cryst Growth 201(202):332–335
3. Liu L, Edgar JH (2002) Substrates for gallium nitride epitaxy. Mater Sci Eng R 37:61–127
4. Li G, Wang W, Yang W, Lin Y, Wang H, Lin Z, Zhou S (2016) GaN-based light-emitting diodes on various substrates: a critical review. Rep Prog Phys 79:056501

5. Wei T, Yang J, Wei Y, Huo Z, Ji X, Zhang Y, Wang J, Li J, Fan S (2016) Cross-stacked carbon nanotubes assisted self-separation of free-standing GaN substrates by hydride vapor phase epitaxy. Sci Rep 6:28620
6. Kum H, Lee D, Kong W, Kim H, Park Y, Kim Y, Baek Y, Bae S, Lee K, Kim J (2019) Epitaxial growth and layer-transfer techniques for heterogeneous integration of materials for electronic and photonic devices. Nat Electron 2:439–450
7. Chen Y, Hong S-K, Ko H-J, Kirshner V, Wenisch H, Yao T, Inaba K, Segawa Y (2001) Effects of an extremely thin buffer on heteroepitaxy with large lattice mismatch. Appl Phys Lett 78:3352–3354
8. Morelli DT, Heremans JP, Slack GA (2002) Estimation of the isotope effect on the lattice thermal conductivity of group IV and group III-V semiconductors. Phys Rev B 66:195304
9. Geim AK, Grigorieva IV (2013) van der Waals heterostructures. Nature 499:419–425
10. Ci L, Song L, Jin C, Jariwala D, Wu D, Li Y, Srivastava A, Wang ZF, Storr K, Balicas L, Liu F, Ajayan PM (2010) Atomic layers of hybridized boron nitride and graphene domains. Nat Mater 9:430–435
11. Zhao W, Ghorannevis Z, Amara KK, Pang JR, Toh M, Zhang X, Kloc C, Tan PH, Eda G (2013) Lattice dynamics in mono- and few-layer sheets of WS_2 and WSe_2. Nanoscale 5:9677–9683
12. Koma A, Sunouchi K, Miyajima T (1984) Fabrication and characterization of heterostructures with subnanometer thickness. Microelectron Eng 2:129–136
13. Chen Q, Yang K, Liang M, Kang J, Yi X, Wang J, Li J, Liu Z (2023) Lattice modulation strategies for 2D material assisted epitaxial growth. Nano Convergence 10:39
14. Choi JH, Kim J, Yoo H, Liu J, Kim S, Baik C-W, Cho C-R, Kang JG, Kim M, Braun PV, Hwang S, Jung T-S (2016) Heteroepitaxial growth of GaN on unconventional templates and layer-transfer techniques for large-area, flexible/stretchable light-emitting diodes. Adv Opt Mater 4:505–521
15. Kouno T, Kishino K, Yamano K, Kikuchi A (2009) Two-dimensional light confinement in periodic InGaN/GaN nanocolumn arrays and optically pumped blue stimulated emission. Opt Express 17:20440
16. Kishino K, Sekiguchi H, Kikuchi A (2009) Improved Ti-mask selective-area growth (SAG) by rf-plasma-assisted molecular beam epitaxy demonstrating extremely uniform GaN nanocolumn arrays. J Cryst Growth 311:2063–2068
17. Chung K, Lee CH, Yi GC (2010) Transferable GaN layers grown on ZnO-coated graphene layers for optoelectronic devices. Science 330:655–657
18. Kobayashi Y, Kumakura K, Akasaka T, Makimoto T (2012) Layered boron nitride as a release layer for mechanical transfer of GaN-based devices. Nature 484:223–227
19. Saidi WA (2015) Trends in the adsorption and growth morphology of metals on the MoS_2(001) surface. Cryst Growth Des 15:3190–3200
20. Ai R, Guan X, Li J, Yao K, Chen P, Zhang Z, Duan X, Duan X (2017) Growth of single-crystalline cadmium iodide nanoplates, CdI_2/MoS_2 (WS_2, WSe_2) van der Waals heterostructures. ACS Nano 11:3413–3419
21. Qi Y, Wang Y, Pang Z, Dou Z, Wei T, Gao P, Zhang S, Xu X, Chang Z, Deng B, Chen S, Chen Z, Ci H, Wang R, Zhao F, Yan J, Yi X, Liu K, Peng H, Liu Z, Tong L, Zhang J, Wei Y, Li J, Liu Z (2018) Fast growth of strain-Free AlN on graphene-buffered sapphire. J Am Chem Soc 140:11935–11941
22. Chang H, Chen Z, Liu B, Yang S, Liang D, Dou Z, Zhang Y, Yan J, Liu Z, Zhang Z, Wang J, Li J, Liu Z, Gao P, Wei T (2020) Quasi-2D growth of aluminum nitride film on graphene for boosting deep ultraviolet light-emitting diodes. Adv Sci 7:2001272
23. Xiao N, Dong X, Song L, Liu D, Tay Y, Wu S, Li L-J, Zhao Y, Yu T, Zhang H, Huang W, Hng HH, Ajayan PM, Yan Q (2011) Enhanced thermopower of graphene films with oxygen plasma treatment. ACS Nano 5:2749–2755
24. Wang BB, Gao D, Levchenko I, Ostrikov K, Keidar M, Zhu MK, Zheng K, Gao B (2016) Self-organized graphene-like boron nitride containing nanoflakes on copper by low-temperature N_2+H_2 plasma. RSC Adv 6:87607–87615

25. Singh G, Sutar DS, Divakar Botcha V, Narayanam PK, Talwar SS, Srinivasa RS, Major SS (2013) Study of simultaneous reduction and nitrogen doping of graphene oxide Langmuir-Blodgett monolayer sheets by ammonia plasma treatment. Nanotechnology 24:355704
26. Lim T, Kim D, Ju S (2013) Direct deposition of aluminum oxide gate dielectric on graphene channel using nitrogen plasma treatment. Appl Phys Lett 103:013107
27. Gokus T, Nair RR, Bonetti A, hmler MB, Lombardo A, Novoselov KS, Geim AK, Ferrari AC, Hartschuh A (2009) Making graphene luminescent by xxygen plasma treatment. ACS Nano 3:3963–3968
28. Duan L, Zhao L, Cong H, Zhang X, Lu W, Xue C (2019) Plasma treatment for nitrogen-doped 3D graphene framework by a conductive matrix with sulfur for high-performance Li-S batteries. Small 15:1804347
29. Dai XJ, Chen Y, Chen Z, Lamb PR, Li LH, du Plessis J, McCulloch DG, Wang X (2011) Controlled surface modification of boron nitride nanotubes. Nanotechnology 22:245301
30. Choi JK, Huh JH, Kim SD, Moon D, Yoon D, Joo K, Kwak J, Chu JH, Kim SY, Park K, Kim YW, Yoon E, Cheong H, Kwon SY (2012) One-step graphene coating of heteroepitaxial GaN films. Nanotechnology 23:435603
31. Song YQ, Gao YQ, Liu XT, Ma J, Chen BH, Xie Q, Gao X, Zheng LM, Zhang Y, Ding QJ, Jia KC, Sun LZ, Wang WD, Liu ZT, Liu BY, Gao P, Peng HL, Wei TB, Lin L, Liu ZF (2022) Transfer-enabled fabrication of graphene wrinkle arrays for epitaxial growth of AlN films. Adv Mater 34:2105851
32. Kim J, Bayram C, Park H, Cheng CW, Dimitrakopoulos C, Ott JA, Reuter KB, Bedell SW, Sadana DK (2014) Principle of direct van der Waals epitaxy of single-crystalline films on epitaxial graphene. Nat Commun 5:4836
33. Al Balushi ZY, Miyagi T, Lin Y-C, Wang K, Calderin L, Bhimanapati G, Redwing JM, Robinson JA (2015) The impact of graphene properties on GaN and AlN nucleation. Surf Sci 634:81–88
34. Lin Y-C, Lu N, Perea-Lopez N, Li J, Lin Z, Peng X, Lee C H, Ce Sun, Calderin L, Browning P N, Bresnehan MS, Kim MJ, Mayer TS, errones MT, Robinson JA (2014) Direct synthesis of van der Waals solids. ACS Nano 8:3715–3723
35. Feng Y, Yang X, Zhang Z, Kang D, Zhang J, Liu K, Li X, Shen J, Liu F, Wang T, Ji P, Xu F, Tang N, Yu T, Wang X, Yu D, Ge W, Shen B (2019) Epitaxy of single-crystalline GaN film on CMOS-compatible Si(100) substrate buffered by graphene. Adv Funct Mater 29:1905056
36. Wang H, Maiyalagan T, Wang X (2012) Review on recent progress in nitrogen-doped graphene: Synthesis, characterization, and its potential applications. ACS Catal 2:781–794
37. Liang D, Jiang B, Liu Z, Chen Z, Gao Y, Yang S, He R, Wang L, Ran J, Wang J, Gao P, Li J, Liu Z, Sun J, Wei T (2024) Quasi van der Waals epitaxy of single crystalline GaN on amorphous SiO_2/Si(100) for monolithic optoelectronic integration. Adv Sci 11:2305576
38. Fernandez-Garrido S, Ramsteiner M, Gao G, Galves LA, Sharma B, Corfdir P, Calabrese G, de Souza SZ, Pfuller C, Trampert A, Lopes JMJ, Brandt O, Geelhaar L (2017) Molecular beam epitaxy of GaN nanowires on epitaxial graphene. Nano Lett 17:5213–5221
39. Kong W, Li H, Qiao K, Kim Y, Lee K, Nie Y, Lee D, Osadchy T, Molnar RJ, Gaskill DK, Myers-Ward RL, Daniels KM, Zhang Y, Sundram S, Yu Y, Bae SH, Rajan S, Shao-Horn Y, Cho K, Ougazzaden A, Grossman JC, Kim J (2018) Polarity governs atomic interaction through two-dimensional materials. Nat Mater 17:999–1004
40. Liu F, Wang T, Gao X, Yang HY, Zhang ZH, Guo YC, Yuan Y, Huang Z, Tang JL, Sheng BW, Chen ZY, Liu KH, Shen B, Li XZ, Peng HL, Wang XQ (2023) Determination of the preferred epitaxy of III-nitride semiconductors on wet-transferred graphene. Sci Adv 9:eadf8484
41. Yoo H, Chung K, Choi SK, Kang CS, Oh KH, Kim M, Yi G-C (2012) Microstructures of GaN thin films grown on graphene layers. Adv Mater 24:515–518
42. Li T, Liu C, Zhang Z, Yu B, Dong H, Jia W, Jia Z, Yu C, Gan L, Xu B, Jiang H (2018) Understanding the growth mechanism of GaN epitaxial layers on mechanically exfoliated graphite. Nanoscale Res Lett 13:130
43. Scharifker BR, Mostany J (1984) Three-dimensional nucleation with diffusion controlled growth: part I. Number density of active sites and nucleation rates per site. J Electroanal Chem 177:13–23

44. Oh H, Hong YJ, Kim K-S, Yoon S, Baek H, Kang S-H, Kwon Y-K, Kim M, Yi G-C (2014) Architectured van der Waals epitaxy of ZnO nanostructures on hexagonal BN. NPG Asia Mater 6:e145
45. Wu Q, Yan J, Zhang L, Chen X, Wei T, Li Y, Liu Z, Wei X, Zhang Y, Wang J, Li J (2017) Growth mechanism of AlN on hexagonal BN/sapphire substrate by metal-organic chemical vapor deposition. CrystEngComm 19:5849–5856
46. Wang LL, Yang SY, Gao YQ, Yang JK, Duo YW, Song S, Yan JC, Wang JX, Li JM, Wei TB (2023) Quasi van der Waals epitaxy of a stress-released AlN film on thermally annealed hexagonal BN for deep ultraviolet light-emitting diodes. ACS Appl Mater Interfaces 15:23501–23511
47. Rybin M, Pereyaslavtsev A, Vasilieva T, Myasnikov V, Sokolov I, Pavlova A, Obraztsova E, Khomich A, Ralchenko V, Obraztsova E (2016) Efficient nitrogen doping of graphene by plasma treatment. Carbon 96:196–202
48. Chen Z, Liu Z, Wei T, Yang S, Dou Z, Wang Y, Ci H, Chang H, Qi Y, Yan J, Wang J, Zhang Y, Gao P, Li J, Liu Z (2019) Improved epitaxy of AlN Film for deep-ultraviolet light-emitting diodes enabled by graphene. Adv Mater 31:1807345
49. Chang H, Chen Z, Li W, Yan J, Hou R, Yang S, Liu Z, Yuan G, Wang J, Li J, Gao P, Wei T (2019) Graphene-assisted Quasi van der Waals epitaxy of AlN film for ultraviolet light emitting diodes on nano-patterned sapphire substrate. Appl Phys Lett 114:091107
50. Heilmann M, Munshi AM, Sarau G, Gobelt M, Tessarek C, Fauske VT, van Helvoort AT, Yang J, Latzel M, Hoffmann B, Conibeer G, Weman H, Christiansen S (2016) Vertically oriented growth of GaN nanorods on Si using graphene as an atomically thin buffer layer. Nano Lett 16:3524–3532

Chapter 4
Quasi van der Waals Epitaxy of Nitride Films on Single Crystal Substrates

Due to the lack of cost-effective homogeneous substrates, nitride films are generally grown on various heterogeneous substrates [1, 2]. Whether sapphire or Si substrates, nitrides have a large lattice mismatch and thermal mismatch with them. Although the mismatch is smaller for SiC substrates, SiC substates are less used due to their high price, and nitrides especially AlN, grown on SiC, are prone to cracking. On the whole, the large mismatch in lattice and thermal expansion coefficients between nitrides and conventional crystalline substrates results in high interfacial energy at the interfaces [3, 4], and thus direct heteroepitaxial growth of nitride on them will produce a large number of natural dislocations and cause the epitaxial layer to crack.

In recent years, with the development of vdWE technology on 2D materials, using graphene [5, 6], h-BN [7, 8] and TMDC [9, 10] as insertion layers has become a promising method for epitaxial high-quality nitride films on large mismatch substrates. It is expected to break the influence of substrates in the large mismatch epitaxy process. As mentioned in the previous section, the process of epitaxial 3D nitride films on 2D layered materials is called qvdWE process [11–13]. This chapter focuses on the formation mechanism and dislocation annihilation rules of qvdWE nitride films, and systematically explains the influence of the qvdWE process on nitride films based on the characteristics of 2D materials.

T. Wei et al., *Quasi van der Waals Epitaxy of Nitride Semiconductor and Its Applications*, Wide Bandgap Semiconductors,
https://doi.org/10.1007/978-981-95-3469-2_4

4.1 Quasi van der Waals Epitaxy of Nitride Films on Two-Dimensional Materials/Planar Sapphire Substrate

4.1.1 Quasi van der Waals Epitaxy of Nitride Films on Graphene/Planar Sapphire Substrate

Sapphire substrate has high chemical stability, low price and very high full-spectrum transmittance, so heteroepitaxial growth of high-quality nitrides on (0001) sapphire substrate has become the mainstream technology in the current industry. However, due to the large lattice and thermal expansion coefficient mismatch between nitrides and sapphire, a large number of dislocations are usually generated in the epitaxial layer. Therefore, how to improve the growth mode of nitride epitaxy on sapphire and further alleviate the impact of large mismatch substrates on the epitaxial layer has become a key core issue.

To overcome the above problems, heteroepitaxial nitrides on sapphire currently usually adopt a "two-step" growth strategy: first, a LT nucleation layer with an amorphous structure is grown, and columnar crystals are formed by annealing and solid-phase recrystallization at high temperatures. Then, the initial HT layer continues to nucleate on these columnar structure buffer layers, forming small islands of columnar structures. The islands continue to grow laterally and longitudinally and then merge to form a thin film [14], as shown in Fig. 4.1. In 1986, Japanese scientists Hiroshi Amano and Isamu Akasaki first used LT AlN nucleation layer technology on sapphire substrate, and then obtained high-quality GaN epitaxial films by HT growth. In 1991, Nakamura [15] used LT GaN nucleation layer to replace AlN buffer layer, which can also help to obtain high-quality GaN epitaxial layers with bright surface. In a short period of time, this two-step GaN epitaxy method is widely adopted, and the research on GaN-based materials and devices enters into a rapid development stage. These three Japanese scientists also won the 2014 Nobel Prize in Physics due to realizing high quality GaN and blue LEDs. Currently, AlN epitaxy on sapphire also follows the epitaxy method of GaN, and there will generally be a LT nucleation layer, and then HT AlN epitaxy is performed. In such a two-step process, the LT nucleation layer will inevitably accumulate a large number of dislocations in the initial epitaxial layer, which are difficult to eliminate in the subsequent growth process, and the complex pre-treatment steps and temperature conversion process will also increase the epitaxial growth time.

2D materials represented by graphene, due to their weak vdW force connection between the layers, are expected to alleviate the lattice mismatch and thermal mismatch between the substrate and the epitaxial layer, thereby possibly avoiding the generation of a large number of dislocations in heteroepitaxy. Therefore, the growth of high-quality nitrides on sapphire covered with 2D materials has received widespread

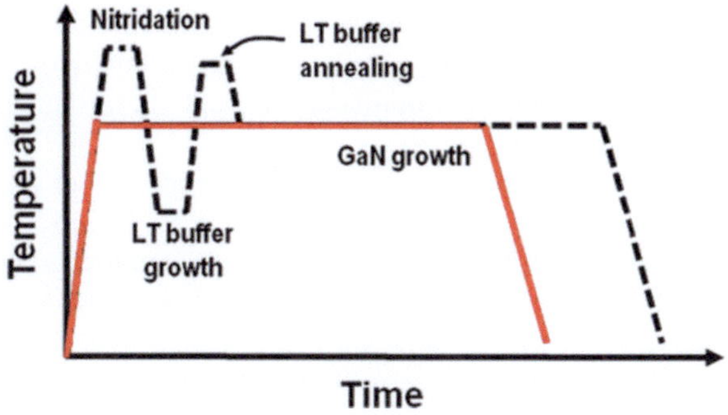

Fig. 4.1 Schematic diagram of conventional two-step growth process on sapphire substrate (black dashed line) and one-step growth on graphene buffer (red line)

attention in recent years. In 2010, Kunook Chung et al. from Seoul National University transferred graphene to a sapphire substrate, and then performed plasma treatment on the graphene surface, realizing GaN single-crystal thin film epitaxy on graphene for the first time [16]. Jae-Kyung Choi et al. used graphene as an insertion layer and achieved one-step HT heteroepitaxial growth on sapphire through the MOCVD method, also obtaining high-quality flat GaN thin films [17]. The introduction of a graphene buffer layer can eliminate complex steps in the MOCVD process, such as substrate surface pretreatment, buffer layer deposition, and buffer heat treatment, thereby significantly simplifying the growth process of nitrides, as shown in Fig. 4.1.

Therefore, the method of inserting a graphene interlayer to avoid the use of a LT buffer layer in the traditional two-step method has been deeply studied for achieving one-step vdWE of nitrides on 2D materials. Qi et al. [18] transferred single-layer graphene grown on Cu (111) foil to a sapphire substrate, ensuring that the graphene was undamaged during the transfer process. Then, AlN growth was carried out on the obtained graphene/sapphire substrate using the MOCVD method. After initial nucleation, the 3D nucleation islands gradually merged into a continuous film. When the growth duration was extended to about 60 min, AlN would fully aggregate on the graphene/sapphire to form a flat epitaxial layer, which was in stark contrast to the rough AlN film obtained on sapphire in Fig. 4.2a. At the same time, for the AlN thin film grown on sapphire, the addition of a graphene interlayer effectively eliminated the residual stress caused by the mismatch, approaching a stress-free state. The strains in AlN and sapphire could impose profound effects on their Raman spectra. For example, the E_2 (high) mode of AlN around 657 cm^{-1} could provide a clear signature for the biaxial strain within the basal plane [19]. As shown in Fig. 4.2b, the strain-sensitive E_2 (high) peak corresponding to AlN/graphene/sapphire was located at 658.9 cm^{-1} at a lower wave number, compared to the peak of AlN/sapphire (661.7 cm^{-1}). It was closer to the peak of stress-free single crystal AlN (657.4 cm^{-1}), indicating that the compressive strain in AlN with graphene buffer was significantly released. The same conclusion can be drawn from the analysis of the A_1 (LO) peak (AlN/graphene/sapphire was about 890.4 cm^{-1}, bulk single crystal AlN was about 890.0 cm^{-1}, AlN/sapphire was about 896.7 cm^{-1} (Fig. 4.2c). For sapphire in Fig. 4.2d, the E_2 peak from AlN/graphene/sapphire was 752.0 cm^{-1} (black), which was closer to the stress-free state (green), while in the AlN/sapphire structure (red) it moved to a lower wave number (748.3 cm^{-1}), further confirming that the tensile strain of sapphire in the AlN/graphene/sapphire structure had decreased. At the same time, the 2D and G peaks of graphene also showed higher stress sensitivity. When graphene was present between AlN and sapphire, its G and 2D peaks moved to higher wave numbers (1648.3 cm^{-1} and 2810.3 cm^{-1}) relative to the intrinsic state (1581.6 cm^{-1} and 2676.9 cm^{-1}, respectively) in Fig. 4.2e, showing a compressive strain state. The reason for the large peak shift in graphene was that it bore the compressive stress originally applied to the AlN epitaxial layer, and the stress was relaxed through its own torsional deformation.

Figure 4.2f–g showed the Raman mapping (5 μm × 5 μm) of the E_2 (high) peak of AlN on AlN/sapphire and AlN/graphene/sapphire, respectively. The peak of E_2 (high) for AlN shifted to a higher wave number, while the peak of E_2 (high) for AlN in AlN/graphene/sapphire was closer to the original AlN, proving that the AlN layer on graphene had effectively released stress over a wide range and exhibited a strain-free state. With the help of weak vdW force, graphene can also effectively release the strain of the epitaxial GaN and other epitaxial films [20]. These reports demonstrated the strain relaxation mechanism by graphene, but there was no detailed discussion or rigorous verification of this statement.

At the same time, conventional heteroepitaxy of lattice-mismatched systems produces lots of dislocations above a critical strain energy to release the accumulated strain energy as the film thickness increases. The formation of dislocations, which severely degrade the device performances, is fundamentally unavoidable in highly lattice-mismatched epitaxy, such as nitride growth on sapphire. Sang-Hoon Bae have observed that heteroepitaxy on graphene coated substrates allowed for spontaneous relaxation of misfit strain owing to the slippery graphene surface while achieving single-crystalline films by reading the atomic potential from the substrate [21]. To track intermediate relaxation, they investigated a system with a small lattice mismatch (0.74%) by growing $In_{0.4}Ga_{0.6}P$ films on GaAs substrates, thus avoiding immediate strain relaxation. As shown in Fig. 4.3a, b, $In_{0.4}Ga_{0.6}P$ grown on graphene/GaAs was substantially relaxed by 78%, whereas $In_{0.4}Ga_{0.6}P$ grown directly on GaAs is relaxed by 30%. In Fig. 4.3c, $In_{0.4}Ga_{0.6}P$ grown directly on GaAs showed a gradual increase in relaxation as a function of thickness. However, $In_{0.4}Ga_{0.6}P$ films grown on graphene/GaAs relaxed spontaneously, signifying that the degree of relaxation was greater than that of heteroepitaxial $In_{0.4}Ga_{0.6}P$ on GaAs even when the films are 40 nm thick. It is well known that the relaxation pathway of strained epitaxial films for heteroepitaxy is by the introduction of a dislocation. However, here, the slippery graphene surface can help to relax the misfit strain via spontaneous relaxation. Figure 4.3d–g revealed that high-density dislocations with significant strain fringe were present near the conventional heteroepitaxial interface of GaP-GaAs and substantial reduction of dislocations at the interface with the heteroepitaxial interface of GaP-graphene-GaAs. Furthermore, the EBSD map reveals the out-of-plane single crystallinity of GaP in Fig. 4.3h and the four-fold symmetry revealed by HRXRD phi scan also confirmed the in-plane single crystallinity of GaP in Fig. 4.3i.

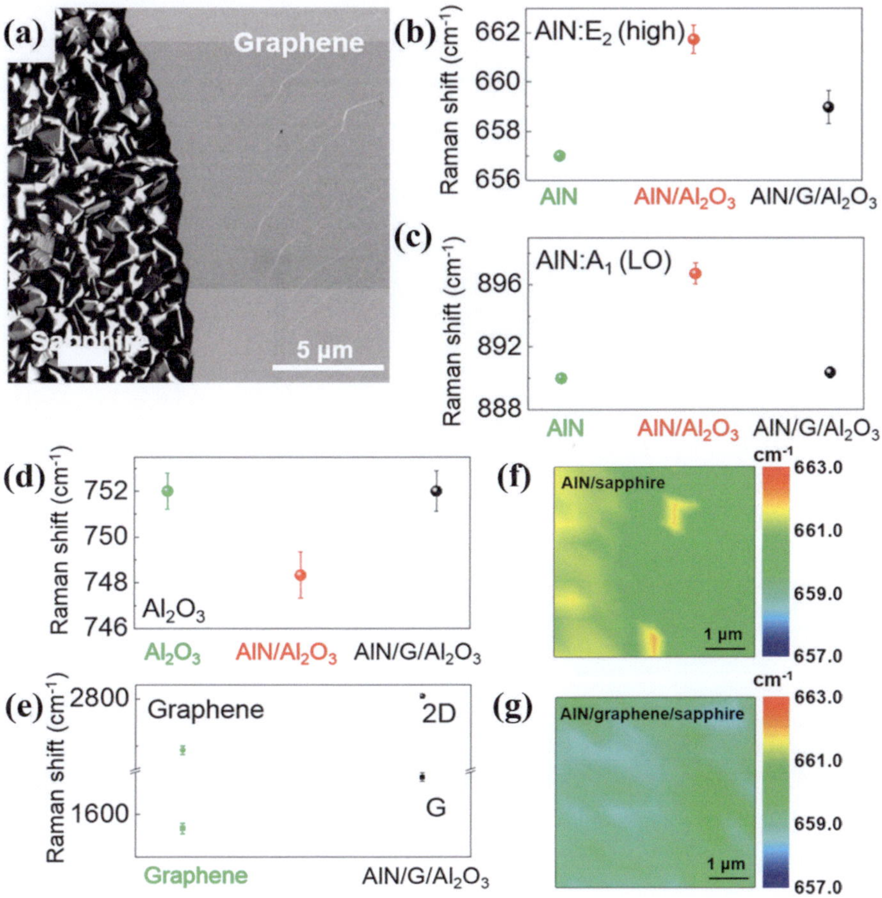

Fig. 4.2 Surface morphology of AlN grown on graphene/sapphire and sapphire substrates of (**a**). Relative Raman shifts of the E_2 (high) (**b**) and A_1 (LO) peaks (**c**) of AlN in pristine AlN, AlN/sapphire, and AlN/graphene/sapphire, respectively. **d** Relative Raman shifts of sapphire in pristine sapphire, AlN/sapphire, and AlN/graphene/sapphire. **e** Relative Raman shifts of G and 2D peaks of graphene in pristine graphene and AlN/graphene/sapphire. **f–g** Raman mappings (5 μm × 5 μm) of the E_2 (high) peak of AlN in AlN/sapphire and AlN/graphene/sapphire, respectively [18]

The spontaneous relaxation by graphene could also enable largely lattice-mismatched systems with minimized dislocation density, which could eventually broaden the material spectrum for advanced electronics and photonics. Bo Shi et al. proposed to reduce the edge TD density of the GaN epilayer on sapphire by nearly 1 order of magnitude via graphene-assisted remote heteroepitaxy [22]. The cross-sectional STEM of the GaN/graphene/Al_2O_3 interface and corresponding GPA e_{xx} mapping image were shown in Fig. 4.4a, b, respectively. Strain cores induced by geometrical misfit dislocations (GMDs) with the typical tension–compression contrast were not observed at the interface with graphene. Moreover, the fast Fourier

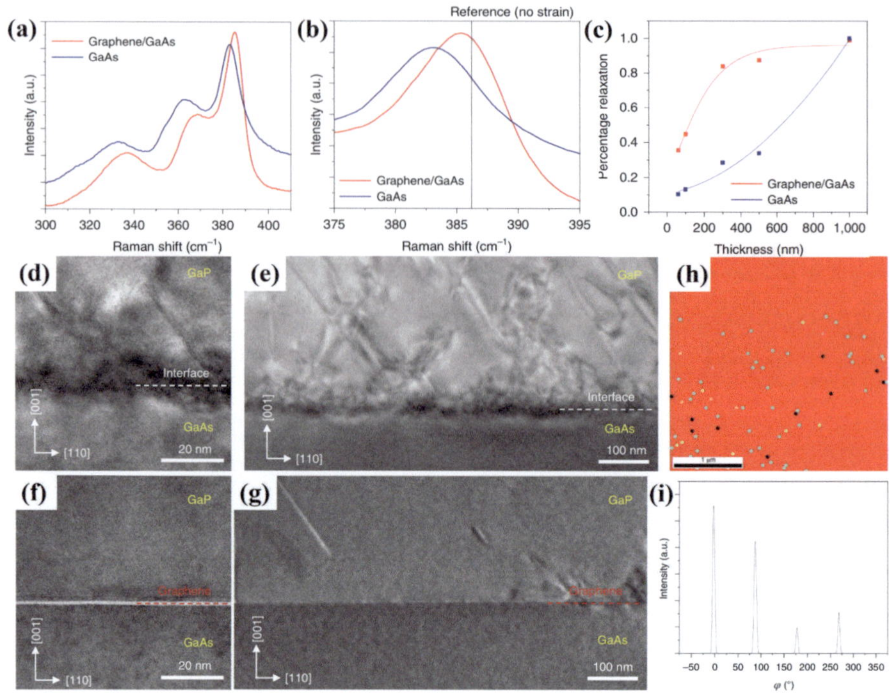

Fig. 4.3 Highly mismatched system of GaP on GaAs. **a**, **b** Raman spectra on InGaP with 0.74% tensile strain, grown on graphene/GaAs and on bare GaAs. **c** Strain-relaxation efficiency of InGaP grown on graphene/GaAs and on GaAs. **d–e** Cross-sectional view of GaP grown on bare GaAs and **f**–g cross-sectional view of GaP grown on graphene/GaAs. (h) EBSD plot of the exfoliated GaP epilayer. **i** HRXRD azimuthal off-axis φ scan of the GaP epilayer [21]

filtered image (Fig. 4.4c) also illustrated that the insertion of graphene significantly reduced the periodic edge dislocation cores at the interface.

As shown in Fig. 4.4d, DFT calculations revealed a significantly reduced energy barrier required for interface displacement on a graphene surface compared to that on a bare sapphire substrate. The maximum energy barrier for the GaN epilayer displacement on graphene-coated sapphire was calculated to be 323 mJ m^{-2}, while the calculated elastic energy required for the introduction of a GMD was 764 mJ m^{-2} by using dislocation theory [23]. In contrast, the energy required for displacing the GaN epilayer directly on a bare sapphire substrate was 1026 mJ m^{-2}, as depicted in Fig. 4.4e. The mechanism of remote epitaxy in large mismatched systems to reduce the density of edge TDs was schematically exhibited in Fig. 4.4c, d. Here, spontaneous lattice relaxation of the GaN films effectively accommodated the substantial mismatch at the interface, thereby reducing the formation of GMDs.

In addition to the novel pathway of relaxing misfit strain in heteroepitaxial films via interface displacement, graphene-driving strain-pre-store engineering had been presented by Chang et al. [24] As shown in Fig. 4.5a, after the graphene film was

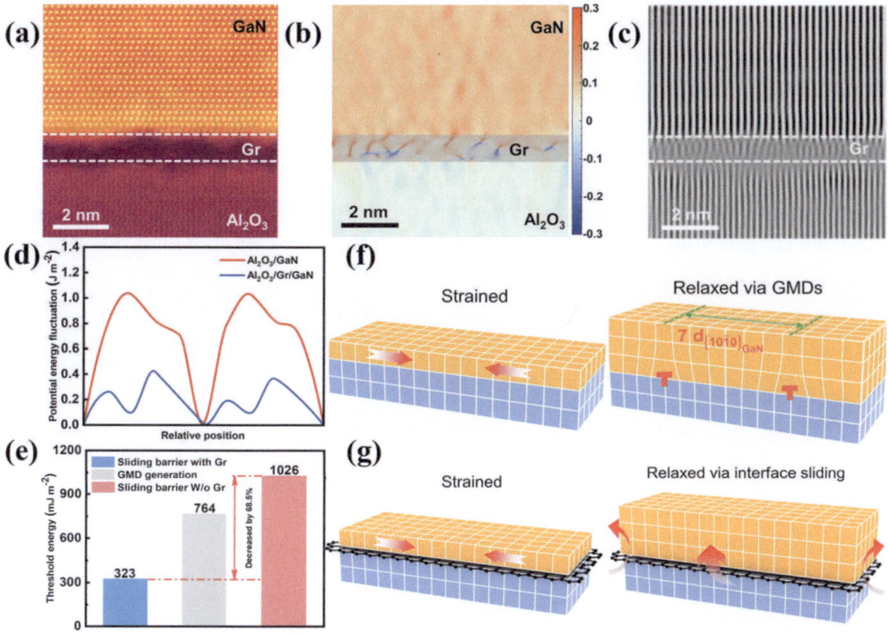

Fig. 4.4 **a** Cross-sectional STEM image of the GaN/graphene/Al_2O_3 interface, viewed along the $\langle 11\ \bar{2}\ 0\rangle_{GaN}/\langle 11\ \bar{2}\ 0\rangle_{GaN}/\langle 1\ \bar{1}\ 00\rangle Al_2O_3$. **b** GPA e_{xx} mapping images based on STEM images. **c** Inverse fast Fourier transform image of the GaN/graphene/Al_2O_3 interface. **d** Energy barrier required for the interface sliding of GaN/Al_2O_3 and GaN/graphene/Al_2O_3. **e** Threshold energy for the situations of interface sliding on graphene (323 mJ m^{-2}), the introduction of a GMD (764 mJ m^{-2}), and interface sliding on bare sapphire (1026 mJ m^{-2}). **f** Schematic diagram of strain relaxation via the introduction of the periodic geometrical misfit dislocation network. **g** Schematic diagram of strain relaxation via interface sliding [22]

grown directly on the sapphire, the defects were introduced into the Gr film by N_2 plasma treatment, and then the epitaxy of the AlN film was performed. In Fig. 4.5b–d, the hexagonal AlN nucleation islands grown on the bare sapphire were quite large and widely distributed with an average diameter and an island density of 84 nm and 4.2 × 10^9/cm^2, respectively. On the contrary, graphene clearly changed the above-mentioned growth mode. Higher-density and smaller-size nucleation islands had been formed on the graphene-buffered sapphire (Fig. 4.5c). It can be estimated that the AlN nucleation islands reached a density of 1.28 × 10^{11}/cm^2 and an average diameter of 17 nm (only about one-fifth of AlN nucleation islands on bare sapphire). Obviously, the threading dislocations originated from the interface between AlN and graphene/sapphire and first formed the dislocation-dense region, followed by sharp bending and annihilation within a thickness as shown in Fig. 4.5e. Figure 4.5f demonstrated the schematic growth model of graphene to affect the crystallinity and strain state of the AlN epilayer. It was revealed that the small-size AlN nucleation islands induced by N_2 plasma-treated graphene may pre-store enough high tensile strain in epilayer during the coalescence, which compensated the compressive strain caused by hetero-mismatch and enabled the epitaxy growth of strain-free AlN films. Therefore, the graphene treatment condition can also regulate the strain of nitride epilayers.

To avoid the size limitation and repeatability issue brought about by graphene transfer from the metal substrate to the target substrate, as well as to reduce the risk of graphene film contamination during the transfer process, Chen et al. [25] directly grew high-quality graphene on c-plane sapphire using a metal-free catalyst atmospheric pressure CVD. The as-grown graphene film fully covered the sapphire

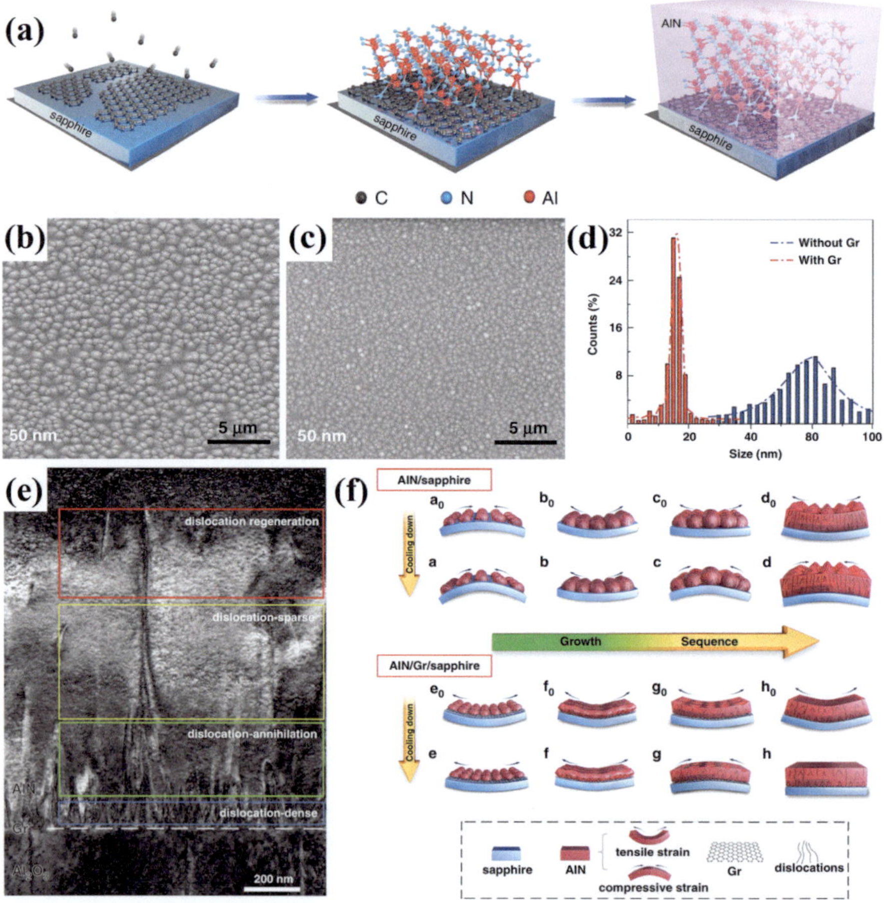

Fig. 4.5 **a** Schematic diagram of the key steps involved in the growth of high-quality AlN films on N_2-plasma-treated graphene/sapphire. SEM images of AlN nucleation on the sapphire substrate **b** without and **c** with graphene. **d** Size distribution analysis of AlN nucleation on sapphire and Gr/sapphire substrate. **e** Dark field images of epitaxial AlN/graphene/sapphire with g = [0002]. **f** Schematic diagrams of the epitaxial process of AlN film grown on Gr/Sapphire. The growth development of AlN films without and with graphene during the processes from nucleation growth to nucleation island coalescence to completion of coalescence and final film formation [24]

substrate, with the number of layers adjustable between 1 and 6. Here, the as-grown AlN on sapphire formed randomly 3D oriented AlN clusters with grain sizes of several hundred nanometers as shown in Fig. 4.6a. However, continuous and smooth AlN film was obtained on plasma treated graphene in Fig. 4.6b and the directly grown AlN film on plasma treated graphene showed a step-like flat surface with RMS roughness of ~ 0.27 nm in Fig. 4.6c. Moreover, large-scale view of an EBSD mapping showed (0001) orientation single-crystal feature of as-grown AlN epilayer.

The qvdWE growth of AlN on graphene was also expected to release the stress. Figure 4.6d showed the Raman spectrum of E_2 phonon mode of as-grown AlN films on sapphire and graphene. The E_2 peak of the AlN film grown on the graphene sapphire was located at 657.8 cm^{-1}, almost overlapping with stress-free AlN (657.4 cm^{-1}). However, the AlN epilayer grown on sapphire showed a larger frequency (569.2 cm^{-1}), facing much higher compressive strain. The residual stress in AlN epilayer can be estimated according to $\sigma_{xx} = \Delta\omega/K$, where $\Delta\omega$ was the E_2 peak shift from stress-free AlN crystal value of 657.4 cm^{-1} and K was the stress coefficient ~ 3.7 cm^{-1}/GPa. The residual stress of AlN was significantly reduced from 0.49 GPa to 0.11 GPa by graphene. Notably, the crystalline quality of as-grown AlN on graphene was comparable or higher with that typically obtained via conventional low temperature buffer-assisted epitaxy on sapphire [26]. The epitaxial relationship between the epilayer and c-plane sapphire through the graphene was explored by

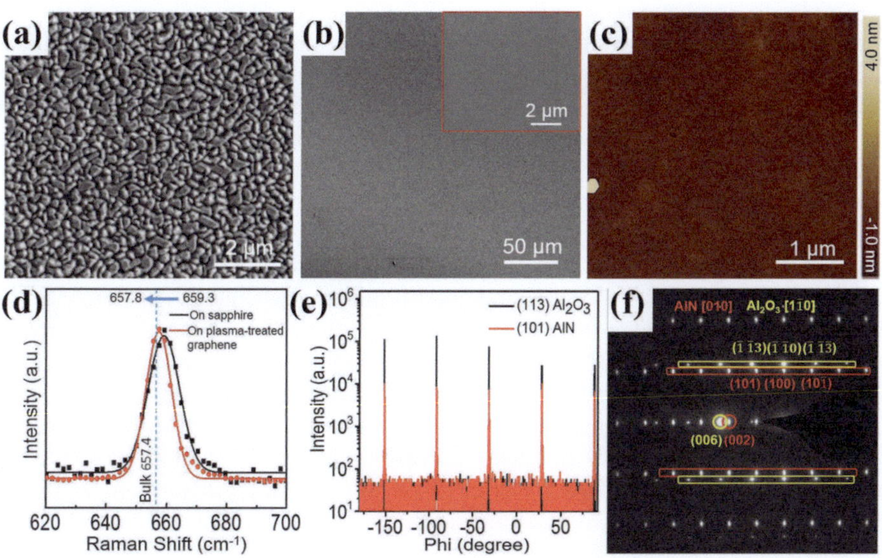

Fig. 4.6 AlN epilayer on sapphire and plasma treated graphene. SEM images of as-grown AlN films on sapphire **a** and plasma treated graphene **b** with increase growth time to 1 h. **c** AFM height image of as-grown AlN film on plasma treated graphene, showing atomically-flat feature with RMS roughness ~ 0.27 nm. **d** Raman spectra of AlN films grown on sapphire and plasma treated graphene. **e** Grazing incidence XRD azimuthal off-axis phi scan through Al_2O_3 (113) and AlN (101). **f** Selected area electron diffraction patterns taken at a AlN/graphene/sapphire interface [25]

performing grazing incidence XRD and STEM. The XRD φ scan in Fig. 4.6e and selected area electron diffraction (SAED) patterns in Fig. 4.6f revealed that the AlN unit cells were rotated by 30° with respect to the sapphire through the graphene, still in good agreement with previous reports of epitaxial wurtzite III-Nitride on sapphire.

4.1.2 *Quasi van der Waals Epitaxy of Nitrides on h-BN/ Planar Sapphire Substrate*

h-BN belongs to the hexagonal crystal system, also known as white graphene. h-BN is the most common BN crystal structure, and is composed of B atoms and N atoms with SP^2 hybridization. Compared to graphene, h-BN exhibits better stability and has a higher decomposition temperature, exceeding 1500 °C. Therefore, using h-BN as a buffer layer or insertion layer can also achieve nitride growth. However, the strong polarity of h-BN significantly attenuates electrostatic potential propagation, thereby affecting the nucleation and orientation of nitride epilayer. Un like graphene, monolayer h-BN may disrupt the transmission of the crystallographic information of the substrate, resulting in the growth of an GaN epilayer with two intermixed phases [27]. Furthermore, as the number of layers of h-BN increases, the effects of qvdWE through the h-BN template become more dominant than those of RE.

Similar to graphene, due to the lack of dangling bonds on the sp^2 bonded BN surface, it is easy to cause uneven distribution of low-density GaN crystal islands on h-BN/sapphire, which cannot merge into a continuous layer with good morphology. Even if a continuous layer can be obtained randomly, the crystal structure quality is relatively poor. Therefore, to grow uniform films on h-BN, oxygen plasma treatment is generally used to increase the surface reactivity of the inert surface. O atoms can replace some N atoms in h-BN material, and the binding of O atoms may change the sp^2 hybrid orbit of B atoms to sp^3 hybrid orbit [28, 29].

To understand the different growth modes of GaN and AlN buffer layers on h-BN interlayer, Lulu Wang et al. performed first principles calculations to compare the adsorption of Ga and Al atoms on h-BN monolayer [30]. They found the adsorption of Al atom on pristine h-BN was rather weak, with adsorption energies around − 0.34 to − 0.48 eV and the adsorption of Ga on pristine h-BN was even weaker. The O defects introduced by O_2 plasma treatment was needed to enhance the adsorption and nucleation of both AlN and GaN on h-BN interlayer. The Al atom bonded to both the O defect and the N atom in h-BN, resulting in a strong adsorption energy up to − 4.9 eV. Such strong adsorption indicated that the O defects on h-BN can serve as effective nucleation center for AlN. On the other hand, the Ga atom could only bond to the O atom in the B–O–N and B–O defects, while the Ga adsorption energy was calculated to be − 3.5 eV, less than that of Al atom by more than 1 eV. The smaller adsorption energy of Ga meaned the O defects were much less effective to enhance Ga nucleation on h-BN. Therefore, the GaN nucleation density was less than AlN on h-BN/sapphire. Figure 4.7 demonstrated the comparison of GaN and AlN nucleation on sapphire and h-BN/sapphire [31]. Obviously, due to the low adhesion coefficient of Ga atom, it was difficult for GaN to adsorb and nucleate on h-BN surface during the LT growth process. Afterwards, the sparse nucleation islands could not coalesce during HT growth process and the GaN film showed a discontinuous surface, as shown in Fig. 4.7a0–c0. Unlike the GaN, AlN nuclei presented markedly different growth mode on the h-BN/sapphire substrate. AlN nucleation islands with high-density and small-volume were uniformly distributed on h-BN/sapphire in Fig. 4.7a1–c1. The rapid lateral coalescence of AlN epilayer with the introduction of h-BN interlayer was attributed to the fact that h-BN can strongly promote the 2D growth tendency during the AlN growth. The quasi-2D growth of AlN on h-BN enabled the rapid coalescence of dense AlN islands in a short time, showing a surprising continuous surface morphology. In contrast, for the growth on bare sapphire, the HT-AlN buffer layer was initiated with sparse nucleation islands at HT. As the nuclei continued to grow to 12 min, the AlN nucleation islands swelled in a typical 3D-dominated Volmer-Weber growth mode that was unfavorable for merging, thus resulting in a rough surface as shown in Fig. 4.7a2–c2.

Qingqing Wu et al. had also studied the growth of AlN on sapphire, untreated h-BN/sapphire, and oxygen plasma-treated h-BN-O_2/sapphire, and compared the effects of substrate treatment on the quality of AlN epitaxial thin films [32]. After

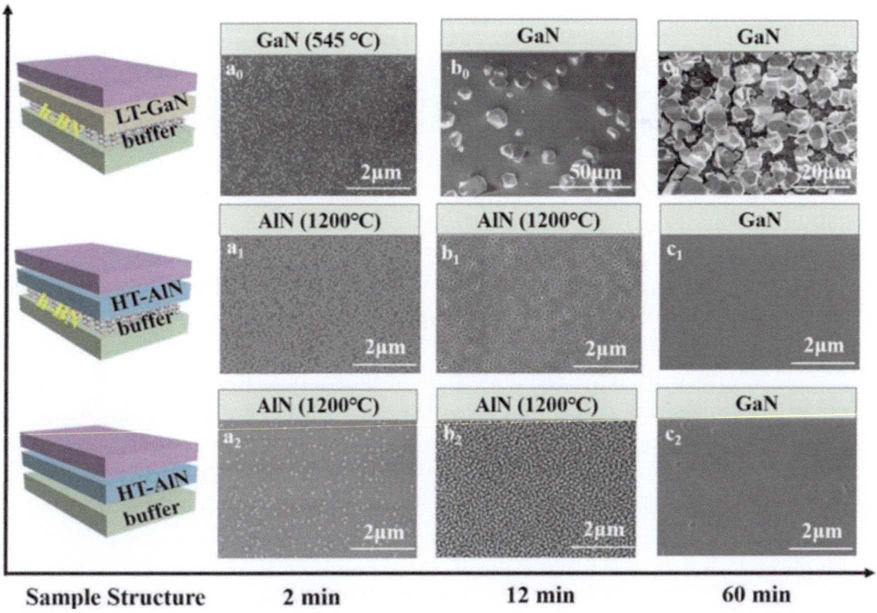

Fig. 4.7 (a_0) SEM images of the LT GaN buffer layer grown for 2 min. SEM images GaN epilayer grown on LT buffer layer for (b_0) 12 min, (c_0) 1 h. SEM images of HT-AlN interrupted buffer layer with (a_1) 2 min and (b_1) 12 min grown on ≈ 2-nm-thick h-BN/sapphire, and (a_2) 2 min and (b_2) 12 min grown on bare sapphire, respectively. GaN epilayer grown for 1 h on the HT-AlN buffer layer on (c_1) h-BN/sapphire and (c_2) bare sapphire [31]

O_2 plasma treatment, the atomic connection structure of h-BN material changed. The chemical state of B atoms with N-B-O rose slightly to follow the principle of minimum energy. In addition, some B-O chemical bonds may deviate from the c-axis perpendicular to the sapphire (0001) plane and tilt. In this structure, a small amount of N atoms were replaced by O atoms, and most O atoms were suspended on B atoms, and the B-O chemical bonds suspended on h-BN can be regarded as hanging bonds. Therefore, due to the increase in dangling bonds on the h-BN-O_2 surface, this treatment can benefit the subsequent AlN nucleation. Figure 4.8 showed the characterization results of AlN and DUV LEDs grown on two types of h-BN. The AlN grown on hBN without oxygen plasma treatment showed a porous structure, with a pore density of about 1.5×10^{-8} cm^{-2} in Fig. 4.8a. The lack of dangling bonds on h-BN made it difficult for AlN to merge, resulting in a porous surface. Figure 4.8b, c demonstrated the AlN surface grown on h-BN-O_2 was continuous and atomically flat, with the RMS of only 0.25 nm. The XRD rocking curve results in Fig. 4.8d, e showed that the FWHMs of the (0002) and (10–12) crystal plane rocking curves of AlN on h-BN-O_2/sapphire were 409 arcsec and 963 arcsec, respectively. The FWHMs of the (0002) and (10–12) crystal plane rocking curves of AlN on h-BN/sapphire were 227 arcsec and 1148 arcsec, respectively. The total dislocation density of AlN on h-BN/sapphire was obviously higher than that of AlN on h-BN-O_2/sapphire. Further, a conventional DUV LED device structure was grown on this AlN template, as shown in the inset of Fig. 4.8f. When the injection current was 90 mA and the integration time was 100 ms, a strong EL emission peak at 290 nm with an intensity of 12,000 could be clearly observed. This work confirmed that the dangling bonds on the surface of h-BN treated with oxygen plasma were significantly increased, which can promote the nucleation and merging of AlN and enabled the luminescence of DUV LED devices.

In addition to oxygen plasma treatment, Fang Liu et al. proposed a chemical activation method to generate dangling bonds on the surface of h-BN and thus to improve the nucleation ability of sp^3-hybrided GaN. To activate the surface, h-BN was treated by hydrochloric acid (HCl) and deionized water, which was expected to introduce the hydroxide (OH) and thus the oxygen bonds so that the nucleation sites can be created. The artificially added N–O bonds not only created more sites for following GaN nucleation but also modified the lattice polarity of GaN to be uniform Ga-polarity one, leading to high-quality material with flat surface [33]. They illustrated the nucleation mechanism of GaN on h-BN in Fig. 4.9. During HCl treatment, N–OH and B–O–N bonds were generated through the help of hydroxide ions, in which H atoms were then desorbed during the temperature ramping of substrate in MOCVD growth. When the Ga and N atoms were supplied, N adatoms preferred to bind with N–O bonds and formed the N–O–N bonds. And then, three Ga adatoms bound with each N–O–N bond and formed N–O–N–Ga bond and thus initiated the GaN epitaxy. This kind of atomic bond configuration at the interface leaded to the epitaxy of Ga-polarity GaN, which was experimentally confirmed after the growth by chemical etching effect.

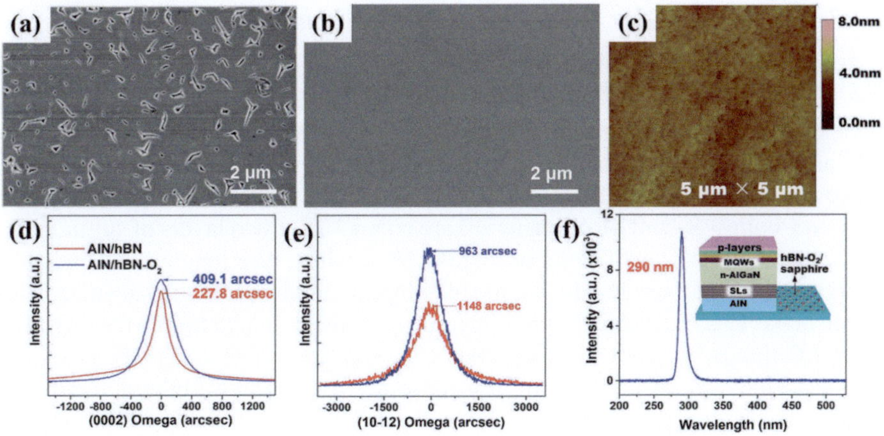

Fig. 4.8 SEM images of **a** AlN on h-BN/sapphire and **b** AlN on h-BN-O_2/sapphire, **c** AFM image of AlN on h-BN-O_2/sapphire, X-ray rocking curves of **d** (0002) and **e** (10–12) AlN grown on h-BN/sapphire and h-BN-O_2/sapphire (the numbers are the FWHM values), **f** schematic diagram of an AlGaN-based DUV LED structure on the h-BN-O_2/sapphire substrates and the EL spectrum of the DUV LED structure [32]

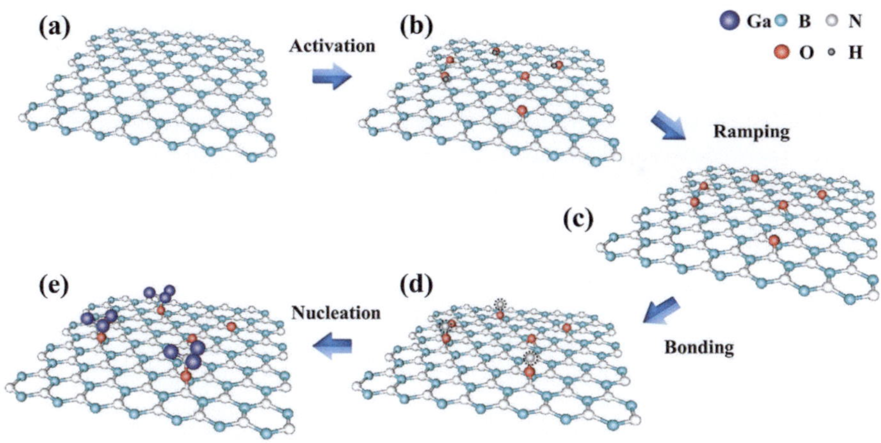

Fig. 4.9 Schematic diagrams of activation of h-BN and nucleation mechanism of GaN. **a** Surface of h-BN. **b** Formation of B-O-N and N–O bonds on h-BN surface by trapping OH- in HCl solution. **c** Desorption of H atoms from activated h-BN surface (i.e., from N–O–H bonds) during the temperature ramping of substrate. **d** Generation of N–O–N bonds on N–O bonds by trapping N atoms. **e** Formation of N–O–N–Ga structures by binding three Ga atoms on every N–O–N, indicating the start of Ga-polarity GaN epitaxy [33]

4.1.3 Quasi van der Waals Epitaxy of Nitrides on Disulfide/Planar Sapphire Substrate

Another natural choice for a 2D buffer layer lies in TMDs. TMDs, represented by WS_2 and MoS_2, stand out among many 2D materials owing to their extremely similar lattice constants to group III-nitride. Compared to the single carbon atom layer of graphene, both WS_2 and MoS_2 materials are composed of two layers of sulfur atoms sandwiched by one layer of metal atoms [34]. Meanwhile, their longitudinal thickness is larger, which is more conducive to shielding the influence of the substrate. This should make TMDs capable of providing a proper potential field for nitrides' growth.

Priti Gupta et al. from India pioneered the transfer of mechanically exfoliated WS_2 and MoS_2 2D materials to SiO_2/Si substrates and successfully grew strain-free single crystal islands of GaN materials on these 2D materials using MOCVD [35]. They formed stress-free single-crystal GaN islands on WS_2 and the PL peaks of GaN showed strong band-edge emission. Moreover, they found that MoS_2 played a role in the early stages of GaN nucleation, although there was a certain degree of MoS_2 degradation at 1140 °C in the presence of hydrogen and ammonia. This work confirmed that TMDs could be used as a novel buffer layer for lattice approximation matching of nitride growth. Subsequently, Tangi [13] used a single layer of MoS_2 with only 0.8% lattice mismatch with GaN as the interlayer, and grew GaN on the CVD-deposited single-layer MoS_2/sapphire via the MBE method. The existence of the single-layer MoS_2 and the quality of the epitaxial GaN were characterized by Raman spectroscopy. Figure 4.10a–c showed the Raman spectra of MoS_2/sapphire, GaN/MoS_2/sapphire, and GaN/sapphire. The black and red curves in the figure showed the phonon modes at 385.5 and 405.0 cm^{-1}, which corresponded to the in-plane vibrations of Mo and S atoms ($E^1{}_{2g}$) and out-of-plane S atoms ($A^1{}_g$) in MoS_2, respectively. In addition to the MoS_2 phonon modes, the red curve also contained the low-intensity $E_2{}^H$ phonon feature of the 3 nm thick GaN layer grown on MoS_2. The blue curve showed the Raman peak obtained on the GaN epitaxial film with the $E_2{}^H$ phonon mode at 568.5 cm^{-1}, which matched the $E_2{}^H$ phonon mode of the GaN thin layer in the red curve. This suggested that the GaN thin layer had achieved the same degree of stress relaxation as the sample corresponding to that on sapphire, due to the small lattice mismatch (0.8%) between GaN and MoS_2. Figure 4.10d showed a typical cross-sectional HRTEM image obtained at the GaN/single-layer MoS_2/sapphire heterojunction, where the thickness of the MoS_2 at the heterojunction was 0.8 nm, consistent with the thickness of the S–Mo–S single layer, confirming the stable existence of the single-layer MoS_2 during the growth of GaN.

Yue Yin et al. also grew high-quality single-crystalline AlN thin film on sapphire substrate with an intrinsic WS_2 overlayer by MOCVD, which had surface roughness and defect density similar to that grown on conventional sapphire substrates as shown in Fig. 4.11a, b. The curves of XRD-ϕ scan revealed that the AlN (0002) facets were rotated by 30° with sapphire (0006) facets through WS_2, describing the epitaxial relationship was [1100] AlN// [11–20] sapphire in Fig. 4.11c. The EBSD mapping provided the evidence that most of the area of the AlN film displayed

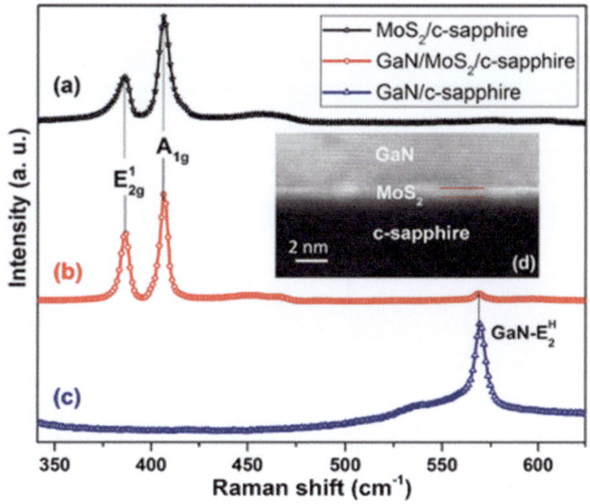

Fig. 4.10 The micro-Raman spectra of samples A-C acquired on MoS_2, GaN/MoS_2, and epi-GaN samples, respectively. The inset (d) shows the HRTEM cross-sectional image obtained at typical GaN/SL-MoS_2/c-sapphire heterojunction [34]

almost (0001) single crystallinity, as demonstrated in red by the inverse pole figure color triangle (Fig. 4.11d). These results all strongly suggested that single-crystalline AlN film was grown on WS_2 film, and the atomically resolved STEM image showed clearly distinguishable line between AlN and sapphire as the signal of WS_2 existed in Fig. 4.11f. The DUV LEDs were also deposited on the AlN/WS_2/sapphire template. LOP increased simultaneously with the injection current, revealing that the EL emission was generated from the carrier injection and radiative recombination at MQW layers in Fig. 4.11g. The faint peak-position shift should be attributed to the low-stress property of AlN thin film. These results demonstrated that conventional DUV LEDs could be fabricated on the WS_2/sapphire substrate.

Finally, using 2D materials as buffer/release layers for nitride qvdWE or RE becomes a promising solution for pursuing high-quality and transferrable nitride epitaxial layers. Furthermore, this new epitaxial approach facilitates heterogeneous integration between nitride materials and other material systems easier, which is important for advanced electronics and photonics applications. Compared to graphene, polar 2D materials such as h-BN and TMDs behave differently in qvdWE. Since their lattice potential can better screen the lattice potential field of the substrate, polar 2D materials can reduce the dependence of the subsequent growth process on the substrate. For this reason, polar materials, especially TMDs, are more suitable for guiding the deposition of nitride films on different substrates, especially amorphous substrates or metal substrates. These results have the potential to circumvent the requirement of substrate release for nitride growth and provide a pathway for the integration of different functional materials [37].

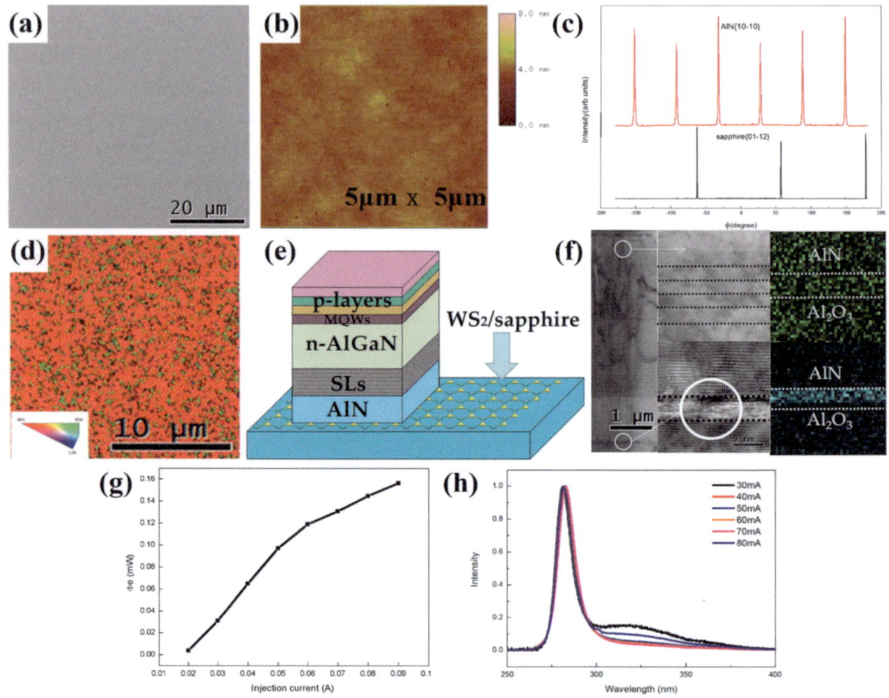

Fig. 4.11 **a** SEM image and **b** AFM image of AlN thin film growth on WS_2/sapphire substrate. **c** X-ray powder diffraction φ scan curve. **d** EBSD mapping of AlN film. **e** Schematic illustration of the DUV LED structure. (f) Cross-sectional STEM image and EDX mapping of heterojunction DUV LEDs. **g** LOP of the fabricated LEDs at various injection currents. **h** The normalized EL spectra of fabricated LEDs with currents ranging from 30 to 80 mA [36]

4.2 Quasi van der Waals Epitaxy of Nitride Films on Two-Dimensional Materials/Pattern Sapphire Substrate

Due to the large lattice mismatch and thermal expansion mismatch between nitrides and sapphire substrates, significant residual stress and high dislocation density are always induced, which in turn affect the performance of the devices grown on them. To address this issue, various methods have been developed to achieve high-quality nitride preparation, such as multi-layer buffer layer, superlattice layer, and gradient composition layer, etc. From the perspectives of stability and repeatability, pattern sapphire substrate (PSS) technique is widely recognized as an effective method that can stably and effectively obtain high-quality nitride materials. On the one hand, during the lateral merging process of nitrides on the PSS, the dislocations generated at the interface will bend and annihilate each other, thereby achieving the purpose of improving crystal quality. On the other hand, due to the high refractive index of nitride materials, the light emitted by the optoelectronic device is severely totally reflected at the interface between the device and the air. However, the epitaxial thin film on the PSS will form an air gap, increasing the scattering of light in the LED, thereby improving the light extraction efficiency (LEE) of the device [38]. In 2013, Dong et al. [39] had improved the quality of AlN materials by epitaxially growing AlN on a nano-patterned sapphire substrate (NPSS) in Fig. 4.12a, compared to the conventional flat sapphire substrate (FSS). A 4 μm AlN film was epitaxially grown on NPSS at HT, and the surface of the AlN template reached atomic-level flatness, as shown in Fig. 4.12b. Figure 4.12c–e showed the bright-field cross-sectional TEM image of AlN epilayers on NPSS and FSS. High density of threading dislocations (TDs) originating from the AlN/sapphire interface were observed above the un-etched sapphire regions. The TDs in the vicinity of the voids bent towards the voids sidewalls and terminated because of the lateral over-growth of AlN, decreasing the TD density in AlN above the etched sapphire regions. Compared to that on FSS, the crystal quality of AlN grown on NPSS was significantly improved, which in turn improved the material quality of the 283 nm DUV MQWs structure epitaxially grown on the AlN template, achieving a 43% increase in internal quantum efficiency (IQE).

However, the epitaxial growth of nitrides on NPSS is also facing great challenge, especially for AlN epilayer [40]. Due to the high surface stickiness coefficient of Al atoms on sapphire, the surface migration rate of Al atoms is much lower than that of Ga atoms. At the same time, the higher bond energy of AlN makes it tend to grow in a 3D island mode [41], which is not conducive to 2D film growth. Therefore, even

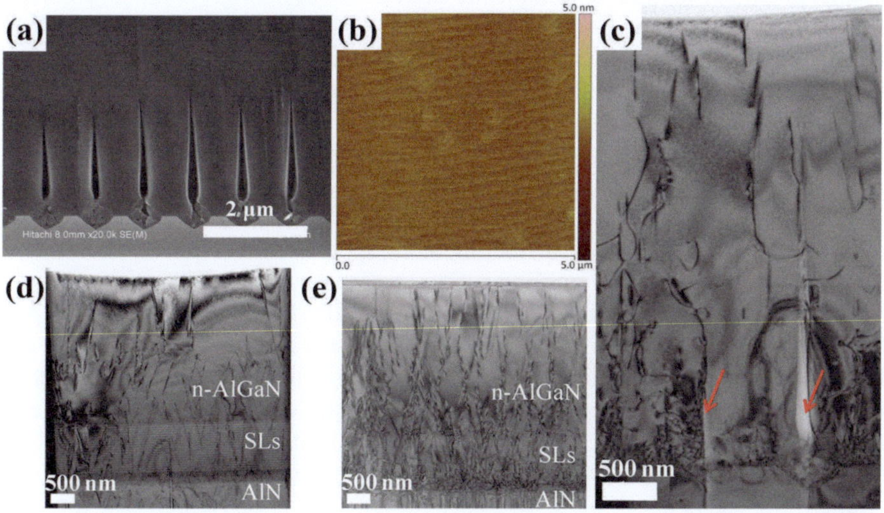

Fig. 4.12 **a** A cross-sectional SEM image of AlN grown on NPSS. **b** An AFM image of the AlN grown on NPSS (5 × 5 μm^2). Bright-field cross-sectional TEM images of: **c** AlN grown on NPSS with g = [11–20], **d** the epilayers above AlN template of NPSS with g = [− 2110], and **e** the epilayers above AlN template of FSS with g = [0–110] [39]

on NPSS, AlN films still require a higher coalescence thickness, usually over 3 μm as shown in Fig. 4.12a. This means that AlN epilayer requires a longer growth time and higher material costs on NPSS than that on flat sapphire substrate. Fortunately, by using 2D materials as the insertion layer and epitaxially growing AlN films on the patterned substrate, not only the stress-relieving effect of 2D materials can be fully utilized, but also the many problems encountered in epitaxial growth on patterned substrates can be effectively solved.

In order to solve the above problems, Hongliang Chang et al. prepared NPSS using nanoimprint lithography, introducing graphene inserting layer for AlN growth. To avoid complex and technically demanding graphene transfer, they directly grew a high-quality graphene layer on NPSS using a catalyst-free CVD method [41, 42]. To enhance the reactivity of graphene for AlN nucleation, they first treated graphene covered NPSS by N_2 plasma and then directly grew AlN epilayers by conventional two-step and one-step methods. In Fig. 4.13a_1–a_3, the epitaxy growth of AlN on FSS with graphene coalesced quickly after the nucleation at high density, resulting in quasi-flat surface morphology with a thickness of only 150 nm and a mirror-smooth surface with 500 nm. The rapid lateral coalescence of AlN in the presence of graphene might be attributed to the fact that graphene can effectively promote a strong 2D growth tendency of AlN growth, consistent with the expected quasi-2D results of the first-principles calculations. Obvious effects of graphene happened on AlN grown on NPSS. The LT AlN buffer grown for 6 min had dense but uniform smaller nucleation islands (Fig. 4.13b1), and a flat film surface was obtained after continued HT growth for 2 h (Fig. 4.13b2). In contrast, the QvdWE growth of AlN with graphene buffer at 1270 °C for 6 min exhibited a surprisingly continuous surface morphology in the flat region of NPSS (Fig. 4.13c1), further proving that AlN did indeed grow on the graphene surface with quasi-2D mode. After growing for 2 h, AlN film demonstrated an extremely smooth surface (Fig. 4.13c2). It was noted that the thickness of the lateral coalescence of AlN grown on NPSS with LT AlN buffer was 2.90 μm, as depicted in Fig. 4.13b_3. In sharp contrast, the AlN film with graphene buffer showed a lateral coalescence thickness of only 1.08 μm from the SEM images of the cross section in Fig. 4.13c_3, attributed to the graphene-driven quasi-2D growth mode of AlN film. For a comparison, the HT AlN growth on NPSS without buffer demonstrated rough and nonuniform surface morphology after 2 h growth.

The model to elucidate the effect of different buffer layers on the growth of AlN from cluster coalescence to covering pattern until the film was presented. As was shown in Fig. 4.13b_4, at the initial stage of AlN growth on NPSS at LT, Al atoms were easily adsorbed above the as-grown AlN cluster due to the minimum energy principle. At the same time, the LT condition reduced the diffusion length of the Al adatoms so that the Al atoms tended to be stably adsorbed in the lowest energy position. It meaned above the grown AlN cluster (Fig. 4.13b4), there was a typical 3D-dominated Volmer–Weber growth mode of AlN (Fig. 4.13b5). Therefore, the lateral coalescence tendency of AlN clusters was relatively slow and it was hard to quickly cover the concave-cone pattern of NPSS, as is shown in Fig. 4.13b_6, b_7. In contrast, the introduction of the graphene buffer avoided the problems encountered in the above growth process. On the surface of the graphene treated by N_2 plasma,

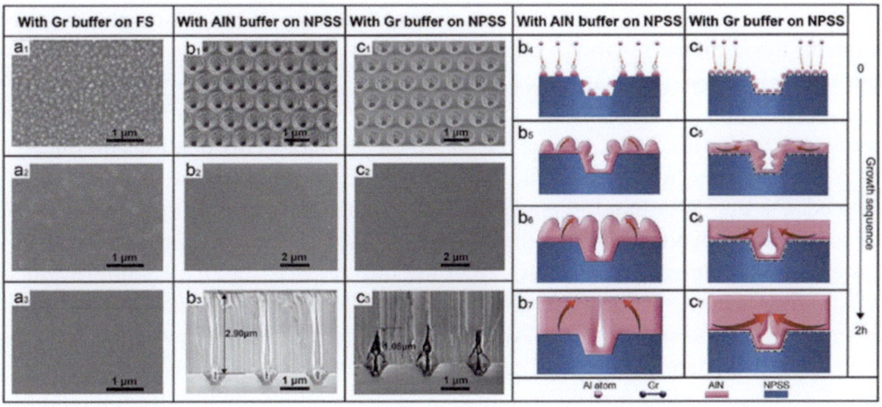

Fig. 4.13 SEM characterization and schematic diagrams of AlN films grown with different buffer layers. SEM images of a_1) 30 nm, (a_2) 150 nm, and (a_3) 500 nm AlN grown at 1270 °C on flat sapphire with graphene buffer. (b_1) SEM image of AlN grown on NPSS for 6 min at 650 °C. (c_1) SEM image of AlN directly grown for 6 min at 1270 °C on NPSS with graphene buffer. The SEM images of AlN grown on NPSS at a HT (1270 °C) for 2 h with (b_2) LT (650 °C) AlN buffer and (c_2) graphene buffer, respectively. The cross-sectional SEM images of AlN films grown on NPSS at a HT (1270 °C) for 2 h with (b_3) LT (650 °C) AlN buffer and (c_3) graphene buffer, respectively. Schematic illustration of morphology evolution for AlN films grown with (b_4–b_7) LT AlN buffer and (c_4–c_7) graphene buffer, respectively [41]

AlN can nucleate relatively densely and uniformly at HT condition. In the presence of the graphene buffer, Al adatoms achieved high enough mobility at HT to move to energetically favorable sites in Fig. 4.13c_4. Meanwhile, compared to sapphire, Al adatoms had a lower migration barrier on the graphene surface, which meaned that Al adatoms could easily diffuse on the graphene surface and had a longer lateral diffusion length. Therefore, the growth mode of the epitaxial growth of AlN on graphene/NPSS was not the 3D-dominant Volmer–Weber mode, but a quasi-2D growth mode with lateral expansion as the main trend from the initial stage (Fig. 4.13c5). Thus, it enabled AlN to more rapidly coalesce and cover the concave-cone pattern (Fig. 4.13c6) and finally formed an extremely smooth film (Fig. 4.13c7). Therefore, qvdWE method on graphene may effectively help to shorten the growth time and cost of AlN epilayer during MOCVD.

At the same time, the graphene covered on NPSS can effectively help the AlN epitaxial layer improve the crystal quality and release the biaxial stress in the epitaxial layer. From the TEM analysis, the dislocations of AlN films were mostly located on the flat surface of the NPSS between the concave-cone patterns rather than above the pattern. As for the AlN film grown with LT AlN buffer in Fig. 4.14a, b, a dislocation-intensive layer on the flat surface may be seen and the dislocation-dense region had a thickness of about 60 nm. Meanwhile, lots of dislocations in the polycrystalline layer were not annihilated in the subsequent lateral growth but spreaded upward. In contrast, graphene buffer can significantly improve this situation. HRTEM image in Fig. 4.14c exhibited a sharp interface of AlN/graphene/sapphire between the concave-cone patterns, which confirmed the stable existence of layered graphene throughout the growth process. The cross-sectional atomic structure of AlN film QvdWE on N_2 plasma-treated graphene/sapphire was shown schematically in Fig. 4.14d. From Fig. 4.14e, f, one may recognize that the existence of graphene introduced multiple effects. On the one hand, the number of dislocations produced at the AlN/graphene/sapphire interface between the concave-cone patterns was greatly reduced. On the other hand, the growth of graphene-driven AlN presented a strong 2D trend, which leaded to some dislocations to incline and stacking faults to annihilate without extending to the surface of the film. Hence, the quality improvement of AlN film should be attributed to the fact that graphene alleviated the mismatch effects between the substrate and AlN film to a certain extent. More importantly, the graphene-driven quasi-2D growth caused few defects generated at the AlN/graphene/NPSS interface, preventing the propagation of dislocations toward the epitaxial film surface. Furthermore, compared with the residual stress in AlN with LT AlN buffer (0.42 GPa), the residual compressive stress of AlN with Gr buffer (0.14 GPa) was significantly reduced from the E2 (high) phonon mode shift in Fig. 4.14g. Meanwhile, the 2D and G peaks of Raman spectra of graphene also exhibited high stain sensitivity. As shown in Fig. 4.14h, G and 2D peaks of the Raman spectra of Gr sandwiched between AlN and NPSS shifted to higher wave numbers, with regard to that of the pristine ones before growth, which implied that Gr was subjected to compressive stress. Therefore, it was inferred that graphene relaxed the stress caused by the mismatch between epilayer and substrate after the combination of graphene distortion of itself as shown in Fig. 4.14d. Therefore, AlN with graphene exhibited a weak residual stress state.

Usually, the compressive strain in MQWs generates a piezoelectric field which tilts the energy band diagram, resulting in a decreasing radiation recombination probability and a red-shift emitting wavelength. The graphene buffer could reduce the compressive strain in MQWs and thus weaken QCSE resulting in higher output power and IQE of LEDs. Wen-Cheng Ke et al. had grown graphene interlayer on 4 inch PSS substrate as shown in Fig. 4.15a [43]. Before loading in MOCVD for InGaN LEDs growth, a 10 nm thick AlN NL using a sputtering system was deposited on the graphene/PSS. From Raman spectra and EL peak, it was evident the embedded graphene interlayer decreased the compressive strain in the GaN template. In Fig. 4.15b, the relationship between the peak energy and injection current for Ref-LED showed a typical blue-shifted behavior of blue InGaN LED. The blue-shifting of 8.5 meV for Ref-LED was due to screening of the polarization field by increasing the applied field generated by the injection electrons and holes in the InGaN well layer. Compared to Ref-LED, the peak energy only had a slight blue shift of 1.3 meV for Gr-LED. According to Fig. 4.15c, the optimized fitting values of piezoelectric fields were 1.60 and 1.16 MV/cm for Ref-LED and Gr-LED, respectively. The fitting results suggested that the small blue-shift of the EL peak energy was caused by a weak QCSE due to the introduction of graphene.

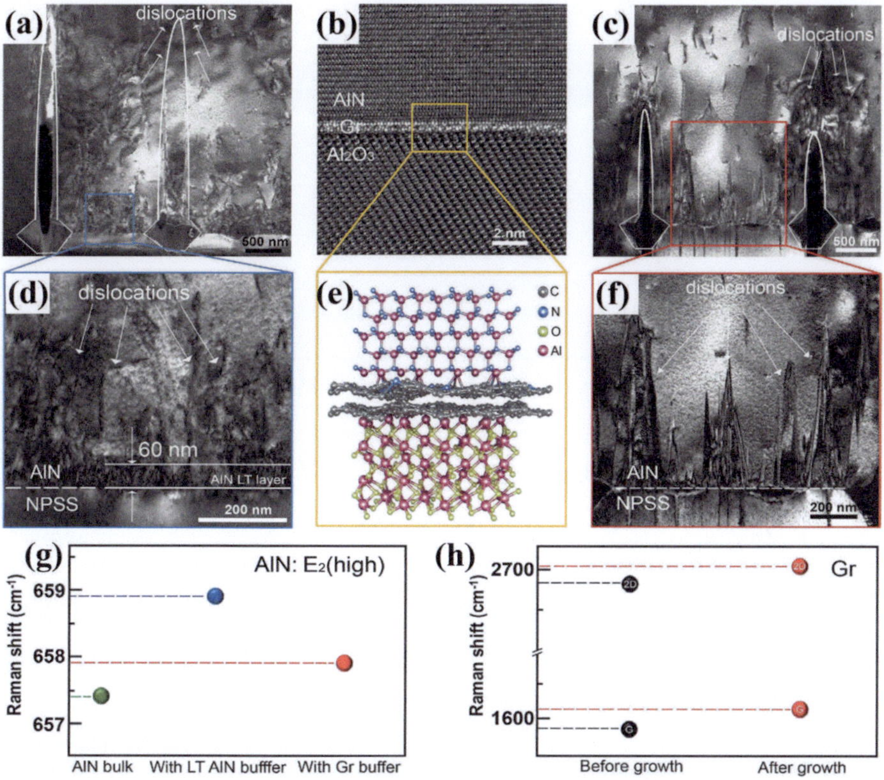

Fig. 4.14 Cross-sectional TEM and Raman characterizations of AlN films grown on NPSS with the different buffer layers. Dark field images of epitaxial AlN on NPSS with **a** LT AlN buffer and **e** graphene buffer with g = [0002]. Magnified dark field images of the AlN/NPSS interface in the blue rectangular region marked in **a** and the AlN/graphene/NPSS interface in the red rectangular region marked in (**e**). **c** The HRTEM image of the AlN/graphene/NPSS interface. **d** The schematic diagram of AlN/graphene/sapphire sandwich structure. **g** Relative Raman shifts of E_2 (high) of AlN in stress-free AlN bulk, AlN/NPSS, and AlN/graphene/NPSS, respectively. **h** Relative Raman shifts of G and 2D peaks of graphene before and after AlN growth [41]

4.3 Quasi van der Waals Epitaxy of Nitride Films on 2D Materials/SiC Substrate

Due to the similar crystal structure of SiC and nitrides, the lattice mismatch can be almost ignored, so high-quality nitride films can be grown on SiC substrates. However, growing crack-free nitride epitaxial layers on SiC is still a daunting task [44]. During the growth process, when nitride islands merge together, strain will accumulate in the epitaxial layer, making it prone to produce crack at a certain thickness. If graphene is used as a buffer layer, its surface lacks dangling bonds. Therefore, the epitaxial layer can grow according to its inherent lattice, to some extent avoiding dislocations introduced by lattice mismatch, and thus relieve the

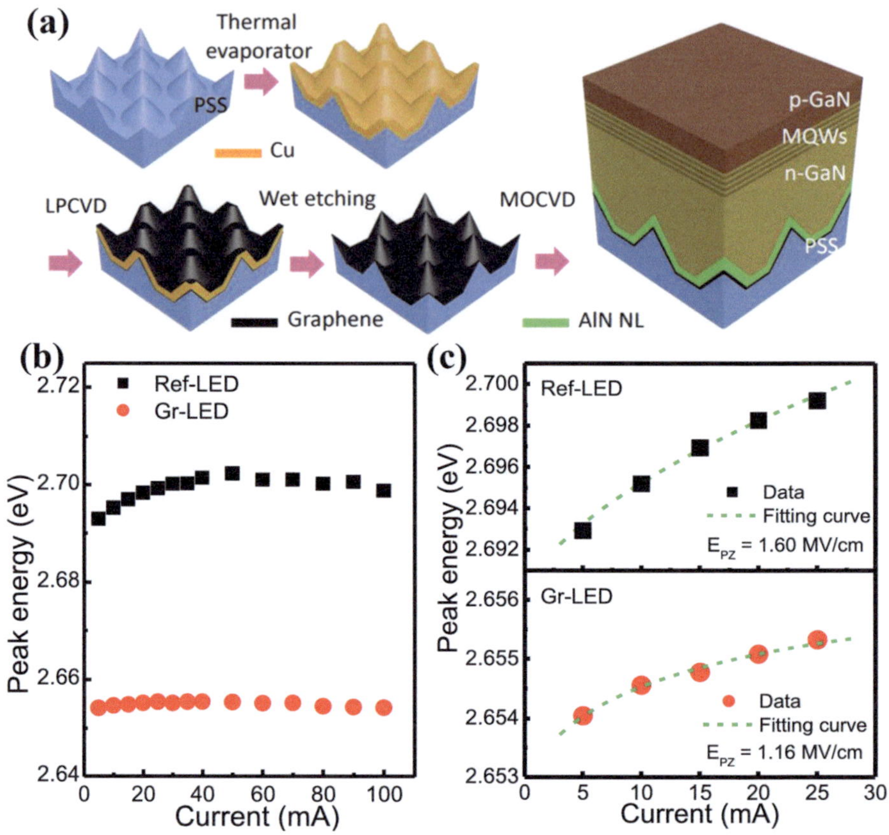

Fig. 4.15 **a** A flow chart of graphene interlayer grown on PSS. **b** Revolutions of the peak energy as a function of injection current. **c** The estimated piezoelectric field by fitting the peak energy in the current range of 5–25 mA for both samples [43]

accumulation of stress in the vertical direction. At the same time, the migration rate of metal atoms of III-nitride on the graphene surface is greatly increased, allowing the metal atoms to migrate to lower energy positions for the nucleation. Combined with the absence of surface dangling bonds, this forms large and sparse nucleation islands, reducing the boundary density, and thus relieving the strain generated during lateral merging. Therefore, many researchers use 2D graphene materials as a buffer layer to improve the epitaxial quality of nitride thin films on SiC substrates.

There are usually two methods to cover graphene on the previously introduced sapphire substrate: one is to grow graphene on a metal substrate and then transfer it to the target epitaxial substrate, and the other is to directly grow graphene on the target substrate through CVD. For transferred graphene, despite its high quality and large area, the complexity of the entire process is greatly increased due to the intermediate wet transfer process. For graphene directly grown on traditional substrates such as sapphire and Si through CVD, this method does not use a metal catalyst

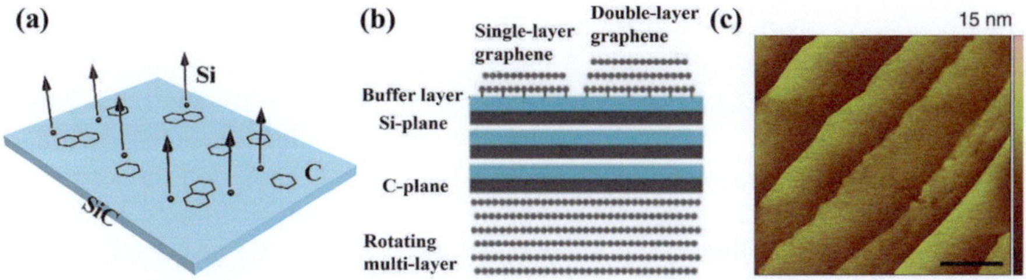

Fig. 4.16 Growth of graphene on SiC substrate. **a** Schematic diagram of the evaporation process of Si atoms on the surface of the SiC substrate. **b** Schematic diagram of the structural differences of graphene obtained on the Si face and C face SiC. **c** AFM image of the surface of the graphene-converted SiC substrate

during the growth process, and only relies on carbon source cracking and recombination to generate graphene. Therefore, the prepared graphene material has smaller domains and poorer quality. A lot of amorphous carbon exists on graphene surface, which usually cannot completely cover the substrate surface [45]. However, high-quality graphene films can be obtained on SiC substrate through high-temperature sublimation. At high temperatures, the Si atoms in the SiC substrate will sublime first, allowing the remaining C atoms to recombine to form graphene [46], as shown in Fig. 4.16a, b. The graphene obtained by this method is of high quality and can completely cover the substrate surface. At the same time, the graphene produced on the SiC substrate is usually accompanied by the formation of the steps [47], as shown in Fig. 4.16c. SiC substrate has two different polar faces, namely the silicon face (0001) and the carbon face (000–1). Graphene prepared on these two faces has different growth mechanisms, and under the same conditions, the carbon face sublimes the atoms much faster than the silicon face. As can be seen from Fig. 4.16 (b), for graphene materials grown on the silicon face, there is a buffer layer structure between the bottom layer and the substrate, and it is AB-type stacking. This type of stacking has a period every two layers, with an interlayer distance of 0.667 nm, and a 60° rotation around the c-axis. For graphene materials grown on the carbon face, there is no buffer layer between the bottom layer and the substrate. The interaction between the layers is very weak, and it is non-AB filled graphene.

4.3.1 Quasi van der Waals Epitaxy of GaN on Two-Dimensional Materials/SiC

In 2014, Kim et al. [47] from IBM Watson research center demonstrated direct growth of high quality single crystalline GaN films on epitaxial graphene (EG) on SiC substate by careful control of growth kinetics. Figure 4.17a–c showed SEM images of GaN films epitaxially grown on graphene under different conditions via MOCVD. The conventional two-step growth resulted in polycrystalline GaN clusters, while one-step growth at 1100 °C led to the formation of continuous GaN stripes arranged along adjacent SiC steps. It can be speculated that these polycrystalline clusters formed by the traditional two-step growth method were due to the limited atomic migration rate at 580 °C and the suppressed nucleation on the graphene surface, leading to the random formation of low-density nuclei on the steps. The surface stripes formed by one-step growth at 1100 °C suggested that the increased atomic migration rate due to the temperature rise allowed the adsorbed atoms to nucleate at the energetically favorable step edges. The improved two-step growth (nucleation at 1100 °C and growth at 1250 °C) can form continuous and smooth GaN films, attributed to the faster lateral growth of GaN nuclei formed along the step edges at 1100 °C under the high temperature conditions of 1250 °C. Typical step-flow GaN morphology was realized with a RMS roughness of 3 Å by modified two-step method in Fig. 4.17d.

The EG formed on the SiC surface via Si sublimation retained its unique orientation throughout an entire SiC substrate. The low-magnification TEM image in Fig. 4.17e showed that a uniform 2.5-μm-thick GaN film was grown on graphene/SiC. The HRTEM image showed well aligned, ordered crystal lattices of GaN, graphene and 4H-SiC in Fig. 4.17f. The XRD spectrum with the 2theta/omega scan represented only two peaks corresponding to single-crystalline wurtzite (0002) GaN and (0004) 4H-SiC (Fig. 4.17g). The full-width half maximum of the GaN (0002) peak (= 0.06°) was in the similar range to that measured from GaN grown on sapphire or SiC.

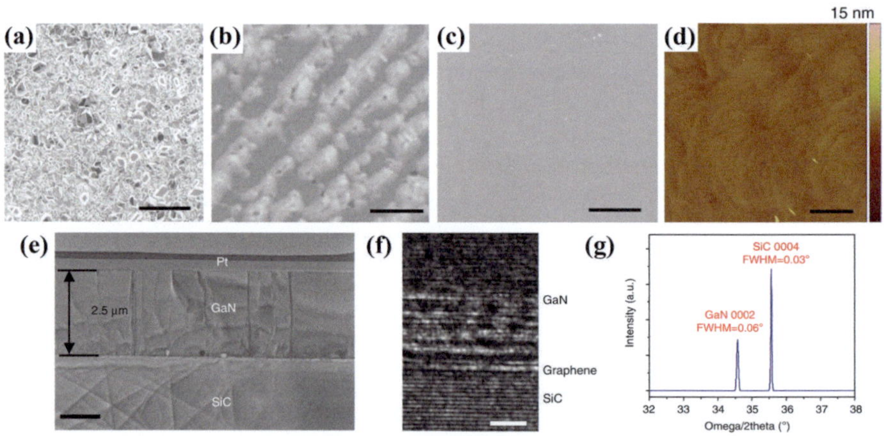

Fig. 4.17 Plan-view SEM images of GaN films grown on EF by **a** the conventional two-step growth (nucleation at 580 °C and growth at 1150 °C), **b** one-step growth at 1100 °C, and **c** modified two-step growth (nucleation at 1100 °C and growth at 1250 °C). **d** AFM image of GaN films grown by modified two-step growth showing atomistically smooth surface with RMS roughness of 3 Å. **e** Low-magnification cross-sectional TEM image GaN grown on graphene/SiC substrate. **f** HRTEM taken at the interface of GaN/graphene/SiC. **g** HRXRD omega/2theta scan [47]

However, to achieve high-quality crystalline thin films on 2D materials, an enhanced seeding effect from the substrate is desirable. Meanwhile, to ensure the substrate is reusable without damaging the van der Waals surface, strong adhesion of the 2D material to the substrate is needed in harsh material growth environments. To solve these problems, Kuan Qiao et al. presented the modification of a graphitized SiC (0001) wafer which consisted of EG on top of an interfacial graphene buffer layer (GBL) on the SiC wafer [48]. The GBL contained a graphene-like honeycomb lattice with lateral sp_2-bonded carbon atoms, while it bonded to the SiC substrate by sp_3-hybridized covalent bonds. They obtained the GBL by mechanically removing the EG from a graphitized SiC surface with a metal-induced 2D material transfer process in Fig. 4.18a. The EG was exfoliated by a layer of tensile-stressed Ni. Because the adhesion of Ni to EG is greater than that of EG to the substrate, the EG can be repeatedly exfoliated by the Ni stressor with a yield of over 95%. Nucleation density was substantially enhanced on the GBL/SiC substrate compared to that on the EG/GBL/SiC substrate, as shown in Fig. 4.18b, c. AFM image showed a smooth as-grown surface of the GaN grown on GBL/SiC with a RMS roughness of 0.18 nm in Fig. 4.18d. The electron channeling contrast imaging (ECCI) image of the GaN grown on the GBL showed a total TD density of 2.1×10^7 cm^{-2}. The triple-axis HRXRD ω-scan of GaN (002) and (105) peaks for the GaN epilayer on the GBL were reduced from 328 to 141 arcsec and from 279 to 193 arcsec in comparison to those of GaN grown on EG/GBL/SiC, as shown in Fig. 4.18f, g. The corresponding screw dislocation density and edge dislocation density estimated from the FWHM were reduced from 2×10^8 to 4×10^7 cm^{-2} and from 5×10^8 to 2×10^8 cm^{-2}, respectively, correlating well with the HRXRD analysis. Until now, this was the lowest dislocation density reported for GaN directly grown on 2D materials or other van der Waals surfaces. They also demonstrated the complete release of GaN membranes from the surface of the GBL, as shown in Fig. 4.18i. These findings indicated that the covalent adhesion of the GBL to the substrate prevented the etching of sp^2-bonded carbon on the surface in the harsh MOCVD growth environment.

4.3.2 Quasi van der Waals Epitaxy of AlN on Two-Dimensional Materials/SiC

Compared to GaN, the growth of thick and crack-free AlN layers on SiC is still a difficult task [49]. When AlN islands merge together, the tensile stress in the film will accumulate, and cracks are easily generated at a certain thickness of below 1 μm. Here, qvdWE growth on graphene could be a potential way to obtain high-quality and crack-free AlN films that are desirable for fabrication of high-performance DUV devices. Yunyu Wang et al. reported the qvdWE growth of crack-free AlN films on graphene/SiC substrate enabled by lattice engineering [50]. The processes of graphene preparation and high-quality AlN films epitaxy are schematically shown in Fig. 4.19. Briefly, the graphene preparation process on semi-insulating 4H-SiC (0001)

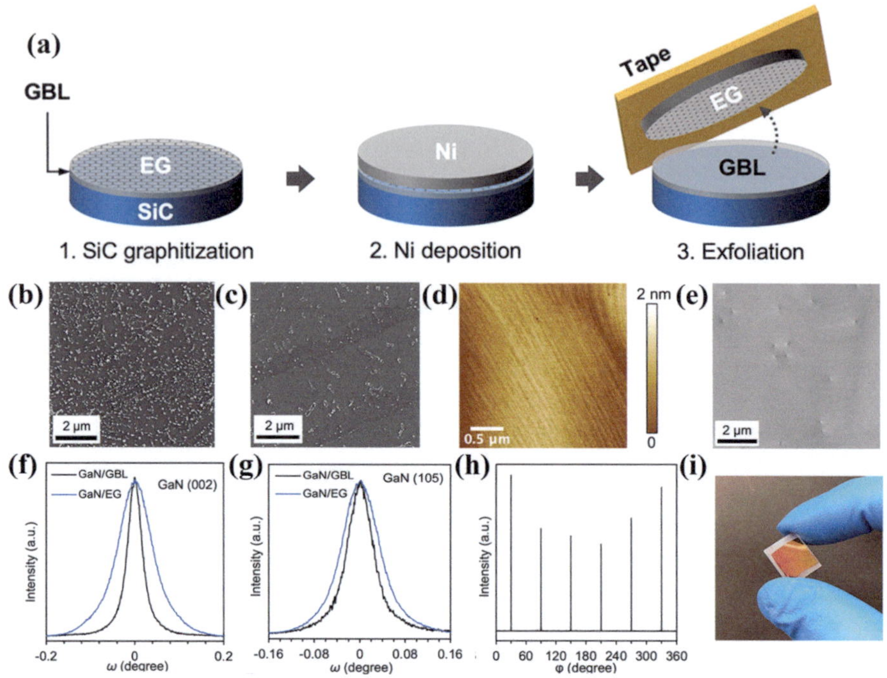

Fig. 4.18 **a** Schematic process of EG exfoliation. **b** SEM image of GaN nucleation grown on EG. **c** SEM image of GaN grown on the GBL. **d** AFM image of the surface of GaN grown on the GBL. **e** ECCI image of GaN grown on the GBL **f** HRXRD rocking curves of GaN (002) on the GBL and EG. **g** HRXRD rocking curves of GaN (105) on the GBL and EG. **h** HRXRD ϕ-scan of GaN grown on the GBL. **i** Photograph of exfoliated GaN on the GBL thin film on flexible substrate [48]

is carried out by thermal decomposition. Si atoms sublimate under high temperature at 1600–1700 °C, leaving the C atoms to recombine to form graphene in Fig. 4.19a. During this process, the atoms on the top surface recombine to be a new lattice structure when Si atoms partially sublimate, which is still connected with the substrate and is called a buffer layer. As the growth continues, Si atoms on the top surface sublimate completely to form a graphene layer while the second layer begins to repeat the sublimation process. Finally, graphene obtained on SiC is a bilayer structure with an intermediate buffer layer, as shown in Fig. 4.19b. After that, AlN epitaxy growth is carried out using MOCVD, as shown in Fig. 4.19c, d. Al atoms migrate on the surface to find a nucleation site and combine with N atoms to form AlN nuclei, then lots of nuclei merge together to form continuous AlN films.

After N_2 plasma treatment, the nuclei plasma-treated graphene/SiC (PG-SiC) became uniformly distributed with higher nucleation density but smaller size, due to the increased nucleation sites which constrained the movement of precursors. It was interesting that many cracks were formed in the AlN epilayers grown on bare SiC substrate (B-SiC), while fewer on PG-SiC, as shown in Fig. 4.20a, b. Figure 4.20c, d showed the AlN films grown on both B-SiC and PG-SiC substrates suffer from

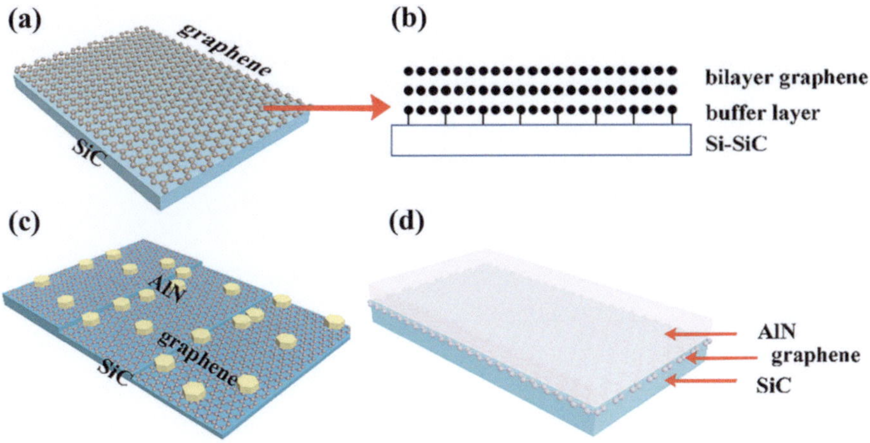

Fig. 4.19 Schematic diagrams of AlN films grown on graphene/SiC. **a** The graphitization process of the SiC substrate to form EG by sublimating Si atoms. **b** The detailed graphene structure, which is bilayer with a buffer layer connected with the SiC substrate. **c** AlN nucleation on graphene/SiC with a short time, indicating the prior nucleation near step edge areas. **d** AlN films epitaxy on graphene/SiC [50]

tensile strain. The E_2-high peak located at 651.5 cm^{-1} and 655.4 cm^{-1} for AlN films grown on B-SiC and PG-SiC, respectively, indicating a decrease in strain on PG-SiC. The calculated strain decreases from 1.486 GPa to 0.432 GPa with the introduction of graphene. Similar phenomenon of strain release in other vdWE systems had also been found, such as AlN on transferred-graphene/sapphire [18]. This may be because the epilayer and substrate were interacted mainly by van der Waals attractions, which was much weaker than chemical bonds.

To better understand the stress release mechanism of the nitride-graphene system, researchers use first-principles calculations to systematically analyze and compare the differences between the SiC/AlN and graphene/AlN interfaces. The adsorption energy of the AlN epitaxial layer can be calculated by the following formula

$$E_{ads} = -(E_{sub/AlN} - E_{sub} - E_{AlN}) \tag{4.1}$$

where E_{sub}, E_{AlN}, $E_{sub/AlN}$ represent the energy of the substrate, AlN epitaxial layer, and the interface of the substrate and AlN, respectively. Positive binding energy corresponds to a stable adsorption state. In the SiC/AlN system, the 4H-SiC substrate is represented by a model of bilayer (BL) thickness, and the bottom layer is passivated by hydrogen atoms. Considering the small lattice mismatch (1%) between SiC and AlN, the lattice constant of their interface is fixed as the lattice constant of SiC. In the graphene/AlN system, the graphene unit cell (composed of 14 C atoms) is made to match the AlN 2 × 2 unit cell by the rotation, which can greatly reduce the stress at the interface. The effect of its thickness is studied by designing AlN of 4 BL and 8 BL thickness, and the top atoms are passivated by pseudo-hydrogen atoms

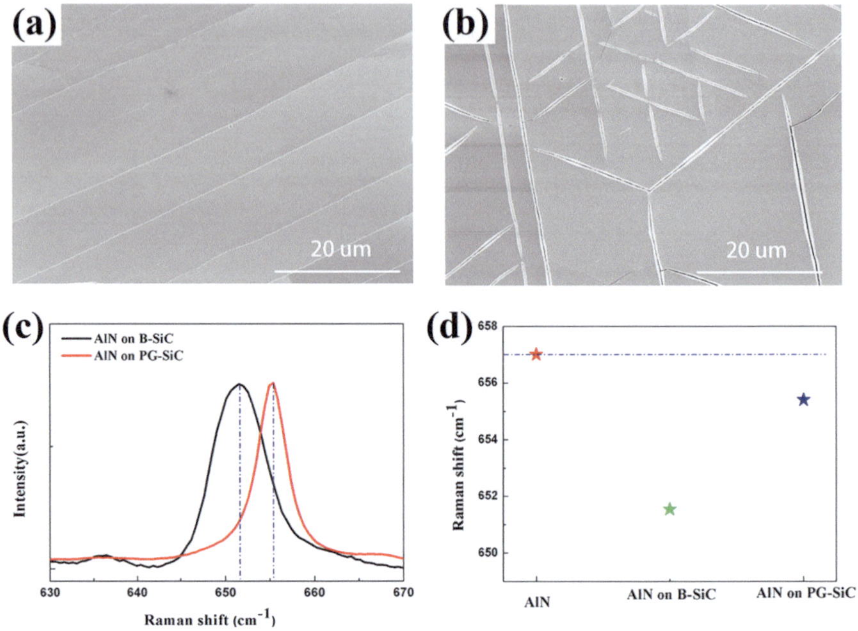

Fig. 4.20 Stress relaxation analyses of AlN films. SEM images of AlN films grown on different substrate: **a** PG-SiC and **b** B-SiC, revealing the reduction of cracks with the help of graphene. **c** Raman shifts of E_2-high peak of AlN films on B-SiC and PG-SiC substrates. **d** Relative Raman shifts of bulk AlN, AlN on B-SiC, and AlN on PG-SiC [50]

to remove the interference of dangling bonds. For 4H-SiC bulk material, AlN bulk material and single-layer graphene, their lattice constants are set to 3.09 Å, 3.12 Å, and 2.47 Å, respectively, consistent with experimental values [51]. In the experiment, double-layer graphene grown on a SiC substrate was used as a buffer layer for AlN thin film growth. Therefore, the main focus was on the interaction between AlN and graphene, using the AlN/graphene interface to study the qvdWE of AlN thin films, while ignoring the influence of the SiC substrate. The calculation model was shown in Fig. 4.21. Figure 4.21a showed the atomic structure of the AlN/SiC interface, in which the AlN layer was directly connected to the SiC substrate. In contrast, the atomic structure of the AlN/graphene interface was demonstrated in Fig. 4.21b–d, in which the graphene lattice rotated 10.9° relative to the AlN lattice plane. For the 4 BL and 8 BL AlN thin films grown on bare SiC, their binding energies were 0.423 eV/Å^2 and 0.420 eV/Å^2, respectively. Such a huge binding energy could be attributed to the strong Si–N bond and the extremely small lattice mismatch between SiC and AlN, which was also the reason for the high-quality AlN growth on the SiC substrate. At the same time, with the increase of AlN thickness, the binding energy decreased to a certain extent, mainly because the stress energy in the AlN thin film increased. When the AlN thin film was very thick, the stress energy can accumulate to a significant level, and dislocations and defects would occur in the AlN thin film to

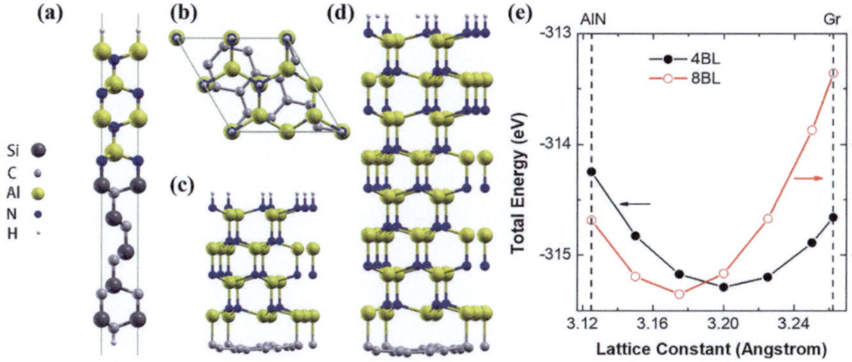

Fig. 4.21 First-principle calculations of AlN/SiC and AlN/graphene interfaces. **a** The geometry of AlN/SiC interface with 4 BL AlN on B-SiC. **b** Top view of the AlN/graphene interface. The side views of AlN/graphene interfaces with **c** 4 BL AlN and **d** 8 BL AlN on graphene. **e** The total energies of AlN/graphene interface with different AlN thicknesses as a function of lattice constant. The large dark gray, small light gray, yellow, blue, and small white balls represent Si, C, Al, N, and H atoms, respectively [50]

release stress. For the 4 BL and 8 BL AlN thin films grown on graphene, first kept the graphene lattice constant unchanged, and stretched the AlN layer in-plane biaxially, with a stretch of about 5%, to make its lattice constant reach 3.26 Å (matching with graphene).

It was found that the binding energy of the AlN thin film grown on graphene was very low, and the adsorption energy of 4 BL AlN on graphene was only 0.017 eV/Å^2, which was an order of magnitude lower than the AlN/SiC system. In addition, the binding energy in this structure significantly decreased with the increase in AlN thickness. When the thickness of the AlN layer was 8 BL, the binding energy decreased to – 0.028 eV/Å^2, where the negative sign indicated that this state was unstable. The low binding energy of AlN on graphene was related to the weak interaction between the two and the large strain in the AlN layer. To separate the effect of stress, the binding energy was decomposed into two parts: one part was the binding energy between the strained AlN layer and graphene, which was calculated to be 0.076 eV/Å^2, and the other part was the strain energy of the AlN layer, estimated to be 0.013 eV/(BL·Å^2). The two had opposite effects on the total binding energy. For 4 BL AlN, the former was still larger than the strain energy, and the total energy was positive, indicating a stable state. However, the strain energy would significantly increase with the increase in AlN thickness. For 8 BL AlN, the strain energy increased to 0.105 eV/Å^2, exceeding the binding energy between AlN and graphene, so the total energy became negative, that was, an unstable state. That was to say, the AlN layer grown on graphene was very unstable due to its weak interface interaction and large stress in the AlN layer. Based on this, the quality of AlN thin films grown on graphene/SiC would be particularly poor, but this was not consistent with experimental results.

Considering the contradictory results between calculations and experiments, there should be some mechanisms that can effectively reduce the stress in the AlN layer.

One possible way was that the lattice constant of AlN decreased continuously with the growth, but this would lead to a large number of bond breaks and reorganizations at the interface, which violated the laws of energetics and kinetics. Another possible way was to reduce the lattice constant of the entire AlN/graphene system. The interaction between the graphene layer and the substrate was weak, and it was easier to compress the thin graphene layer than to expand the thick AlN layer in terms of energetics. The total energy of the AlN/graphene system changed with the lattice constant as shown in Fig. 4.21e. The most suitable lattice constant for epitaxially growing 4 BL AlN on the graphene buffer layer was 3.20 Å. The axial stress in the AlN epitaxial layer was 2.4%, which was much lower than the stress when the inherent lattice constant of graphene was maintained. For 8 BL AlN, the most suitable lattice constant droped to 3.175 Å, making the stress in the AlN epitaxial layer 1.6%. At the same time, calculations had found that the changes in the lattice constant of graphene can increase its strain energy, but it would effectively reduce the strain energy of the AlN layer and the total energy of the AlN/graphene system. As the thickness of AlN increased, the lattice constant of the AlN/graphene system would be closer to the lattice constant of AlN bulk material, thereby reducing the stress of the epitaxial layer [52]. Therefore, the lattice constant of graphene would gradually decrease with the increase in the thickness of the AlN layer. In the calculation process, the effects of the SiC substrate and the underlying graphene (the graphene used was a bilayer structure, only considering the topmost layer) were ignored, and these two would suppress the change of the graphene lattice constant to a certain extent. However, the vdW interaction between different graphene layers was about 0.025 eV/Å^2, which was less than the interaction between graphene and the AlN layer (0.076 eV/Å^2). Therefore, it can be reasonably speculated that the change in the lattice constant of graphene in the experiment should also have a similar rule, although it may be smaller than the calculated value. The speculation was also confirmed by the the displacement of the Raman peak of the graphene layer under different thickness AlN films. When the AlN film was not grown, the 2D peak of graphene was at 2712 cm^{-1}, which was considered to be an intrinsic state without expansion. As the thickness of AlN increased, the 2D peak of graphene began to move to a higher wave number, indicating that graphene was in a compressive strain state, which was consistent with the theoretical conjecture. When the thickness of the AlN layer increased to 0.6 μm, the Raman peak shifted to 2737 cm^{-1}, indicating that the compressive stress received by graphene had reached its maximum. When the thickness of the AlN layer reached 0.9 μm, the wave number of the Raman peak of graphene decreased, indicating that under greater stress, the graphene layer broke and became smaller domain fragments to restore part of the lattice constant and release stress. Based on the above analysis, it was concluded that the graphene layer can act as a flexible sheet that could change its lattice constant to minimize the strain of the epilayer during the growth of AlN. This lattice engineering was of considerable significance to the qvdWE growth of III-nitrides on the graphene layer.

4.4 Summary

In recent years, 2D materials, especially classic graphene and h-BN, as the intermediate layer to assist nitrides in qvdWE, have attracted great attention due to their many advantages. This chapter focuses on the various technical routes of nitride thin film qvdWE assisted by 2D materials on different types of substrates, showing the good crystalline quality of the obtained films, mainly divided into sapphire substrate, PSS, and SiC substrate. The 2D material as a buffer layer can effectively improve the lateral migration rate of nitride metal atoms on the epitaxial substrate surface, thereby promoting the lateral coalescence of the film. At the same time, by shielding the influence of the large mismatch substrate on the epitaxial layer, 2D materials can also effectively improve the epitaxial quality of nitrides on different substrates, and effectively help the epitaxial layer release residual stress. The high-quality low-strain nitride films obtained with the assistance of 2D material engineering provide a high-quality template layer for further growth of III-nitride-based optoelectronic devices, thereby enabling the preparation of devices with stable optoelectronic properties.

References

1. Zhang J, Wang H, Sun W, Adivarahan V, Wu S, Chitnis A, Chen C, Shatalov M, Kuokstis E, Yang J (2003) High quality AlGaN layers over pulsed atomic layer epitaxially grown AlN templates for deep ultraviolet light emitting diodes. J Electron Mater 32:364–370
2. Banal RG, Funato M, Kawakami Y (2008) Initial Nucleation of AlN grown directly on sapphire substrates by metal organic vapor phase epitaxy. Appl Phys Lett 92:241905
3. Wu H, Zhao W, He C, Zhang K, He L, Chen Z (2019) Growth of high quality AlN/sapphire templates with high growth rate using a medium temperature layer. Superlattices Microstruct 125:343–347
4. Walde S, Hagedorn S, Weyers M (2019) Impact of intermediate high temperature annealing on the properties of AlN/sapphire templates grown by metal organic vapor phase epitaxy. Jpn J Appl Phys 58:SC1002
5. Soldano C, Mahmood A, Dujardin E (2010) Production, properties and potential of graphene. Carbon 48:2127–2150
6. Jo G, Choe M, Lee S, Park W, Kahng YH, Lee T (2012) The application of graphene as electrodes in electrical and optical devices. Nanotechnology 23:112001
7. Lin Y, Connell JW (2012) Advances in 2D boron nitride nanostructures: Nanosheets, nanoribbons, nanomeshes, and hybrids with graphene. Nanoscale 4:6908–6939
8. Zhang K, Feng Y, Wang F, Yang Z, Wang J (2017) Two dimensional hexagonal boron nitride (2d-h-BN): Synthesis, properties and applications. J Mater Chem C 5:11992–12022
9. Huang J-K, Pu J, Hsu C-L, Chiu M-H, Juang Z-Y, Chang Y-H, Chang W-H, Iwasa Y, Takenobu T, Li L-J (2014) Large-area and highly crystalline WSe_2 monolayers: From synthesis to device applications. ACS Nano 8:923–930
10. Jo S, Costanzo D, Berger H, Morpurgo AF (2015) Electrostatically induced superconductivity at the surface of WS_2. Nano Lett 15:1197–1202
11. Zhao C, Ng TK, Tseng C-C, Li J, Shi Y, Wei N, Zhang D, Consiglio GB, Prabaswara A, Alhamoud AA, Abdulrahman MA, Alyamani AY, Zhang XX, Li L-J, Ooi BS (2017) InGaN/GaN nanowires epitaxy on large-area MoS_2 for high-performance light-emitters. RSC Adv 7:26665–26672

12. Liang DD, Jiang B, Liu ZT, Chen ZL, Gao YQ, Yang SY, He R, Wang LL, Ran JX, Wang JX, Gao P, Li JM, Liu ZF, Sun JY, Wei TB (2024) Quasi van der Waals epitaxy of single crystalline GaN on amorphous SiO_2/Si(100) for monolithic optoelectronic integration. Adv Sci 11:202305576
13. Han M, Ryu BD, Ko KB, Hong C-H (2023) Quasi van der Waals epitaxial growth of GaN on hexagonal boron nitride via metal-organic chemical vapor deposition. Phys Status Solidi B 260:2200496
14. Sasaki T, Matsuoka T (1995) Analysis of two-step-growth conditions for GaN on an AlN buffer layer. J Appl Phys 77:192–200
15. Nakamura S (1991) GaN growth using GaN buffer layer. Jpn J Appl Phys 30:L1705–L1707
16. Chung K, Lee CH, Yi GC (2010) Transferable GaN layers grown on ZnO-coated graphene layers for optoelectronic devices. Science 330:655–657
17. Choi JK, Huh JH, Kim SD, Moon D, Yoon D, Joo K, Kwak J, Chu JH, Kim SY, Park K, Kim YW, Yoon E, Cheong H, Kwon SY (2012) One-step graphene coating of heteroepitaxial GaN films. Nanotechnology 23:435603
18. Qi Y, Wang Y, Pang Z, Dou Z, Wei T, Gao P, Zhang S, Xu X, Chang Z, Deng B, Chen S, Chen Z, Ci H, Wang R, Zhao F, Yan J, Yi X, Liu K, Peng H, Liu Z, Tong L, Zhang J, Wei Y, Li J, Liu Z (2018) Fast growth of strain-free AlN on graphene-buffered sapphire. J Am Chem Soc 140:11935–11941
19. Prokofyeva T, Seon M, Vanbuskirk J, Holtz M, Nikishin SA, Faleev NN, Temkin H, Zollner S (2001) Vibrational properties of AlN grown on (111)-oriented silicon. Phys Rev B: Condens Matter Mater Phys 63:125313
20. Mun D-H, Bae H, Bae S, Lee H, Ha J-S, Lee S (2014) Stress relaxation of GaN microstructures on a graphene-buffered Al_2O_3 substrate. Phys Status Solidi-R 8:341–344
21. Bae S-H, Lu K, Han Y, Kim S, Qiao K, Choi C, Nie Y, Kim H, Kum HS, Chen P, Kong W, Kang B-S, Kim C, Lee J, Baek Y, Shim J, Park J, Joo M, Muller DA, Lee K, Kim J (2020) Graphene-assisted spontaneous relaxation towards dislocation-free heteroepitaxy. Nat Nanotech 15:272–276
22. Shi B, Liu ZT, Li Y, Chen Q, Liu JX, Yang KL, Liang M, Yi XY, Wang JX, Li JM, Kang JJ, Gao P, Liu ZQ (2024) Atomic evolution mechanism and suppression of edge threading dislocations in nitride remote heteroepitaxy. Nano Lett 24:7458–7466
23. Lebedev V, Cimalla V, Pezoldt J, Himmerlich M, Krischok S, Schaefer JA, Ambacher O, Morales FM, Lozano JG, Gonzalez D (2006) Effect of dislocations on electrical and electron transport properties of InN thin films. I. Strain relief and formation of a dislocation network. J Appl Phys 100:094902
24. Chang HL, Liu ZT, Yang SY, Gao YQ, Shan JY, Liu BY, Sun JY, Chen ZL, Yan JC, Liu ZQ, Wang JX, Gao P, Li JM, Liu ZF, Wei TB (2022) Graphene-driving strain engineering to enable strain-free epitaxy of AlN film for deep ultraviolet light-emitting diode. Light Sci Appl 11:88
25. Chen Z, Liu Z, Wei T, Yang S, Dou Z, Wang Y, Ci H, Chang H, Qi Y, Yan J, Wang J, Zhang Y, Gao P, Li J, Liu Z (2019) Improved epitaxy of AlN film for deep-ultraviolet light-emitting diodes enabled by graphene. Adv Mater 31:1807345
26. Wu Y, Hanlon A, Kaeding JF, Sharma R, Fini PT, Nakamura S, Speck JS (2004) Effect of nitridation on polarity, microstructure, and morphology of AlN films. Appl Phys Lett 84:912–914
27. Kong W, Li H, Qiao K, Kim Y, Lee K, Nie Y, Lee D, Osadchy T, Molnar RJ, Gaskill DK, Myers-Ward RL, Daniels KM, Zhang Y, Sundram S, Yu Y, Bae SH, Rajan S, Shao-Horn Y, Cho K, Ougazzaden A, Grossman JC, Kim J (2018) Polarity governs atomic interaction through two-dimensional materials. Nat Mater 17:999–1004
28. Chung K, Oh H, Jo J, Lee K, Kim M, Yi G-C (2017) Transferable single-crystal GaN thin films grown on chemical vapor-deposited hexagonal BN sheets. NPG Asia Mater 9:e410
29. Zhao Y, Wu X, Yang J, Zeng XC (2012) Oxidation of a two-dimensional hexagonal boron nitride monolayer: A first-principles study. Phys Chem Chem Phys 14:5545–5550
30. Wang LL, Yang SY, Gao YQ, Yang JK, Duo YW, Song S, Yan JC, Wang JX, Li JM, Wei TB (2023) Quasi van der Waals epitaxy of a stress-released AlN film on thermally annealed

hexagonal BN for deep ultraviolet light-emitting diodes. ACS Appl Mater Interfaces 11:23501–23511
31. Wang LL, Yang SY, Zhou F, Gao YQ, Duo YW, Chen RF, Yang JK, Yan JC, Wang JX, Li JM, Zhang YF, Wei TB (2024) Wafer-scale transferrable GaN enabled by hexagonal boron nitride for flexible light-emitting diode. Small 20:2306132
32. Wu QQ, Yan JC, Zhang L, Chen X, Wei TB, Li Y, Liu ZQ, Wei XC, Zhang Y, Wang JX, Li JM (2017) Growth mechanism of AlN on hexagonal BN/sapphire substrate by metal-organic chemical vapor deposition. CrystEngComm 19:5849–5856
33. Liu F, Yu Y, Zhang YT, Rong X, Wang T, Zheng XT, Sheng BW, Yang LY, Wei JQ, Wang XP, Li XB, Yang XL, Xu FJ, Qin ZX, Zhang ZH, Shen B, Wang XQ (2020) Hexagonal BN-assisted epitaxy of strain released GaN films for true green light-emitting diodes. Adv Sci 7:200917
34. Tangi M, Mishra P, Ng TK, Hedhili MN, Janjua B, Alias MS, Anjum DH, Tseng C-C, Shi Y, Joyce HJ, Li L-J, Ooi BS (2016) Determination of Band Offsets at GaN/single-layer MoS_2 Heterojunction. Appl Phys Lett 109:032104
35. Gupta P, Rahman AA, Subramanian S, Gupta S, Thamizhavel A, Orlova T, Rouvimov S, Vishwanath S, Protasenko V, Laskar MR, Xing HL, Jena D, Bhattacharya A (2016) Layered transition metal dichalcogenides: promising near-lattice-matched substrates for GaN growth. Sci Rep 6:23708
36. Yin Y, Ren F, Wang Y, Liu Z, Ao J, Liang M, Wei T, Yuan G, Ou H, Yan J, Yi X, Wang J, Li J (2018) Direct van der Waals epitaxy of crack-free AlN thin film on epitaxial WS_2. Materials 11:2464
37. Song WR, Chen Q, Yang KL, Liang M, Yi XY, Wang JX, Li JM, Liu ZQ (2023) Recent advances in mechanically transferable III-nitride based on 2D buffer strategy. Adv Funct Mater 33:2209880
38. Kim M, Fujita T, Fukahori S, Inazu T, Pernot C, Nagasawa Y, Hirano A, Ippommatsu M, Iwaya M, Takeuchi T, Kamiyama S, Yamaguchi M, Honda Y, Amano H, Akasaki I (2011) AlGaN-based deep ultraviolet light-emitting diodes fabricated on patterned sapphire substrates. Appl Phys Express 4:092102
39. Dong P, Yan J, Wang J, Zhang Y, Geng C, Wei T, Cong P, Zhang Y, Zeng J, Tian Y, Sun L, Yan Q, Li J, Fan S, Qin Z (2013) 282-nm AlGaN-based deep ultraviolet light-emitting diodes with improved performance on nano-patterned sapphire substrates. Appl Phys Lett 102:241113
40. Imura M, Nakano K, Narita G, Fujimoto N, Okada N, Balakrishnan K, Iwaya M, Kamiyama S, Amano H, Akasaki I, Noro T, Takagi T, Bandoh A (2007) Epitaxial lateral overgrowth of AlN on trench-patterned AlN layers. J Cryst Growth 298:257
41. Chang H, Chen Z, Liu B, Yang S, Liang D, Dou Z, Zhang Y, Yan J, Liu Z, Zhang Z, Wang J, Li J, Liu Z, Gao P, Wei T (2020) Quasi-2D growth of aluminum nitride film on graphene for boosting deep ultraviolet light-emitting diodes. Adv Sci 7:2001272
42. Chang H, Chen Z, Li W, Yan J, Hou R, Yang S, Liu Z, Yuan G, Wang J, Li J, Gao P, Wei T (2019) Graphene-assisted Quasi van der Waals epitaxy of AlN film for ultraviolet light emitting diodes on nano-patterned sapphire substrate. Appl Phys Lett 114:091107
43. Ke W, Chiang C, Lin Y, Liao Y, Cheng W, Chang K, Lin J (2023) Improved wavelength stability and heat dissipation of InGaN-based light-emitting diodes using a graphene interlayer on patterned sapphire substrate. Appl Phys Lett 122:203501
44. Ismach A, Druzgalski C, Penwell S, Schwartzberg A, Zheng M, Javey A, Bokor J, Zhang Y (2010) Direct chemical vapor deposition of graphene on dielectric surfaces. Nano Lett 10:1542–1548
45. Virojanadara C, Syvjarvi M, Yakimova R, Johansson LI, Zakharov AA, Balasubramanian T (2008) Homogeneous large-area graphene layer growth on 6H-SiC (0001). Phys Rev B 78:245403
46. Zhang F, Chen X, Zuo Z, Qin X, Xu X, Zhao X (2018) High performance metal-graphene-metal photodetector employing epitaxial graphene on SiC (0001) surface. J Mater Sci-Mater Eletron 29:5180–5185
47. Kim J, Bayram C, Park H, Cheng CW, Dimitrakopoulos C, Ott JA, Reuter KB, Bedell SW, Sadana DK (2014) Principle of direct van der Waals epitaxy of single-crystalline films on epitaxial graphene. Nat Commun 5:4836

48. Qiao K, Liu Y, Kim C, Molnar R, Osadchy T, Li W, Sun X, Li H, Myers-Ward R, Lee D, Subramanian S, Kim H, Lu K, Robinson J, Kong W, Kim J (2021) Graphene buffer layer on SiC as a release layer for high-quality freestanding semiconductor membrane. Nano Lett 21:4013–4020
49. Okumura H, Kimoto T, Suda J (2014) Formation mechanism of threading-dislocation array in AlN layers grown on 6H-SiC (0001) substrates with 3-bilayer-high surface steps. Appl Phys Lett 105:071603
50. Wang Y, Yang S, Chang H, Li W, Chen X, Hou R, Yan J, Yi X, Wang J, Wei T (2020) Flexible graphene-assisted van der Waals epitaxy growth of crack-free AlN epilayer on SiC by lattice engineering. Appl Sur Sci 520:146358
51. Baskin Y, Meyer L (1955) Lattice constants of graphite at low temperatures. Phys Rev 100:544–544
52. Chae SJ, Kim YH, Seo TH, Duong DL, Lee SM, Park MH, Kim ES, Bae JJ, Lee SY, Jeong H, Suh E-K, Yang CW, Jeong MS, Lee YH (2015) Direct growth of etch pit-free GaN crystals on few-layer graphene. RSC Adv 5:1343–1349

Chapter 5
Large Mismatch Substrate Nitrides Quasi van der Waals Epitaxy

As a non-covalent epitaxy, QvdWE can greatly alleviate the lattice mismatch and thermal mismatch between the epitaxial layer and the substrate, which can improve the quality of nitrides epitaxial on the traditional substrate while expanding the range of substrate choices, and become a tool for exploring the nitride epitaxial on the new substrates. Commonly referred to as new substrates mainly include: metals and metal alloys, amorphous glass, amorphous plastics and so on. Compared with the traditional sapphire, Si, SiC, and other more commonly used single-crystal substrates, the above new amorphous substrates have many unique advantages, but there are also some problems.

In this chapter, QvdWE of nitrides on amorphous glass substrates, metal substrates, and other large mismatch substrates will be presented separately, with an exhaustive description of the current progress of research in each section, and the current difficulties encountered in nitride epitaxy on a variety of new substrates will be elaborated.

5.1 Amorphous Glass Substrate for Quasi van der Waals Epitaxy Nitride

Amorphous glass includes quartz glass (pure SiO_2) and ordinary glass (Na_2SiO_3, $CaSiO_3$, SiO_2 or Na_2O–CaO–$6SiO_2$). Compared with other amorphous substrates, glass substrates are mature in process, inexpensive, not subject to size limitations and high transparency, but also compatible with the liquid crystal process. They not only can significantly reduce the cost but also be conducive to the realization of large-area film preparation. Because the softening temperature of ordinary glass is very low, between 500 and 600 °C, at high temperatures, ordinary glass will not only soften but also cause the epitaxial film penetration impurities, resulting in unintentional doping, and deteriorating the quality of nitride films. Therefore, the

T. Wei et al., *Quasi van der Waals Epitaxy of Nitride Semiconductor and Its Applications*, Wide Bandgap Semiconductors,
https://doi.org/10.1007/978-981-95-3469-2_5

traditional MOCVD method for growing GaN materials can only use quartz glass with a very high softening temperature (1300–1500 °C) and a small coefficient of thermal expansion as the substrate.

It is very difficult to grow high-quality single-crystal nitride films by direct epitaxy on amorphous glass. Thus, it is important to choose materials as the buffer layers that have a small lattice mismatch with the nitride and have good wettability and are easy to attach and nucleate on amorphous planes. QvdWE allows the epitaxial layer to have a large lattice mismatch and different lattice symmetry with the substrate by introducing 2D materials, so QvdWE can be used to reduce the difficulty of growing single-crystalline nitride thin films on amorphous quartz glass substrates.

The most commonly used 2D buffer layer material for nitride epitaxy on amorphous quartz substrates is graphene, which has many advantages over other materials: (i) The lattice mismatch requirement is negligible; (ii) The growth temperature can be lowered, and the migration barrier of group III metals on graphene is very low, and adsorbed atoms can easily diffuse on the surface; (iii) The length of the atomic diffusion is larger, which promotes the creation of large 2D islands that can be polymerized into a film of small dislocation density. The planar hexagonal structure similar to that of wurtzite GaN (0001) surface can influence the lattice structure of the epitaxial layer as well as possess a good thermal conductivity that can reduce the temperature of the device. Since its preparation technology is also relatively mature, the QvdWE of nitride on amorphous substrates has mainly focused on the graphene buffer layer in recent years, and the main research progress is as follows:

In 2012, Kunook Chung et al. of Seoul National University [1] drew on their own experience of growing GaN films on sapphire substrates and performed GaN growth directly on SiO_2 substrates as shown in Fig. 5.1. They first transferred the CVD-grown graphene from a copper foil to the surface of a polycrystalline SiO_2 substrate, keeping its morphology flat. Then, they subjected the graphene to oxygen plasma treatment to make a certain number of nucleation sites appear. Subsequently, they grew ZnO nanowalls on graphene as an intermediate layer. Based on this, they prepared GaN films using a LT layer-lateral merger layer and a high-temperature epitaxial layer growth sequence. Compared with the graphene-free region, the GaN films grown in the region with a graphene buffer layer demonstrated a flatter and smoother surface, which showed the possibility of epitaxial nitride films on amorphous substrates, although large cracks appeared.

To further examine the quality of the films, they subsequently performed XRD tests on the grown GaN film samples. The test results showed that the FWHM of the X-ray diffraction rocking curve (002) for GaN films grown on a polycrystalline substrate with CVD graphene films could be as low as 0.8°, which indicated that the GaN films were well oriented on the c-axis. The value of FWHM using a ZnO layer as an intermediate layer was much smaller than the previously reported 3–6° of GaN films grown with mechanically exfoliated graphene layers on SiO_2 substrates. This result strongly suggested that the graphene film played a crucial role in the formation of GaN films with good vertical orientation. They also investigated the in-plane orientation of GaN films by measuring {112} diffraction φ scans. As shown in Fig. 5.2c, the φ scans of GaN films grown without CVD graphene films (blue

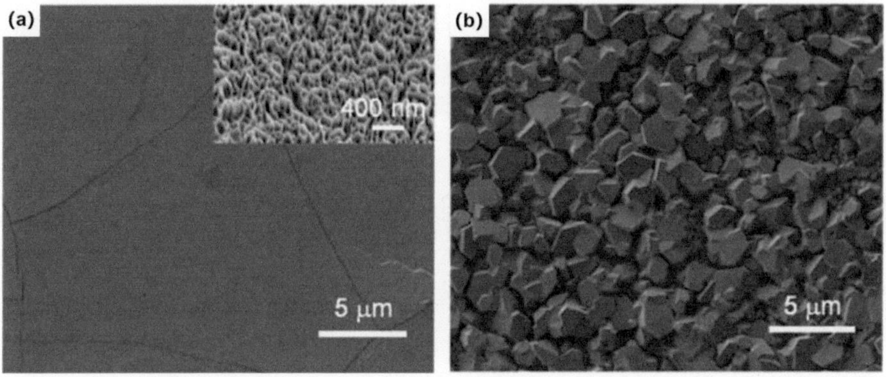

Fig. 5.1 **a** GaN SEM illustration of graphene region on SiO_2 substrate. Inset is SEM illustration of ZnO nanowall. **b** Graphene-free region on SiO_2 substrate [1]

solid line) showed no obvious peaks, and the φ scans of GaN films grown on CVD graphene films (black solid line) contain three sets of lattices. This result indicated that the GaN grains in the films had large grain angle boundaries and preferred in-plane orientations, which were caused by the use of CVD graphene films.

In the following year, Yoo et al. [2] of Seoul National University continued a similar approach using ZnO nanowalls as an intermediate layer to grow GaN films on oxygen plasma-treated CVD graphene for structural analysis, which was subsequently characterized as shown in Fig. 5.3. SEM results showed a certain number of pits and cracks on the surface of the film. The orientation of the film was subsequently characterized using EBSD, and its c-direction growth directions were all in the (001) plane, while the out-plane orientations were not uniform, with the presence of obvious grain boundaries. The EBSD grain boundary dislocation angle mapping images further revealed that both high-angle and low-angle grain boundaries existed in this film, and the grain sizes corresponding to the high-angle grain boundaries ranged from a few micrometers to tens of micrometers. Through the in-depth characterization of this film, a clearer understanding of the GaN films grown by this method had been obtained, which would help in the subsequent work.

Baek et al. [3] of Seoul National University also performed GaN microdisk array growth on amorphous SiO_2/Si substrate using micropatterned graphene dots as localized nucleation layers. Figure 5.4a shows a schematic diagram of the method, in which the graphene film synthesized on Ni film by applying the CVD method was first transferred to the SiO_2/Si substrate using a standard transfer method. Then, the large-size graphene film was formed using electron beam lithography. A resist was used as an etch mask, and the rest of the graphene film was then removed by etching through an oxygen plasma at a plasma power of 30 mA for 30 s to form regular hexagonal arrays with a diameter of 3 μm and a dot spacing of 10 μm in Fig. 5.4b. Before GaN growth, ZnO nanostructures were grown on graphene micropatterns using MOCVD to enhance GaN nucleation and crystallization. As shown in SEM images of Fig. 5.4c, highly networked 600 nm tall and 20 nm thick

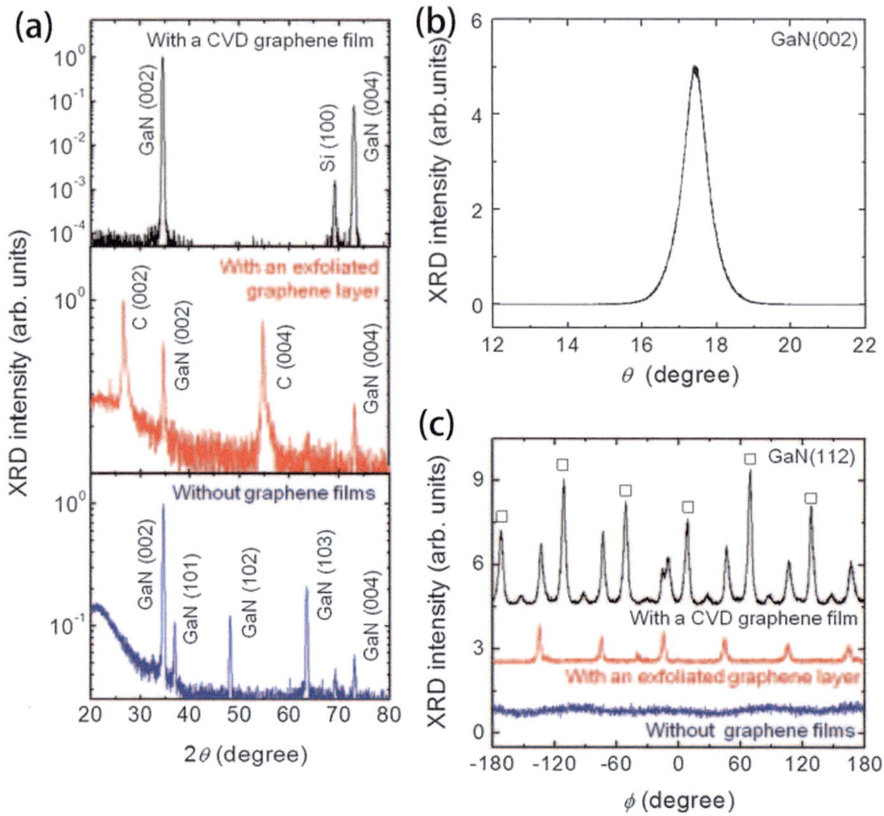

Fig. 5.2 **a** θ–2θ scans of GaN films with CVD graphene film (top), mechanically exfoliated graphene layer (middle) and graphene-free film (bottom). **b** XRD rocking curves of GaN films on CVD graphene films. **c** XRD φ scan curves of GaN films epitaxialized using CVD graphene film (black solid line), mechanically exfoliated graphene layer (red solid line) and graphene-free film (blue solid line) [1]

ZnO nanowalls were grown only inside the graphene micropattern. This intermediate layer played a key role in the heterogeneous epitaxial growth of high-quality GaN films on graphene layers. Finally, GaN microdisks were grown by ELOG on ZnO-coated graphene microdot arrays using a two-step process. Subsequent characterization of the grown GaN microdisks demonstrated hexagonal arrays of GaN microdisks with flat top surfaces and sidewalls grown on micropatterned graphene films, as shown in Fig. 5.4d. The diameter of individual hexagons extended from 3 μm in the original graphene pattern to 8.6 ± 0.4 μm. This indicated that the crystalline GaN microdisks were overgrown on the ZnO nanowalls in both lateral and vertical directions. The natural formation of hexagonal microdisks with flat sides did not require further etching, which corroborated the high quality of the epitaxial GaN microdisks.

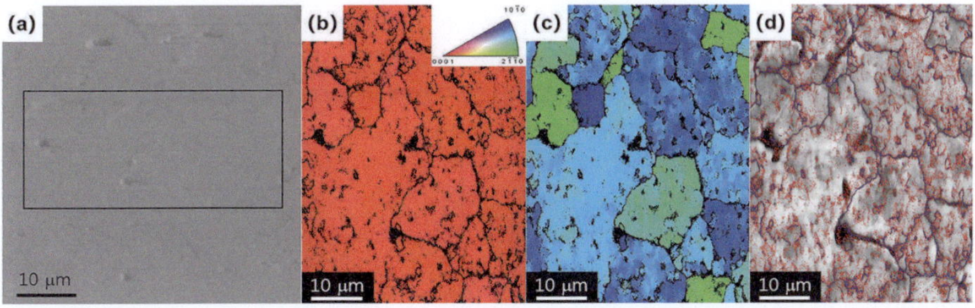

Fig. 5.3 **a** 70° tilt view SEM image of GaN thin film used for EBSD analysis. **b** EBSD normal direction antipodal map of GaN thin film. **c** EBSD transverse direction antipodal map of GaN thin film. **d** EBSD grain boundary misorientation angle mapping image (red > 1°, blue > 3°). The grain boundaries with orientation dislocation angles less than 3° were defined as low-angle grain boundaries, while those with orientation dislocation angles greater than 3° were defined as high-angle grain boundaries [2]

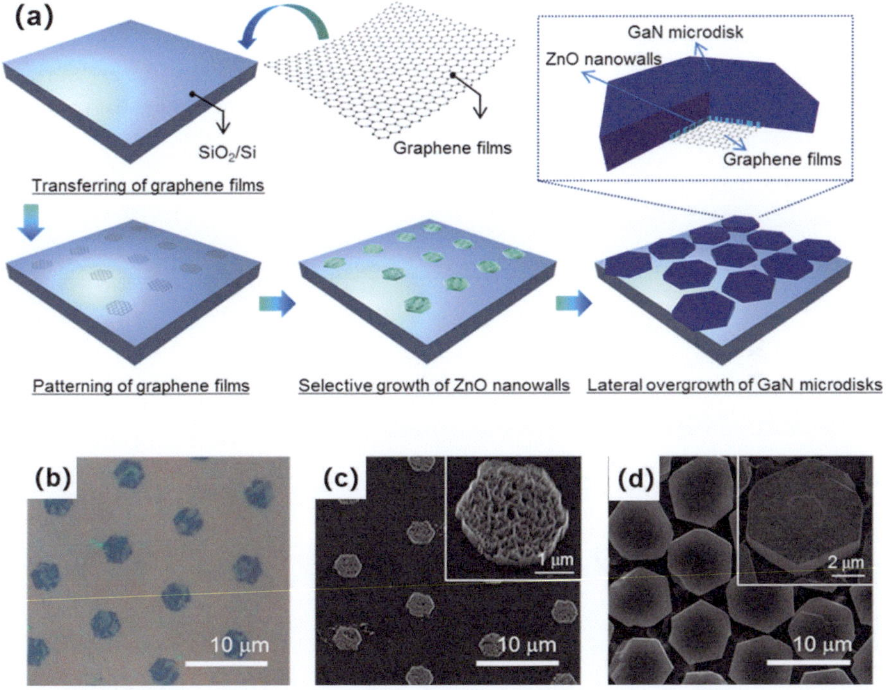

Fig. 5.4 **a** Schematic of the fabrication process used to grow hexagonal GaN microdisk arrays on CVD graphene films. **b** Optical microscopy images of hexagonal patterned CVD graphene films. **c**, **d** Planar SEM images of ZnO nanowalls grown on patterned graphene films and grown hexagonal GaN microdisks, respectively. The insets in **c** and **d** were the corresponding 30° tilted SEM images [3]

To reduce the enhancement of complexity at the interface by microstructures, Chae et al. [4] of Sungkyunkwan University treated the graphene surface transferred to SiO_2 using 50 cycles of ozone plasma, and then directly carried out the MOCVD epitaxial growth of GaN thin films without the aid of microstructures and other buffer layers, as shown in Fig. 5.5. After the growth at high temperature of 3 h (1040 °C), the GaN polycrystalline domains grown on SiO_2/Si substrate still failed to merge into a film, and there were big gaps between them. However, most of the line dislocations in GaN were perpendicular to the growth direction of [0002] under this growth condition, and no corrosion craters were observed after KOH thermo etching.

To further understand the properties of the sample, they performed TEM characterization. Figure 5.6a shows TEM images of GaN near the graphene interface. The inset clearly showed that the crystal structure had no defects such as voids or cracks. The thickness of the multilayer graphene (10 layers) varied slightly and still showed the characteristics of multilayer graphene on nickel substrate, which was also clearly visible at the interface. In addition, the HRTEM images of the GaN crystals in Fig. 5.6b presented well-arranged lattice arrays. The lattice spacing between neighboring planes was 0.52 nm, which matched well with the d-spacing of GaN (0001) planes grown on sapphire. The electron diffraction pattern further verified that the sample was a single crystal material. This indicated that despite the presence of polycrystalline domains that were not fully merged into a film, the sample still

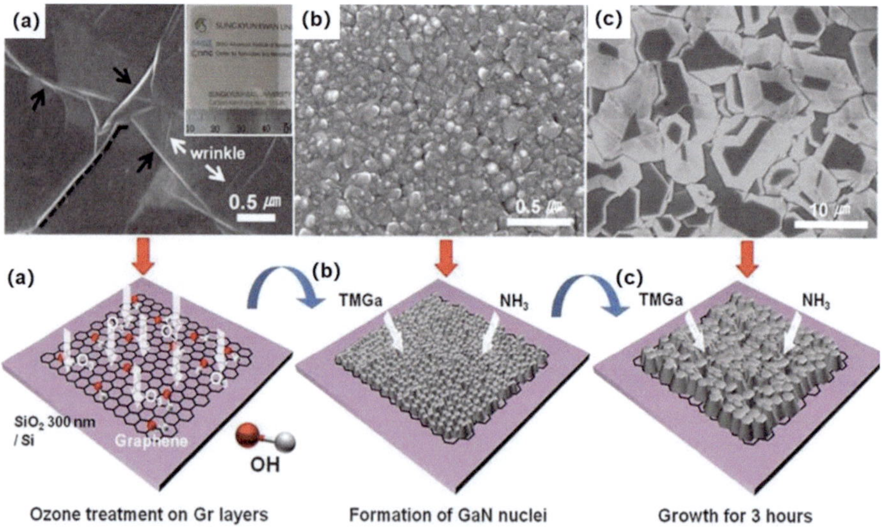

Fig. 5.5 **a** SEM image and corresponding schematic of the graphene surface after synthesis. Black arrows indicated bumpy structures formed during graphene transfer. The inset showed graphene transferred on a PET film (four 3.5 cm) with 80% transmittance at 550 nm. **b** SEM image and corresponding schematic of initial nucleation grown at 530 °C for 5 min. **c** SEM image and corresponding schematic of GaN crystals grown at 1040 °C for 3 h after the initial nucleation step [4]

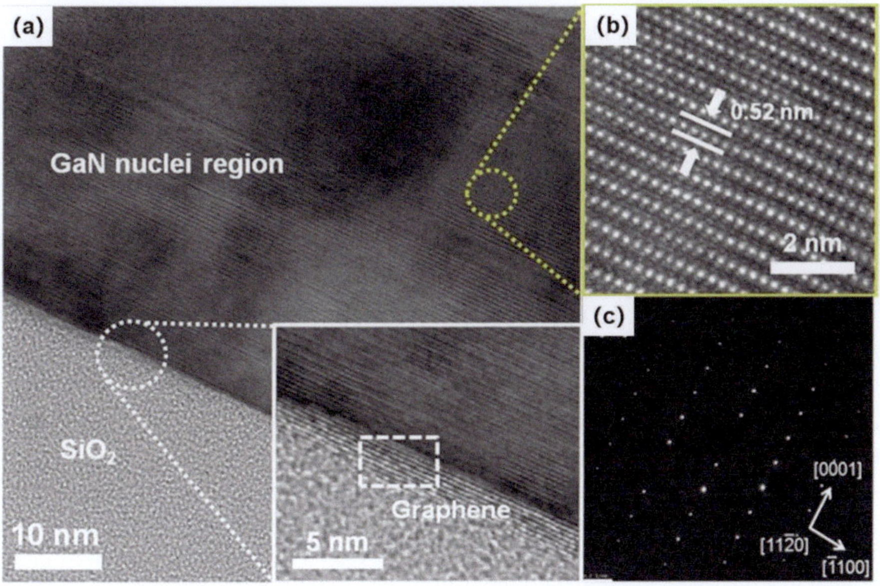

Fig. 5.6 **a** Low magnification cross-sectional TEM image of GaN nucleation region on graphene. The number of graphene layers is shown in the inset. **b** Corresponding high-resolution TEM image. **c** Electron diffraction pattern of GaN crystals grown on Gr/SiO_2/Si [4]

exhibited single-crystal properties in certain regions. This work was a useful attempt for the next step in the direct growth of single-crystal GaN films on 2D materials.

In addition to the MOCVD growth method, the preparation of GaN materials by pulsed sputter deposition (PSD) is also a hot research topic nowadays. Shon et al. [5] of the University of Tokyo prepared GaN thin films with high quality by pulsed deposition method, and the results were shown in Fig. 5.7. They first grew multilayer graphene on Ni substrate by CVD and then transferred it to the amorphous substrate to deposit a 50 nm thick AlN layer and a 1000 nm thick GaN layer by PSD equipment. Smooth and flat GaN films were obtained on the amorphous substrate after the addition of graphene, and the whole epitaxial layer was grown along the c-axis without yellow light bands, and the crystal quality was very good. By continuing the growth of the MQWs layer, red, green and blue three-color LEDs were prepared, in which the IQE of the green LED was 7.4%. During the use of sputtering method for nitride epitaxy, the growth temperature used is very low, which can greatly reduce the production cost and is not limited by the substrate material. Only the quality of the grown film needs to be further improved.

In 2021, the nanorod-assisted vdWE of nearly single-crystalline GaN films was realized for the first time by Liu et al. using graphene-coated amorphous silica glass substrates, and flexible inorganic LEDs were successfully fabricated [6]. The graphene layer effectively guided the lattice arrangement of nitride at the initial growth stage, avoiding the disorderly polycrystalline structure of nitride during its

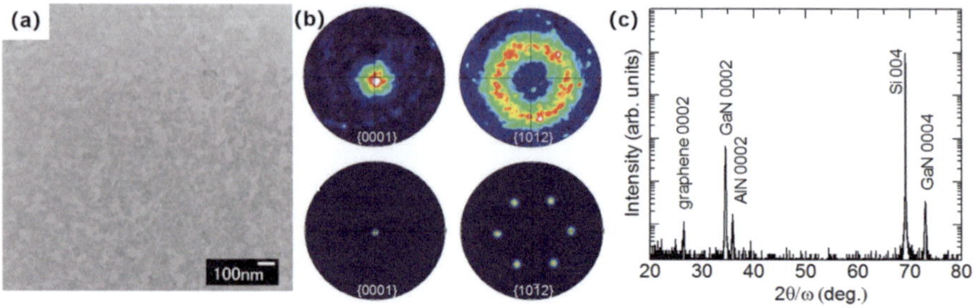

Fig. 5.7 **a** SEM image, **b** {0002} {10–12} EBSD and **c** XRD characterization for GaN materials prepared by PSD [5]

growth on amorphous substrates. Meanwhile, incorporating the nanocolumn buffer layer avoided the excessive accumulation of nitride grains on the graphene surface. Subsequently, ELOG of the nanorods leads to the coalescence of continuous and flat high-quality GaN films. By utilizing the weak van der Waals forces at the nitride/graphene interface, they mechanically released the growing epitaxial layer and transferred it to a flexible conductive polymer film to produce flexible LEDs. To verify the optical characteristics of blue LEDs, the PL spectra were measured. The fabricated flexible LED displayed strong blue luminescence, and the IQE of the blue LED was 48.67% (Fig. 5.8).

Subsequently, the same group first presented the feasibility of single-crystalline nitride epitaxy on an amorphous substrate via an atomic WS2 buffer layer. It is found that the first layer of the nitrogen adatom simultaneously donates electrons to the S atoms performing weak interaction and to the top Ga atoms forming covalent bonds, which plays an important role in the interfacial lattice arrangement. Thereby, successfully performing epitaxy of a single-crystalline GaN film on WS2-glass wafer. This work offers a method to achieve semiconductor epitaxy without the constraint of the substrate lattice, which could also serve as a realistic roadmap for the heterogeneous integration of semiconductors by epitaxy [7].

Wihle active attempts continue to be made by employing graphene, more and more 2D materials are available for scientists to choose from as other 2D material preparation techniques are further improved. Among them, h-BN belongs to the hexagonal crystal system with a layer structure similar to that of graphene, with a hexagonally symmetric honeycomb lattice in the facets. It is formed by alternating arrangements of sp^2-hybridized B and N atoms, with a B-N chemical bond length of 0.144 nm, and an in-face lattice constant of 2.504 nm [8, 9]. In the c-axis direction, the B atoms are arranged alternately against the N atoms, and the layers are connected by vdW forces with a spacing of about 0.333 nm. Compared with graphene, its longitudinal thickness is larger, which is more favorable for shielding the influence of the substrate [10]

In 2017, Chung et al. [11] of Seoul National University realized the direct growth of single-crystal GaN layers on centimeter-scale h-BN. The centimeter-scale h-BN

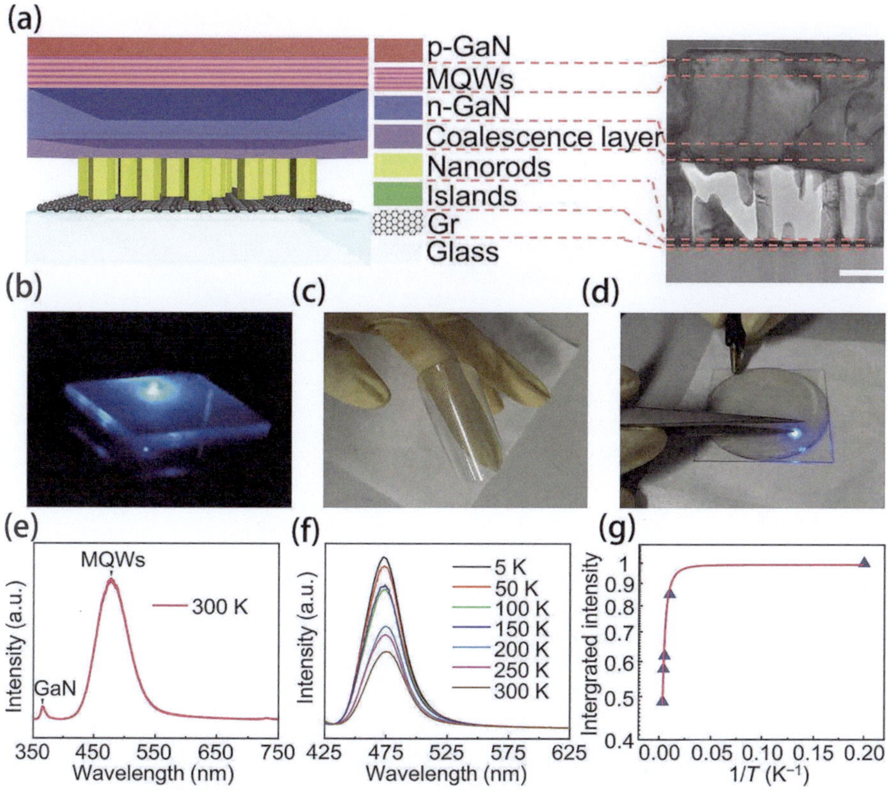

Fig. 5.8 Nitride films grown by vdWE on amorphous graphene-covered glass substrate. **a** Schematic diagram and cross-sectional TEM image of the GaN-based LED. **b** Vertically aligned nanorods with uniform height in a 25° tilted-view SEM image. **c** The SEM image of the continuous and planar GaN films. **d** Photograph of flexible blue LEDs being exfoliated from a glass substrate. RT PL spectrum **e** and temperature-dependent PL spectra **f** of the blue LED. **g** A relatively high IQE is shown by the integrated PL intensity of emissions from MQWs in dependence on temperature [6]

film was first synthesized on single-crystal Ni(111) using CVD, which was transferred to an amorphous fused silica substrate with no epitaxial relationship with GaN. Subsequently, GaN layers with a flat and continuous surface morphology, well-oriented along the c-axis and in-plane crystals were obtained, driven by the atomic steps inherent in the CVD-grown h-BN layers, as shown in Fig. 5.9.

The structural properties of GaN films grown on CVD-grown h-BN layers were examined using XRD measurements. For the θ–2θ scan results, two major XRD diffraction peaks at 34.57° and 72.91°, corresponding to the (0002) and (0004) orientations of the wurtzite GaN, respectively, can be observed in the 20°–80° measurement range in Fig. 5.10a. No other XRD peaks from the very thin h-BN layer and the amorphous fused silica support substrate were measured. The inset in Fig. 5.10a showed the rocking curve of the GaN (0002) peak, where the FWHM of the measured

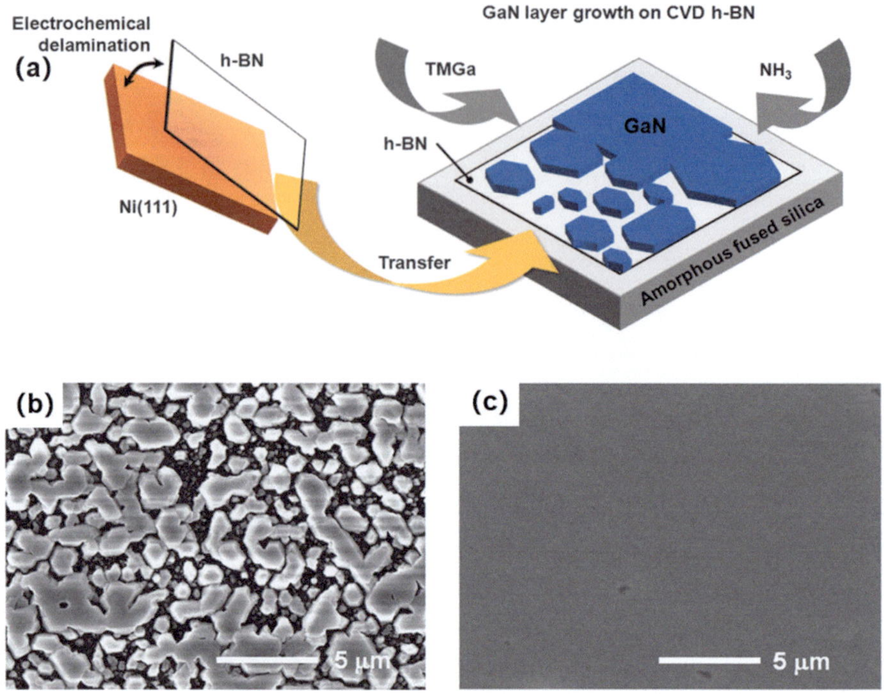

Fig. 5.9 **a** Schematic of CVD-grown h-BN transferred from Ni(111) to an amorphous fused silica substrate and of GaN layer growth using MOCVD. **b** SEM images of high-density discrete GaN islands that almost merged into GaN microstructures of a few micrometers in size and **c** a fully agglomerated GaN film grown on CVD-grown h-BN [11]

rocking curve was 2.37°. Although the FWHM of peak was slightly larger than that of the GaN film grown on the CVD-grown graphene film, both the results of the θ–2θ scan and the rocking curve indicated that the GaN film had a better c-axis orientation grown on the CVD-grown h-BN. In addition, azimuthal scans in the GaN (10–12) direction showed diffraction peaks with sixfold symmetry, as shown in Fig. 5.10b. The monocrystallinity of GaN layers grown on CVD-grown h-BN was also investigated by EBSD. The EBSD results further showed that the GaN films exhibited a single orientation despite the presence of low-angle grain boundaries with 4° intrafacial orientation differences. The (0002) polar plot showed well-aligned crystal orientations along the c-axis, and the (10–10) polar plot clearly demonstrated the sixfold symmetry of the GaN film.

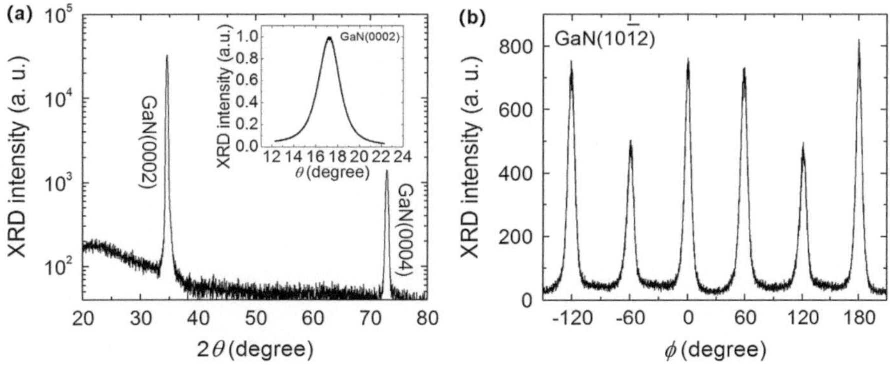

Fig. 5.10 **a** θ–2θ scan, inset showing the rocking curve of the GaN (0002) peak. **b** ϕ-scan in the GaN (10–12) direction for GaN films grown on h-BN grown by CVD [11]

5.2 Metal Substrate Nitride Epitaxy

Great progress has been made in the epitaxy of III-group nitrides on sapphire, SiC, Si, GaAs and GaN substrates. Sapphire is widely used as a substrate for epitaxial growth of nitride devices because it maintains its physical and chemical stability under high-temperature growth conditions of nitride films. However, because (i) the lattice mismatch between sapphire and GaN is as high as 15%; [12] (ii) the thermal mismatch between sapphire and GaN at room temperature is high (about − 18.4%); (iii) sapphire has a low thermal conductivity (0.25 W/cm-K at 100 °C); [13] (iv) sapphire is an insulator and there is a lateral flow of the electric current through the device, therefore the current is not uniformly distributed, generating more heat. These drawbacks largely affect the electrical and optical properties of GaN-based LED devices.

Compared with the traditional sapphire substrate, metal as the substrate material for epitaxial nitride has three unique advantages. First, metals have high thermal conductivity, such as Ag and Cu thermal conductivity of 4.19 and 3.95 W/cm-K, respectively, which can conduct the heat generated within the LED device in a timely manner to reduce the temperature of the device's node area and thus to improve the device's IQW. Second, metal can be used as the substrate material for growing GaN-based vertically structured LED devices, and the current inside the device flows almost entirely vertically through the epitaxial layer of the GaN base, which results in a uniform current distribution and a reduction in the heat generated by the current, which is beneficial to the heat dissipation of the device. In addition, it is not necessary to plate the cathode electrode by corroding the epitaxial layer, which makes full use of the material of the active layer. Third, the metal substrate material is cheaper compared to other substrates, reducing the manufacturing cost of the device.

However, epitaxial growth of III-group nitrides on metal substrates also faces three major challenges. First, the forbidden bandwidth of metal materials is small, which has an absorption effect on light, and the light emitted from the active layer will be strongly absorbed on the surface of the substrate, which reduces the luminous rate of the LED. Secondly, the high coefficient of thermal expansion of metal materials will lead to huge compressive stress in the epitaxial layer, which is more likely to cause cracking of the epitaxial film than sapphire. Finally, metal materials are chemically unstable at high temperatures. When the epitaxial temperature is higher than 700 °C, an interfacial reaction between the epitaxial nitride and the metal substrate occurs, which seriously affects the quality of the epitaxial film. Scientists such as H. Kinoshita et al. have tried to apply traditional MOCVD or MBE technology to epitaxially grow nitrides directly on chemically variable substrates, and found that it was quite difficult to epitaxialize films at high temperature [14].

In order to give the unique advantages of metal substrates and overcome the difficulties of epitaxial growth of nitride on metal substrates, researchers have carried out a lot of research and developed techniques, mainly including LLO and pulsed laser deposition (PLD). This part will discuss how to realize the epitaxial growth of nitride films on metal substrates.

5.2.1 Laser Stripping and Pulsed Laser Deposition Technologies

The laser stripping technology utilizes laser energy to break down the GaN buffer layer at the GaN/sapphire interface, thus realizing the separation of the LED epitaxial wafer from the sapphire substrate. The advantages of the technology are that the epitaxial wafer is transferred to a high thermal conductivity heat sink, which can improve the current expansion in large-size chips. n-side is the light emission side, and thus the light-emitting area is enlarged, and the electrodes block the light less. It is easy to prepare microstructures, and it reduces the number of etching, milling, and scribing, and what's more important is that the sapphire substrate can be reused. Laser stripping technology is mature and commercialized.

PLD is a technique that utilizes a laser to bombard an object and then deposits the bombarded material on a different substrate to obtain a precipitate or a thin film. Because the pulsed laser energy can provide high kinetic energy for the film precursor, it largely reduces the growth temperature of III-group nitride films, thus inhibiting the interfacial reaction between the epitaxial layer and the substrate [15]. In addition, PLD technology is easy to obtain multicomponent films with desired stoichiometric ratios, and has the characteristics of high deposition rate, short experimental period, and arbitrary adjustment of process parameters, which provides a good opportunity for low-temperature growth of high-quality nitride films on metal substrates. This provides the feasibility of low-temperature epitaxial growth of high-quality nitride films on metal substrates. Many research teams have done a lot of research on epitaxy of LEDs and nitride films on metal substrates by applying the above technology, and have achieved fruitful results in improving the luminous rate

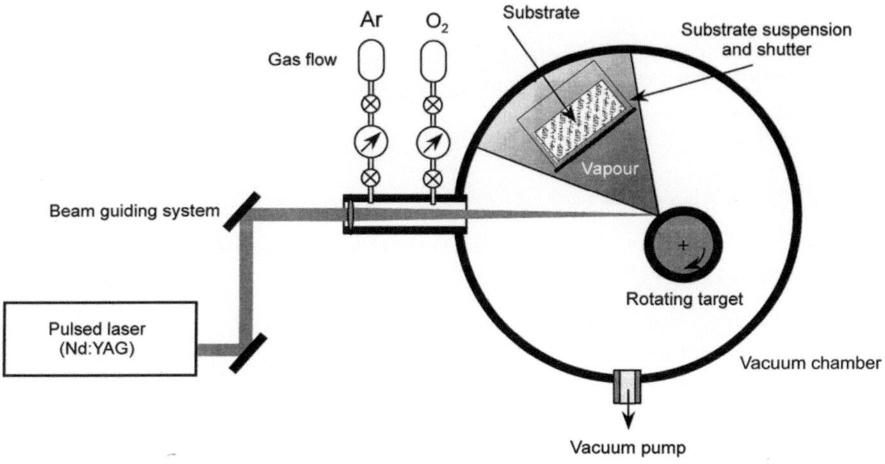

Fig. 5.11 Schematic diagram of the PLD system for thin film deposition [16]

of LEDs and the crystallization quality of epitaxial films. PLD schematic diagrams are shown in Fig. 5.11.

5.2.2 Growing AlN Films

In recent years, AlN has attracted much attention due to its excellent physical properties, such as a wide direct bandgap of 6.2 eV, high thermal conductivity, high breakdown electric field, and large piezoelectric coupling coefficient. These factors make AlN suitable for a wide variety of device applications such as thin-film body acoustic resonators (FBARs), tunneling magnetoresistive (TMR) devices, and short-wavelength optical emitters. Single-crystal AlN films are usually grown by MOCVD or MBE at relatively high temperatures (above 800 °C), which makes it difficult to avoid interfacial reactions between the AlN and the substrate. Recently, it has been found that the substrate temperature required for epitaxial growth can be greatly reduced using PLD and atomically abrupt heterogeneous interfaces can be realized.

Single-crystal Ta substrates with orientations (110), (100) and (111) were used as starting materials for crystal growth by Hirata et al. [17] The Ta substrates were flattened by chemical–mechanical polishing and then cleaned with an organic solvent. The substrates were then placed into a PLD chamber with a background pressure of 5.0×10^{-10} Torr and then annealed in ultrahigh vacuum (UHV) at different temperatures for 1 h to obtain a clean Ta surface. AlN films were then grown using an UHV PLD at a substrate temperature of 450 °C. During the growth process, nitrogen of 99.9999% purity was introduced into the PLD chamber at a pressure of 10 m Torr. The polishing process can deteriorate the crystalline quality of the surface region of the Ta substrate. Therefore, an ultra-high-pressure annealing process was performed

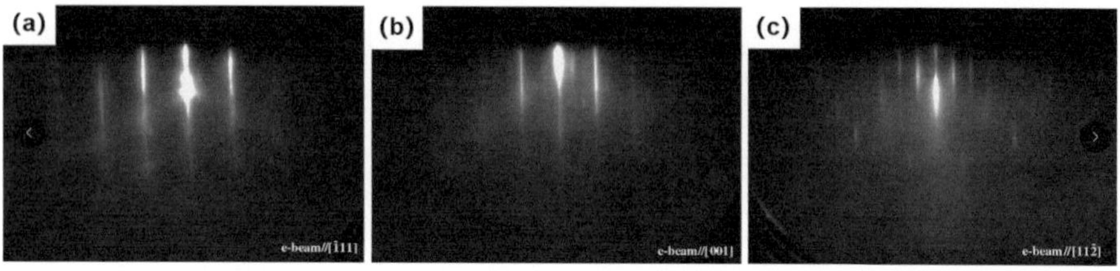

Fig. 5.12 **a** (110), **b** (100) and **c** (111) reflected high-energy electron diffraction images of Ta substrates annealed at 1100 °C for 1 h [15]

to restore its crystallinity. Figures 5.12a–c showed the reflected high energy electron diffraction images of Ta (110), (100) and (111) substrates annealed at 1100 °C for 1 h, respectively. As can be seen from these figures, annealing improved the surface topography of the Ta substrate.

In order to investigate the thermal stability of the heterogeneous interface between AlN and Ta, they performed UHV annealing on 16 nm thick AlN films. Figure 5.13 showed the GIXR variation curves for 10 min of UHV annealing at temperatures of 650 °C, 800 °C, and 1000 °C, respectively. The difference in the intensity profiles of the GIXR was due to the measured settings such as the size of the collimating slit, which had no effects on the shape of the GIXR curves. It can be seen that the shape of the curves did not change in any significant way. In fact, the estimated interfacial layer thickness was constant during the annealing process. These results indicated that the AlN/Ta heterointerface was quite stable at high temperatures (up to 1000 °C), which was important for device fabrication.

A Ta substrate with a clean and flat surface can be obtained by polishing and UHV annealing. GIXR and XPS measurements showed that the AlN/Ta heterointerface was very abrupt, and its abruptness remained unchanged even after annealing at 1000 °C. These results suggested that the PLD low-temperature growth technique was a promising avenue to improve the performance of nitride devices such as FBARs on metal substrates [17]

5.2.3 *GaN on Cu Substrate*

GaN is an important material for the development of short-wave optical devices and high-power electronic devices because of its excellent optical and electrical properties. These devices are typically grown on sapphire substrates, primarily because of the chemical stability of sapphire. However, the performance of these high-power nitride semiconductor devices is often limited by the poor thermal conductivity of the sapphire substrate. The use of high thermal conductivity materials as substrates can effectively solve this problem. Cu substrates are a promising candidate due to their high thermal conductivity of 400 W/mK.

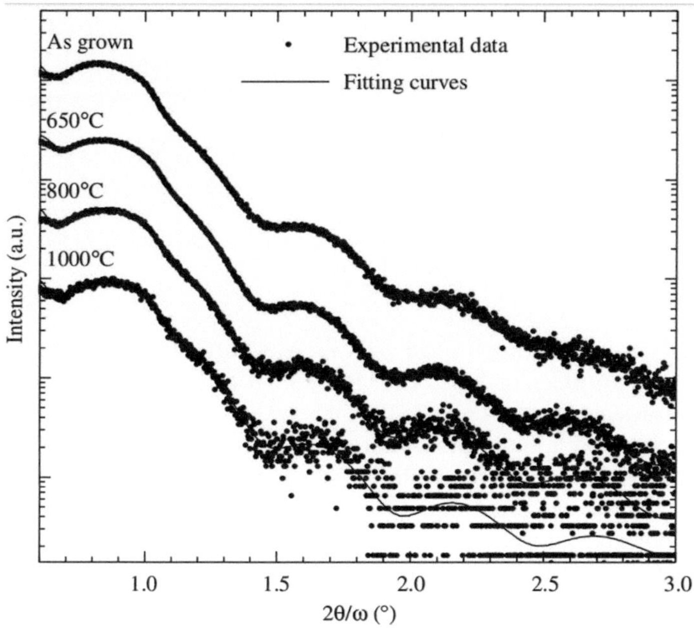

Fig. 5.13 Swept-incidence X-ray diffraction curves of the AlN/Ta structure as a function of temperature [17]

After obtaining high quality AlN films on metal substrates, a number of researchers have made significant progress by attempting to epitaxially grow single-crystal GaN films using AlN as a buffer layer. Shigeyoshi Inoue et al. used PLD to grow GaN thin films on Cu(111) substrates [18]. The epitaxial growth of III-group nitrides on metal substrates using conventional growth techniques such as MOCVD or MBE experiences severe interfacial reactions between the nitrides and the metal substrate. To solve this problem, the reduction of growth temperature of GaN and AlN has been investigated and their epitaxial layer-by-layer growth has been demonstrated using PLD. The low-temperature growth of III-group nitrides using PLD can be attributed to the enhanced kinetic energy of the film precursor, which aids in surface migration.

Figure 5.14a–c showed RHEED patterns of GaN films grown on Cu substrate with a thicknesse of 4, 40 and 130 nm, respectively. The patchy pattern observed at the beginning of the film growth eventually transformed into a sharp stripe pattern, indicating that the surface flatness of GaN films increased with increasing film thickness [18]. These patterns proved the high quality GaN films growth on Cu substrate.

In addition, Li et al. [19] found that Cu can be used as a suitable substrate material for c-plane GaN epitaxy by utilizing MOCVD. By using a LT AlN buffer layer, the alloying of Ga and Cu can be prevented so that GaN layers can be grown on Cu at temperatures of 1000 °C. They used two growth strategies. The first used a standard two-step growth technique involving a LT GaN buffer layer. The second did

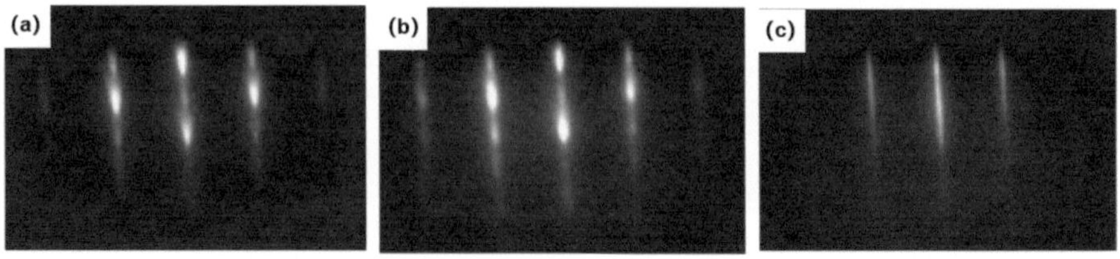

Fig. 5.14 Reflected high-energy electron diffraction patterns of **a** 4 nm, **b** 40 nm and **c** 130 nm GaN films of different thicknesses grown on Cu substrate [18]

not use a LT GaN buffer, but instead grew an AlN buffer layer at 800 °C, 150 Torr, followed by a 2-μm-thick undoped GaN layer at 50 Torr and 1000 °C. At this point, two distinct surface morphologies of the GaN layer can be observed. The first type was characterized by a multidirectional arrangement of GaN grains, as shown in Fig. 5.14a, indicating a polycrystalline morphology of GaN. Using selective area-focused gallium ion beam milling, cross-sectional TEM samples were prepared in the region shown in Fig. 5.15a, and the Cu orientation underneath was found to be single-crystal Cu (111). The SEM images of Fig. 5.15b showed the second type of GaN morphology as positive hexagonal pits. The lateral orientation of these pits was consistent, suggesting a possible single-crystal c-plane GaN exfoliation. A representative z-contrast scan TEM image was given in Fig. 5.15c. From the TEM image, both the AlN buffer layer and the GaN exfoliation layer showed single-crystal structures without grain boundaries.

In summary, single-crystal c-plane GaN can be grown on Cu(110) substrates in MOCVD. The growth relies on a LT AlN buffer layer to isolate the alloying reaction between Ga and Cu. In practice, further optimization of the GaN surface layer is needed, such as improving the GaN surface morphology and reducing the thread dislocation density [18, 19]. Epitaxial growth of GaN on Cu substrates may increase the power limit of future opotoelectronic and electronic devices due to the high thermal conductivity of copper.

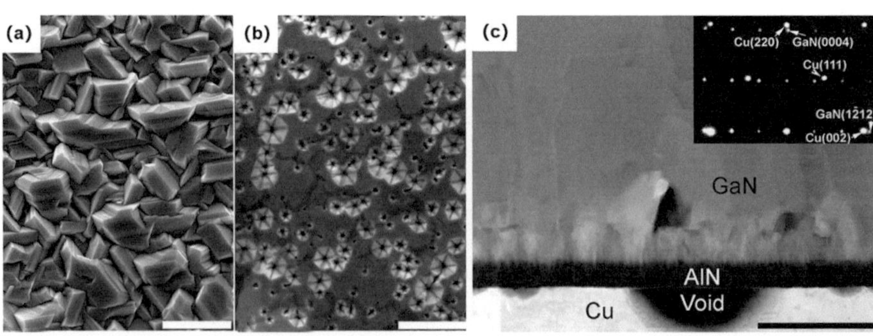

Fig. 5.15 **a** GaN polycrystals grown on Cu substrate. **b** SEM image of GaN single crystals grown on Cu substrate. **c** Cross-sectional STEM image of single crystal GaN grown on Cu [19]

5.2.4 Growth of GaN on Iron Substrates

The advantages of growing GaN and related III-group nitrides on metallic materials have been described above and Fe is considered the most promising substrate due to its excellent magnetic properties and advanced metallurgy. This section describes the epitaxial growth of GaN thin films on single crystal Fe substrates using PLD technology.

Okamoto et al. [20] used a UHV-PLD to grow AlN and GaN thin films on single-crystal Fe (100), (110) and (111) substrates. The Fe substrates were first mechanically polished and cleaned in an organic solvent. The substrates were then placed into a UHV-PLD chamber with a background pressure of 7×10^{-8} Pa and annealed at 700 °C UHV for one hour to remove surface contaminants and oxides. After annealing, the AlN films were grown at 380 °C for 30 min. The LT growth allowed us to suppress interfacial reactions, leading to the formation of polycrystalline films. After polishing, the surfaces of Fe(110), (100) and (111) were smooth. Figure 5.16a–c showed the RHEED plots of Fe(110), Fe(100) and Fe(111) substrates after annealing for 1 h at 700 °C UHV. All patterns showed sharp streaks, which indicated that the crystallinity destroyed by polishing was restored during the annealing process.

It is important to study the thermal stability of the heterogeneous interface between the substrate Fe(110) and AlN(0001). The dependence of the annealing temperature on the thickness of the interfacial reaction layer, obtained by XRR measurements, was given in Fig. 5.17a. The mutant heterogeneous interface remained unchanged at 700 °C annealing. Figure 5.17b showed the in situ XPS spectra of Fe 2p on the AlN surface at annealing to 700 °C. No Fe 2p peaks were observed in these spectra, indicating that there was no diffusion of iron atoms to the AlN surface. These results indicated that the AlN/Fe(110) heterointerface was robust and the AlN grown on Fe(110) can be used as a buffer layer for GaN film growth. PLD epitaxial growth of nitride semiconductors prepared on single crystal iron substrates holds great promise for future applications in optical, electronic and spintoronic devices [20].

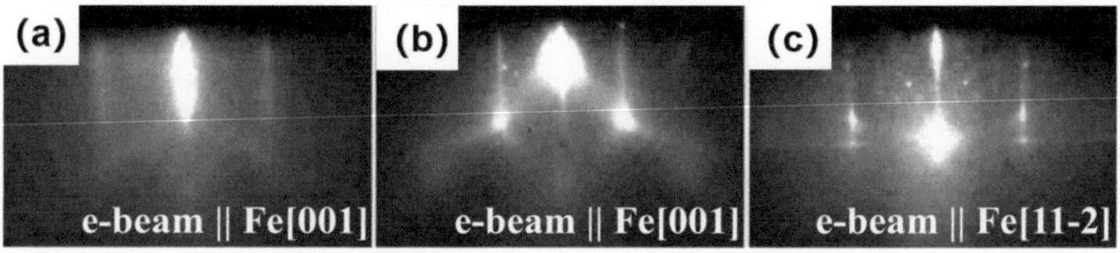

Fig. 5.16 **a** Fe(110), **b** Fe(100) and **c** Fe(111) after annealing at 700 °C UHP for 1 h, and reflected high energy electron diffraction patterns of the substrate [20]

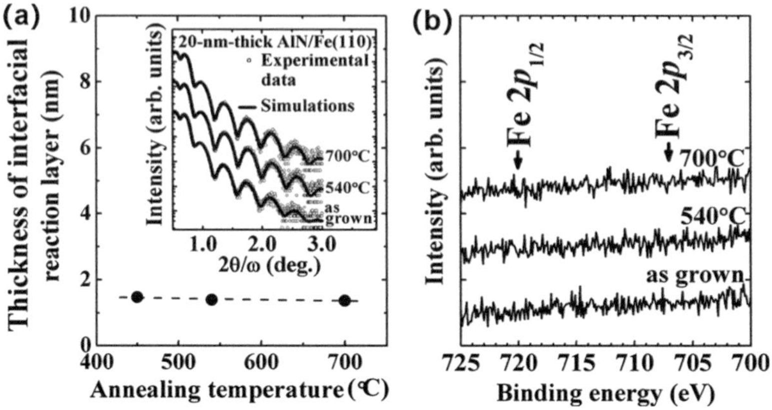

Fig. 5.17 **a** Annealing temperature dependence of the thickness of the interfacial reactive layer obtained by fitting the X-ray reflection curve **b** XPS analysis of Fe 2p spectra from the AlN surface during the annealing process [20]

5.2.5 Epitaxial Growth of LED and Nitride Devices on Metal Substrates

Metal materials with a high thermal conductivity, can help LED chip internal heat quickly conduct out. On the one hand, they reduce the device's section temperature and improve the compound efficiency of the electron–hole pairs, thus increasing the IQE. On the other hand, the film on the metal substrate can be fabricated as vertical structure device, making the device's internal current distribution uniform, and effectively reducing the lateral current due to the device heating. Based on the above excellent properties of metal substrates, they have recently attracted much attention in the preparation of nitride devices. However, the lattice mismatch between the metal substrate and the nitride epitaxial layer is large, and the cooling of the device from high temperature to room temperature is likely to result in cracking of epitaxial layer due to the thermal stress generated.

In order to overcome the shortcomings of nitride growth on metal substrate, many scholars have done a lot of fruitful research in this area. At present, some techniques have been used to improve the quality of epitaxial films on metal substrates, such as laser stripping technology, chemical stripping technology, electroplating technology, electrochemical deposition, etc., as well as PLD, pulsed electron-beam deposition (PED) and other preparation techniques. The applications of these techniques have resulted in high-quality nitride films, which to a certain extent overcomes the shortcomings of epitaxial nitride films on metal substrates, and opens up a broad prospect for the epitaxial growth of nitride thin-film devices on metal substrates.

5.3 Si Substrate Nitride Quasi van der Waals Epitaxy

In the heterogeneous epitaxy process, when the lattice mismatch is relatively small, the first few atomic layers of the epitaxial layer will be one-to-one with the substrate, so that there is a lattice deformation in the epitaxial layer, which will lead to a certain amount of stress in the system. Stress will accumulate as the thickness of the epitaxial layer increases, but the ability of the epitaxial system to store stress is not infinite. When the thickness of the film exceeds a critical value, the stress will be partially released through the formation of defects. The larger the lattice mismatch is, the more defects are formed, mostly in the form of penetration dislocations, edge dislocations, stacked layer dislocations, and so on. Materials containing a 2D layered structure have strong covalent bonds in each monolayer and weak vdW forces between neighboring monolayers. Compared to conventional 3D materials, such as GaN and silicon, 2D materials exhibit excellent chemical and thermal stability [21–23], which allows them to withstand the high growth temperatures of III-nitrides. Meanwhile, the weaker vdW forces on both sides of the 2D materials can shield the stresses due to matrix-to-matrix lattice mismatch and thermal expansion coefficient mismatch, thus improving the crystalline quality of the matrix. More importantly, the 2D materials exhibit a planar lattice arrangement similar to that of III-nitrides. Therefore, they act as buffer layers for III-nitride vdWE and sacrificial layers for stress relief in subsequent processes [24]. One of the 2D materials, graphene, is a monolayer of carbon atoms, sp^2 hybridized and arranged in an ortho-hexagonal honeycomb lattice structure [25]. Due to the absence of dangling bonds, the graphene surface is inert to any foreign atoms, which makes it challenging to grow 3D semiconductor materials on its surface by conventional epitaxy. One possible way to circumvent this problem is to use vdWE [24]. Since the bonding mechanism of vdWE is different from conventional epitaxy with strong chemical bonding, the grown material does not have to match the graphite substrate lattice and is free from strain-induced interfacial defects.

Therefore, the introduction of qvdWE allows the epitaxial layer not to be arranged exactly according to the lattice constants of the substrate, but can maintain its intrinsic parameters, which can avoid the defect problem introduced because of the lattice mismatch. Thus, it bypasses the need of the traditional heterogeneous epitaxy on the substrate and allows the epitaxial layer to have a large lattice mismatch with the substrate, so the use of 2D materials as a buffer layer for nitride epitaxy is now a new breakthrough in solving the lattice large mismatch problem. In the following, we focus on large-mismatch lattice nitride qvdWE.

Although many types of substrates have been attempted for heterogeneous epitaxial growth of III-nitride semiconductors, Si is considered to be one of the most attractive substrates because of its high quality and low cost, ease of obtaining large substrates, and conductivity control [26–33]. So far, c-planar GaN(0001)-based devices such as LEDs and field effect transistors (FETs) grown on Si(111) substrates have been fabricated and commercialized. To realize the integration between mature Si-based electronics on Si(100) and GaN-based optoelectronic devices, it is crucial to grow c-plane GaN(0001) on Si(100) substrates. However, it is still difficult to grow high quality epitaxial GaN films [34–36] due to different symmetries in hexagonal GaN(0001) and cubic Si(100). Therefore, some attempts were made to integrate GaN FETs grown on Si(111) and Si(100) MOSFETs by mechanical bonding [37]. Recently it has been reported that graphene can be used as an intermediate layer for growing c-plane GaN on amorphous substrates [1] Since graphene has hexagonal hexagonal symmetry, it should be able to effectively form an epitaxial relationship with the c-plane of GaN, even on substrates without hexagonal symmetry.

In addition, it was found that the initial deposition of AlN can be used as a buffer layer for epitaxial growth of gallium nitride on sapphire and 6H-SiC to improve the quality of gallium nitride thin films [38]. Si(100) substrates offer the possibility to integrate different functionalities with modern Si device technologies, such as UV detection or emission as well as Si integrated circuits (ICs) on a common substrate. For hexagonal GaN grown directly on Si(100), deposition on this surface is also challenging. Deposition of AlN on Si(100) can be studied by PAMBE [39] and magnetron sputtering [40], where epitaxial relationships, microstructures, and phase evolution are investigated. The 20% mismatch between the lattice constants of AlN and Si with Si(100)'s fourfold symmetry, makes it very difficult to prepare epitaxy of single-phase fibrillar zincite AlN ([0001] sixfold symmetry of the basal plane).

In general, heterogeneous epitaxy of such highly mismatched materials is based on specific interfacial interactions defined by a minimum in the total energy. It is common to observe the growth of 2D- or 3D material textures, the latter depending mainly on the improper cutting of the substrate. Substable cubic AlN-type as well as amorphous interfacial phases are often interspersed. Research has focused on controlling the preferred orientation of the films through deposition conditions and substrate pretreatment. However, the mechanism of the preferred orientation of AlN films has not been explained so far. This raises the need to consider the interaction processes during the growth of the AlN/Si interface on an atomic scale.

The formation of epitaxial structures and the growth of woven structures in nitride thin films was investigated by TEM and XRD on reactive magnetron sputtered substrates by Valcheva et al. [41] The TEM from Fig. 5.18a showed that the lower part of the film consists of single crystal columnar grains with flattened tops and the diameter of the particular grains was between 47 and 85 nm. The film was dense, and no holes were observed at the grain boundaries. Figure 5.18b showed SAED plots of the cross section of the films, indicating that the substrate was oriented along the Si[$1\bar{1}0$] direction, and two alignments parallel to the Si[$1\bar{1}0$] direction were observed in the plots along the [$01\bar{1}0$] and [$11\bar{2}0$] directions, respectively. The diffractograms also exhibited the epitaxial relationship of AlN(0001)$\|$Si(001), and from the two overlapping AlN[$11\bar{2}0$] and [$01\bar{1}0$] fibrillar zincite-type lattices, it can be deduced that at least two of the AlN columnar domains were differently rotated around the c-axis for the in-plane orientations, and that there were no detected spots associated with other polymorphic or amorphous phases.

The epitaxial orientation along the $<110>$ nucleation direction was determined by the orientation that can be efficiently bonded as shown in Fig. 5.19a. The low error from (001) was ~ $\pm 0.5°$ and the width of the step was 50–100 nm, which was

Fig. 5.18 **a** Cross-sectional TEM image, and corresponding **b** SAED image, AlN/Si(001) structure along the Si[1$\bar{1}$0] direction [41]

very well suited to the planar dimensions of the domain. In contrast, the nucleation scheme was insufficient to allow the other two domains oriented along B and D of < 100> to appear. There seemed to be another mechanism involved that may enable its nucleation. Thus, in order to clean the natural oxide from the Si substrate at an initial stage, heating it to 900 °C, or subsequently increasing the temperature to 1000 °C, may lead to decomposition of the Si surface. The temperature treatment may affect the roughening of the platform surface and the step edges. Step dynamics of Si(001) had been investigated using a variety of techniques in different temperature ranges and the dynamics of steps had been found to be very complex with respect to key microscopic processes, such as adsorption or detachment of atoms on the edges of the steps and diffusion of atoms along them [42]. It was well known that after treatment at temperatures of 900 °C or more, the Si(001) surface readily opened by thermal activation and the surface went from 2×1 reconstruction to 1×1 relaxation. Thermal decomposition of Si(001) [43] had been reported to result in surface dimpled punctate features, where the dimples were several atomic layers deep and were randomly distributed across the surface depending on temperature and treatment time. In this case, the observation of interfacial islands can be attributed to the formation of surface roughness and Si surface depressions and/or Si surface diffusion and Si stacking. The two regions of B and D may nucleate at these depressions, where the dimers were likely to be open and more favorable to mismatch due to their orientation along Si(001). The results for the crystal orientations in Figs. 5.19 were shown in Tables 5.1.

It can be seen that the wavy Si surface morphology provided topological conditions for the nucleation and growth of rotational AlN domains. The wavelength of the surface undulations determined the domain size, and the localized state of the surface remodeling determined the epitaxial orientation. Thus, the alignment observed in columnar growth appeared to be the result of a delicate energy balance between eliminating dangling bonds and minimizing surface stresses.

Furthermore, although single-domain GaN can be grown on miscut Si (100) substrates, miscut Si substrates are not compatible with standard CMOS processes due to severe anisotropy and reliability issues [44, 45]. Therefore, the realization of epitaxial single crystals is of great importance, and GaN domains are formed on

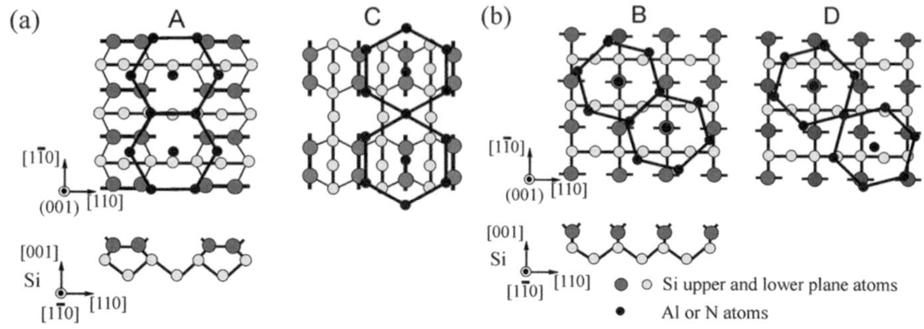

Fig. 5.19 **a** AlN and Si(001) 2 × 1 reconfiguration; **b** Si(001) 1 × 1 unreconstructed surface, with epitaxial relationship between the interfaces schematically shown [39]

Table 5.1 Symbols of the four rotational domains and their in-plane orientations, epitaxial mismatches and volume fractions [42]

AlN domain notation	Epitaxial relation	Nominal epitaxial misfit (%)	Fraction (%)
A	AlN$\langle 11\bar{2}0\rangle$||Si[110]	− 30	24.7
B	AlN$\langle 11\bar{2}0\rangle$||Si[100]	− 1	26.5
C	AlN$\langle 01\bar{1}0\rangle$||Si[110]	− 19	22.3
D	AlN$\langle 01\bar{1}0\rangle$||Si[100]	15	26.5

standard Si (100) substrates in order to integrate GaN-based devices into Si ICs. A potential way to grow single-domain GaN on standard Si(100) substrates is to introduce a buffer layer that shields the asymmetric surface domains of Si(100) and is used as a template for GaN epitaxy. 2D materials are promising candidates due to the weak vdW interactions along the out-of-plane direction. The substrate and epitaxial layers connected by a 2D buffer layer were well controlled [46], which can screen the orientation effect of the Si(100) substrate.

Yuxia Feng et al., reportedly realized epitaxy of single-crystal GaN films with step flow surface by using single-crystal graphene/SiO_2 buffer layer on CMOS-compatible Si(100) substrate [47]. The schematic of the epitaxial process was given in Fig. 5.20. A hybrid buffer layer consisting of a single atom-thick single-crystal graphene and a thin layer of SiO_2 was used for the epitaxy of a single-domain GaN film on Si(100). SiO_2 was used for further screening of asymmetric surface regions of the substrate. Single-crystal monolayers of graphene were grown by CVD and then transferred to the SiO_2/Si(100) surface [48]. This study revealed that the nucleation of group III nitrides on graphene was triggered by the formation of sp^3 hybridized C-N bonds on the graphene surface. In addition, due to the high nucleation density of AlN with single domain structure on graphene folds (preserving the orientation properties of single-crystal graphene), subsequent GaN islands had a higher growth rate and thus can be polycondensed into continuous single-crystal GaN. The successful epitaxy of single-crystal GaN was attributed to the template effect of graphene through the

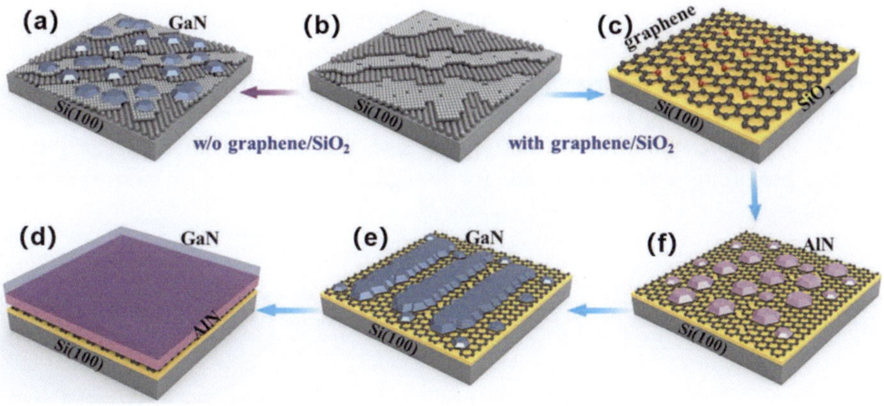

Fig. 5.20 Schematic of epitaxy of single-crystal GaN films on Si(100) with and without single-crystal graphene/SiO_2 interlayer [47]

formation of the C-N covalent bond, which acted as a driving force for the GaN domains as the driving force for the in-plane alignment. These results validated the orientation effect of 2D graphene on III-nitrides, which enabled epitaxy of GaN-based devices on Si(100) and other amorphous or flexible substrates.

Based on the above analysis, the graphene material can be used as a flexible layer to reduce the stress in the epitaxial layer by changing its own lattice constant during the nitride growth process. This stress release mechanism is of great significance for nitride epitaxy on a graphene buffer layer, which ensures high-quality growth of the epitaxial layer in the presence of a large lattice mismatch between the epitaxial layer and the substrate. On conventional heterogeneous substrates, a substrate material with a small lattice mismatch is necessary to obtain high-quality films, while qvdWE on 2D buffer layers, such as graphene, circumvents the need for lattice matching. This effect of stress relief through the addition of a flexible buffer layer is what makes qvdWE superior to conventional heterogeneous epitaxy, especially for substrates with large lattice mismatches and amorphous substrates.

5.4 Summary

This chapter delves deeply into the challenges and applications of nitride qvdWE on large mismatch substrates. The research focuses on the characteristics of nitride qvdWE on amorphous glass substrates and metal substrates, and analyzes the influence of the lattice mismatch between the substrate and the nitride on the growth process. Research indicates that amorphous glass substrates provide a growth environment without the need for lattice matching, but at the same time, they also bring challenges in terms of uniformity and defect control. Metal substrates have attracted much attention due to their high thermal conductivity and good electrical conductivity. However, during the growth process, the problem of chemical reactions between metals and nitrides needs to be addressed. In addition, this study also introduces other large mismatch nitride qvdWE methods, including nitride epitaxy on Si (100) substrates. These methods have successfully overcome the challenges brought by lattice mismatch by optimizing growth conditions and interface engineering, providing a new approach for the growth of nitrides on non-traditional substrates.

References

1. Chung K, In Park S, Baek H et al (2012) High-quality GaN films grown on chemical vapor-deposited graphene films. NPG Asia Mater 4:e24–e24
2. Yoo H, Chung K, In Park S et al (2013) Microstructural defects in GaN thin films grown on chemically vapor-deposited graphene layers. Appl Phys Lett 102:051908
3. Baek H, Lee CH, Chung K et al (2013) Epitaxial GaN microdisk lasers grown on graphene microdots. Nano Lett 13:2782–2785
4. Chae SJ, Kim YH, Seo TH et al (2015) Direct growth of etch pit-free GaN crystals on few-layer graphene. RSC Adv 5:1343–1349
5. Shon JW, Ohta J, Ueno K et al (2014) Fabrication of full-color InGaN-based light-emitting diodes on amorphous substrates by pulsed sputtering. Sci Rep 4:5325
6. Ren F, Liu B, Chen Z et al (2021) van der Waals epitaxy of nearly single-crystalline nitride films on amorphous graphene-glass wafer. Sci Adv 7:eabf5011
7. Yin Y, Liu B, Chen Q et al (2022) Continuous single-crystalline gan film grown on WS_2 -glass wafer. Small 18:2202529
8. Yang Y, Li B, Zhang C et al (2016) The morphology, synthesis, properties, and applications of graphene-like two-dimensional h-BN nanomaterials. Cailiao Daobao 30:143–148
9. Ge L, Yang J, Qiu T (2008) Study progress of preparation methods of hexagonal boron nitride. Electron Comput Mater 27:22–29
10. Wu Q, Yan J, Zhang L et al (2017) Growth mechanism of AlN on hexagonal BN/sapphire substrate by metal–organic chemical vapor deposition. Cryst Eng Comm 19:5849–5856
11. Chung K, Oh H, Jo J et al (2017) Transferable single-crystal GaN thin films grown on chemical vapor-deposited hexagonal BN sheets. NPG Asia Mater 9:e410–e410
12. Shen XQ, Matsuhata H, Okumura H (2005) Reduction of the threading dislocation density in GaN films grown on vicinal sapphire (0001) substrates. Appl Phys Lett 86:021912
13. Liu L, Edgar JH (2002) Substrates for gallium nitride epitaxy. Mater Sci Eng R Rep 37:61–127
14. Li G, Wang W, Yang W et al (2015) Epitaxial growth of group III-nitride films by pulsed laser deposition and their use in the development of LED devices. Surf Sci Rep 70:380–423
15. Ohta J, Fujioka H, Ito S et al (2002) Room-temperature epitaxial growth of AlN films. Appl Phys Lett 81:2373–2375
16. Lackner JM, Waldhauser W, Ebner R et al (2003) Pulsed laser deposition: a new technique for deposition of amorphous SiOx thin films. Surf Coat Technol 163:300–305
17. Hirata S, Okamoto K, Inoue S et al (2007) Epitaxial growth of AlN films on single-crystalline Ta substrates. J Solid State Chem 180:2335–2339
18. Inoue S, Okamoto K, Matsuki N et al (2006) Epitaxial growth of GaN on copper substrates. Appl Phys Lett 88:261910
19. Li Q, Figiel J, Wang G et al (2012) GaN epitaxy on Cu (110) by metal organic chemical vapor deposition. Appl Phys Lett 100:192110

20. Okamoto K, Inoue S, Matsuki N et al (2008) Epitaxial growth of GaN films grown on single crystal Fe substrates. Appl Phys Lett 93:251906
21. Watanabe K, Taniguchi T, Kanda H (2004) Direct-bandgap properties and evidence for ultraviolet lasing of hexagonal boron nitride single crystal. Nat Mater 3:404–409
22. Li D, Müller MB, Gilje S et al (2008) Processable aqueous dispersions of graphene nanosheets. Nat Nanotechnol 3:101–105
23. Geim AK (2009) Graphene: status and prospects. Science 324:1530–1534
24. Yu J, Hao Z, Wang J et al (2019) Study on AlN buffer layer for GaN on graphene/copper sheet grown by MBE at low growth temperature. J Alloy Compd 783:633–642
25. Geim AK, MacDonald AH (2007) Graphene: exploring carbon flatland. Phys Today 60:35–41
26. Fitzgerald EA, Lee KE, Zhang L et al (2016) Novel integrated circuit platforms employing monolithic silicon CMOS+ GaN devices. ECS Trans 75:31
27. Guha S, Bojarczuk NA (1998) Ultraviolet and violet GaN light emitting diodes on silicon. Appl Phys Lett 72:415–417
28. Semond F, Lorenzini P, Grandjean N et al (2001) High-electron-mobility AlGaN/GaN heterostructures grown on Si (111) by molecular-beam epitaxy. Appl Phys Lett 78:335–337
29. Feltin E, Beaumont B, Laügt M et al (2001) Stress control in GaN grown on silicon (111) by metalorganic vapor phase epitaxy. Appl Phys Lett 79:3230–3232
30. Wang W, Li Y, Zheng Y et al (2019) Lattice structure and bandgap control of 2D GaN grown on graphene/Si heterostructures. Small 15:1802995
31. Altman EI (2017) Group III phosphates as two-dimensional van der Waals materials. J Phys Chem C 121:16328–16341
32. Wang W, Zheng Y, Li X et al (2019) 2D AlN layers sandwiched between graphene and Si substrates. Adv Mater 31:1803448
33. Hainey MF Jr, Al Balushi ZY, Wang K et al (2018) Heteroepitaxy of highly oriented GaN films on non-single crystal substrates using a Si (111) template layer formed by aluminum-induced crystallization. Phys Status Solidi (RRL)–Rapid Res Lett 12:1700392
34. Schulze F, Dadgar A, Bläsing J et al (2004) GaN heteroepitaxy on Si (0 0 1). J Cryst Growth 272:496–499
35. Schulze F, Dadgar A, Bläsing J et al (2006) Growth of single-domain GaN layers on Si (0 0 1) by metalorganic vapor-phase epitaxy. J Cryst Growth 289:485–488
36. Rizal U, Swain BS, Swain BP (2015) Synthesis and characterization of Gallium nitride (GaN) thin films deposited by low pressure chemical vapor deposition (LPCVD) technique. Manipal Res Colloq 1–9
37. Lee HS, Ryu K, Sun M et al (2011) Wafer-level heterogeneous integration of GaN HEMTs and Si (100) MOSFETs. IEEE Electron Device Lett 33:200–202
38. Akasaki I, Amano H, Koide Y et al (1989) Effects of ain buffer layer on crystallographic structure and on electrical and optical properties of GaN and $Ga_{1-x}Al_xN$ ($0< x\leqq 0.4$) films grown on sapphire substrate by MOVPE. J Crystal Growth 98:209–219
39. Lebedev V, Jinschek J, Kaiser U et al (2000) Epitaxial relationship in the AlN/Si (001) heterosystem. Appl Phys Lett 76:2029–2031
40. Jang HW, Kang HC, Noh DY et al (2003) Crossover of the preferred growth orientation of AlN/Si (001) films during off-axis radio frequency sputter growth. J Appl Phys 94:2957–2961
41. Valcheva E, Birch J, Persson PO et al (2006) Epitaxial growth and orientation of AlN thin films on Si (001) substrates deposited by reactive magnetron sputtering. J Appl Phys 100:123514
42. Zielasek V, Liu F, Zhao Y et al (2001) Surface stress-induced island shape transition in Si (001) homoepitaxy. Phys Rev B 64:201320
43. Jones DE, Pelz JP, Hong Y et al (1996) Striped phase and temperature dependent step shape transition on highly B-doped Si (001)-(2$\times$ 1) surfaces. Phys Rev Lett 77:330
44. Lebedev V, Jinschek J, Kräußlich J et al (2001) Hexagonal AlN films grown on nominal and off-axis Si (0 0 1) substrates. J Cryst Growth 230:426–431
45. Qin G, Zhou H, Ramayya EB et al (2009) Electron mobility in scaled silicon metal-oxide-semiconductor field-effect transistors on off-axis substrates. Appl Phys Lett 94:073504

46. Pennachio DJ (2019) Extreme epitaxial growth of III-N semiconductor thin films. University of California, Santa Barbara
47. Feng Y, Yang X, Zhang Z et al (2019) Epitaxy of single-crystalline GaN film on CMOS-compatible Si (100) substrate buffered by graphene. Adv Func Mater 29:1905056
48. Xu X, Zhang Z, Dong J et al (2017) Ultrafast epitaxial growth of metre-sized single-crystal graphene on industrial Cu foil. Sci Bull 62:1074–1080

Chapter 6
Flexibility of Quasi van der Waals Epitaxial Nitrides

In recent years, III-group nitride semiconductors have attracted much attention due to their excellent physical properties and wide range of applications in solid-state lighting, flat panel displays, and solar and power electronics. As direct bandgap semiconductors, III-group nitrides offer wide bandgaps ranging from 0.64 eV for InN and 3.4 eV for GaN to 6.2 eV for AlN, spanning the entire UV and visible wavelength bands, as well as the near-infrared range. Thus, III-group nitrides have been widely used to fabricate LEDs, [1] LD [2], photodetectors [3], and solar cells [4]. In addition, HEMT can be fabricated based on 2D electron gases in AlGaN/GaN heterostructures using spontaneous polarization and piezoelectric polarization of wurtzite semiconductors and high electron drift velocity [5–8]. Currently, III-group nitride films with high crystalline quality can be obtained by employing MOCVD [9], MBE [10], or HVPE [11] on c-faceted sapphire [12], Si(111) [13], or 6H-SiC [14] substrates at high growth temperatures, typically above 1000 °C [15]. However, stripping single-crystal group III nitride films from such substrates is difficult because of the strong sp^3-type covalent bonds between the substrate and the epitaxial layer [16]. To overcome this problem, nitrides delaminated from single-crystal substrates by laser radiative heat release [17], transfer methods [18], chemical etching [19], and mechanical stripping [20] have been investigated in numerous studies. However, there are still some bottlenecks for future applications, such as damage during flexible production, size constraints, and cumbersome steps. In addition, flexible amorphous substrates usually cannot tolerate such high growth temperatures due to the amorphous surface atomic arrangement and therefore cannot be used for epitaxial growth of single-crystal films [21]. As a result, it isn't easy to get a large scope for wearable and foldable applications of next-generation (optoelectronic) electronic devices based on III-group nitrides. On the other hand, sp^2-bonded 2D materials (e.g., graphene, h-BN, and transition metal disulfides) show in-plane hexagonal lattice arrangements and weak interlayer bonding properties [22]. If sp^3-bonded III-group nitride films can be grown on sp^2-bonded 2D materials, it is theoretically possible to transfer these functional films onto heterogeneous and/or flexible substrates due to the weak vdW

T. Wei et al., *Quasi van der Waals Epitaxy of Nitride Semiconductor and Its Applications*, Wide Bandgap Semiconductors,
https://doi.org/10.1007/978-981-95-3469-2_6

interactions on both sides of the 2D material layers [23]. In this case, the 2D material not only acts as a buffer layer but also provides a release layer for mechanical stripping of III nitride epitaxial layers. This method may potentially be used to obtain large-scale (up to a few inches) low-cost flexible nitride devices.

6.1 Advantages and Applications of Flexible Devices

The use of flexible electronic and optoelectronic devices in wearable displays, solar cells, sensors, and biomedical devices has recently attracted much attention [24–27]. For bendable and wearable devices, organic films have been widely adopted due to their excellent stretchability and flexibility [28, 29]. At the same time, the realization of flexible devices using inorganic films is still very challenging due to the inherent rigidity and brittleness of inorganic films and single-crystal substrates [30]. However, due to the high radiative complexity and mobility as well as the excellent thermal and mechanical properties of single-crystal inorganic compound semiconductors such as GaN) higher device performance and better long-term stability of the devices are expected [31]. However, for substrates with flexibility (e.g., plastic substrates), the usual growth conditions for high-quality single-crystal films require extremely high temperatures, making it impossible to grow them on flexible substrates with low melting temperatures. Meanwhile, for conventional epitaxial films on rigid substrates, their low tolerance to mechanical deformation makes it difficult to scale up to flexible devices, which also limits their application to flexible devices [18]. However, to prepare flexible optoelectronic devices using nitride films epitaxially grown on rigid substrates, a large number of researchers have attempted to adopt the experimental strategy of stripping high-quality single-crystal nitride films or nitride micro-nanostructures from rigid substrates and transferring them to flexible substrates.

6.1.1 Flexible Detector Devices

The growth of single-crystal high-quality GaN films on multilayer graphene/SiC substrates [32] can be realized using MOCVD. As shown in Fig. 6.1, the surface morphology of GaN films grown on multilayer graphene/SiC substrates was obtained by SEM and AFM, and a delaminateed surface morphology can be seen on the surface of GaN films in Fig. 6.1a. Many clear atomic steps were seen in further observation by highly magnified AFM. The measured GaN step heights and platform widths were 5–8 nm and 150–250 nm, respectively, while the RMS of the film roughness was 1.68 nm, which suggested that GaN films with smooth and flat surfaces had been obtained on multilayer graphene/SiC substrates. Finally, flexible UV photosensitive devices were fabricated by releasing the GaN film from the substrate by direct mechanical peeling and using transparent tape to release the GaN film from

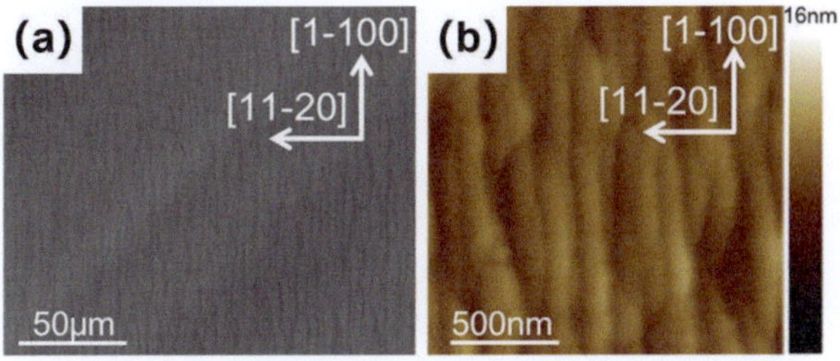

Fig. 6.1 **a** SEM image of GaN film grown on multilayer graphene/SiC. **b** AFM image of surface morphology of GaN film [32]

the graphene layer by direct mechanical peeling, and transferring it to a flexible substrate (on glass rods with different curvatures). Figure 6.2a showed a schematic of a flexible GaN-based UV photodetector. Two indium electrodes were fabricated on the GaN film. Figure 6.2b showed a set of I-V linear curves with bending radii of 1, 2.5, and 4 mm, showing good ohmic contacts. Next, the time response of the UV at 10 V bias voltage and 2.5 mm bending radius was tested. As shown in Fig. 6.2c, the output current increased significantly under illumination conditions, indicating good sensitivity under irradiation with UV light. Figure 6.2d showed the photocurrent (1.22 mA) and dark current (0.73 mA) of the photodetector at 10 V bias voltage and 2.5 mm radius. Thus, the ratio of currents (UV on and off) was about 167%. The rise and decay times of the photodetector were 1.26 and 1.56 s, respectively. It can be concluded that the photodetector had good optoelectronic performance, which can be attributed to the fact that the transport of photogenerated carriers was not affected by bending.

6.1.2 Flexible LED Devices

Figure 6.3 schematically illustrated the basic strategy for epitaxial growth of GaN/ZnO coaxial nanorod heterostructures on graphene-covered heterogeneous substrates and fabrication of flexible LEDs [33]. First, large graphene films were synthesized on a metal foil using CVD and transferred to an arbitrary substrate. In this experiment, a Si substrate with a thin silicon oxide (SiO_2) layer was used (Fig. 6.3a). Although there were some variations in the length and vertical alignment of the nanorods, ZnO nanorods were grown directly on graphene films without an additional seed crystal layer, with densities of 10^8–10^9 cm^{-2}, which was suitable for the fabrication of isolated coaxial nanorod heterostructures. The basic method for fabricating flexible LEDs was shown in Fig. 6.3b. After growing the coaxial nanorod heterostructures on a graphene film, a thin Ni/Au layer was deposited on the p-GaN surface of the nanostructures using electron-beam evaporation and subsequent rapid thermal annealing methods, thus creating isolated ohmic contacts to the individual nanostructures. The gaps between the nanostructures were then filled with an insulating, flexible polymer and an additional Ni/Au metal layer was deposited on top

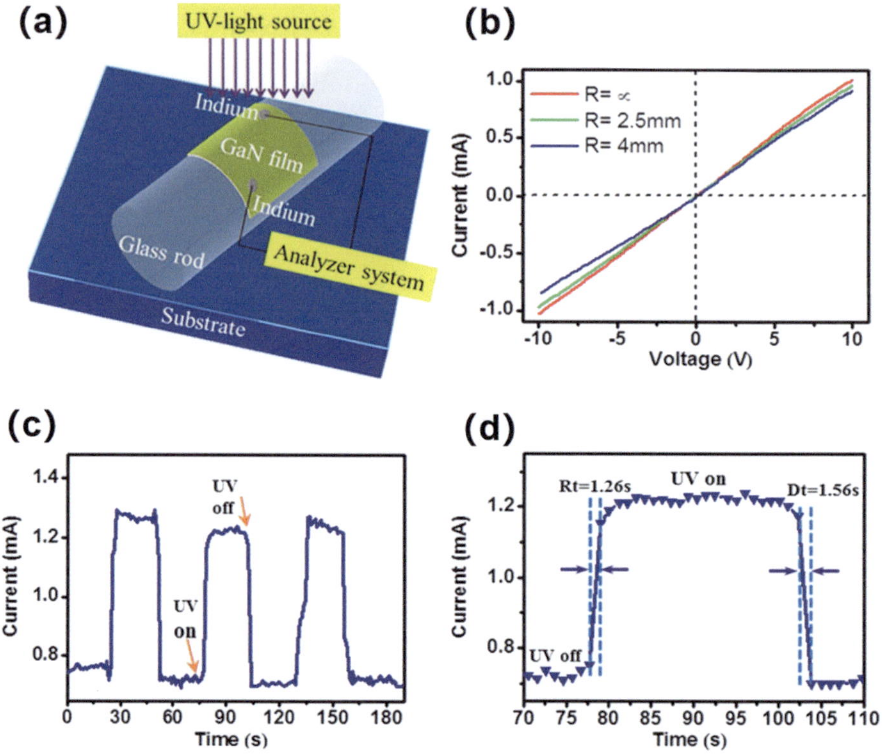

Fig. 6.2 **a** Schematic of GaN-based flexible UV photodetector. **b** I-V characteristic curves for bending radii of 1, 2.5, and 4 mm, respectively. **c** Time response of UV light at 10 V bias voltage and 2.5 mm radius. **d** Zoom-in plot for a single on/off cycle [32]

of the nanostructured embedding layer to enable current diffusion. As an important step in the fabrication of flexible devices, LED devices fabricated on graphene films were transferred to copper-coated polyethylene terephthalate (PET) substrates by wet etching a sacrificial SiO_2 layer underneath the graphene film and using an exfoliation process. The LEDs were then adhered to the Cu-coated PET substrate to create an electrical connection between the graphene film of the device and the insulating plastic substrate.

Nanostructured LEDs transferred to plastic substrates exhibited good EL properties. As shown in the photograph of the LED at 10 mA current in Fig. 6.4a, the emission of blue light was strong enough to be easily observed with the naked eye under indoor-lighting. The light was emitted only from a circular contact area with a diameter of 600 μm and became brighter as the applied current increased. In addition, distinct light-emitting spots were observed in magnified optical microscope images. The emission intensity of each light spot gradually increased while maintaining the number of light spots with increasing applied current. The quantitative characteristics of the LEDs can be further investigated by measuring the power-dependent

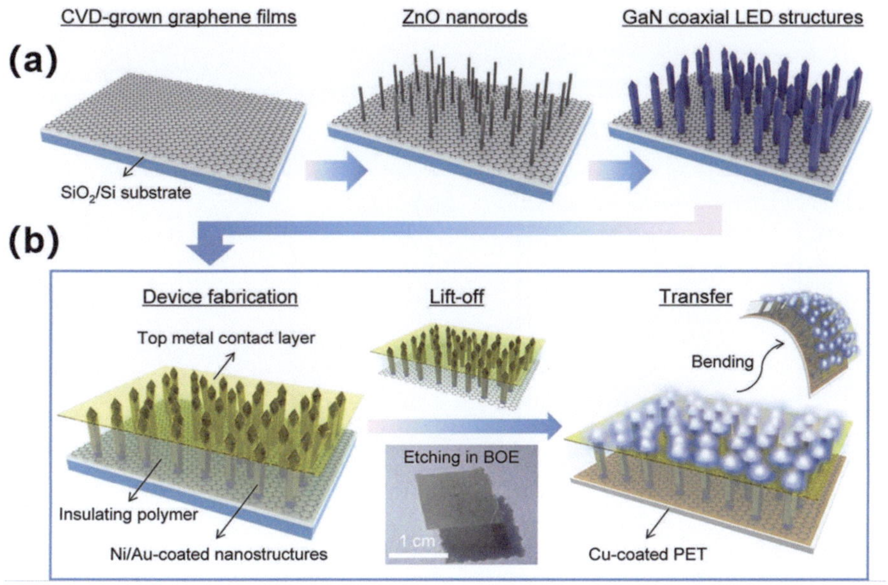

Fig. 6.3 Schematic representation of the process of fabricating flexible LEDs using GaN/ZnO coaxial nanorod heterostructures grown on graphene films. **a** Schematic diagram of GaN/ZnO coaxial nanorod heterostructures grown heterogeneously epitaxially on a large-area graphene film. **b** Schematic diagram of the fabrication of LED devices and their transfer to flexible plastic substrates [33]

electrogenerated spectra and I-V characteristic curves. Figure 6.4b showed the EL spectra at various applied currents from 1 to 10 mA. The major EL peaks were observed at 490–493 nm, which were caused by $In_xGa_{1-x}N/GaN$ MQWs. The nanostructured LEDs exhibited typical rectification behavior in the I-V characteristic curve in Fig. 6.4c, with a turn-on voltage of $\approx$ 5 V, along with a leakage current of about 5×10^{-4} A at a reverse current of − 5 V (solid blue line). However, above the turn-on threshold, the current and the integrated emission intensity (red hollow circles) increased simultaneously with the applied bias voltage. These results suggested that the EL emission originated from the carrier injection and radiation complex with the p–n junction in the MQWs layer of the nanostructured LEDs. Meanwhile, the IQE of the coaxial nanorod heterostructures grown on graphene films was estimated to be $\approx 13\%$ by measuring the temperature-dependent photoluminescence spectra, and the device performance, including the emission efficiency and homogeneity, can be further improved by optimizing the growth of the nanostructures and the device fabrication process.

In order to use the nanostructured LEDs in a flexible form, the device characteristics of the LEDs transferred to Cu/PET substrates were evaluated under substrate bending. Figure 6.5 showed the photos of luminescence, EL and electrical characterization spectra at ∞, 5.5 and 3.9 mm bending radii. As shown in the photograph of Fig. 6.5a, when bending a 13 mm wide substrate to 10 mm and 7 mm widths,

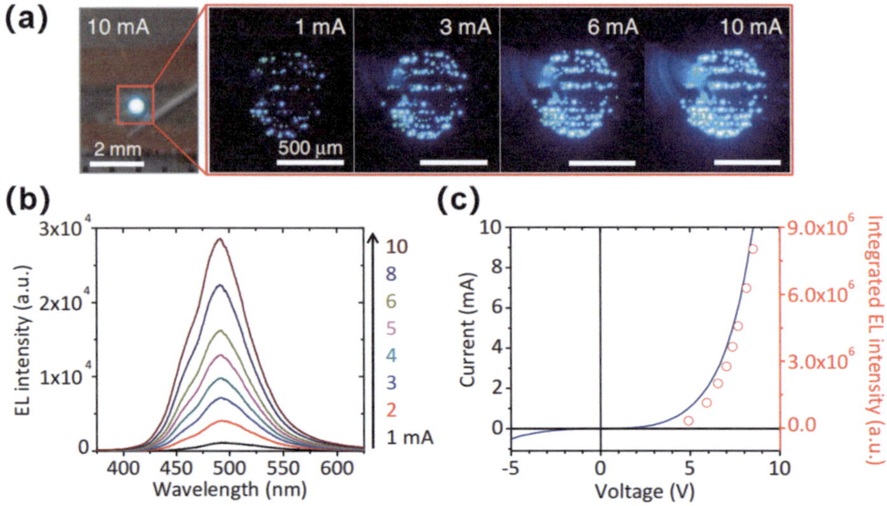

Fig. 6.4 Visible nano-LEDs on graphene films. **a** Photographs and optical microscopy images of the light emitted by the LEDs at different applied currents. **b** Power-dependent EL spectra at applied currents of 1–10 mA. **c** Plot of the I-V characteristic curve of the LED (black solid line) and the integrated EL intensity (red hollow circle) as a function of the applied bias voltage [33]

corresponding to 5.5 and 3.9 mm bending radii, respectively, the LEDs reliably emit blue light even when bending with a fixed current of 10 mA without significant weakening. The EL spectra obtained at different bending radii in Fig. 6.5b showed no significant changes in the position and intensity of the EL peaks, demonstrating that the flexible nanostructured LEDs maintained their optical properties when bent up to a radius of curvature of 3.9 mm. In particular, the lack of displacement of the EL peak during bending indicated that the strain applied to the active region in a single nanostructure was small and negligible. In addition, as shown in Fig. 6.5c, the I-V curves for different bending radii exhibited very similar rectification characteristics with no significant difference in device parameters (e.g., on-state voltage or leakage current). This indicated that no serious mechanical damage or fracture occurred at the top electrode or at the connection between the nanostructure and graphene during the bending test.

In addition to the flexibility, the reliability of the flexible LEDs was also investigated by measuring the EL and electrical characteristics when repeated for up to 100 bending cycles. First, as shown in Fig. 6.5d, under repeated bending, the nanostructured LEDs exhibited almost identical EL spectra at a fixed current of 10 mA, and the integral emission intensity remained almost constant for up to 100 bending cycles. In addition to the luminescence characteristics, the electrical characteristics were also preserved by repeated bending, exhibiting very similar rectified I-V curves, as shown in the inset of Fig. 6.5e. The key electrical device parameters plotted in Fig. 6.5e without significant degradation included the forward bias voltage ($V_{f)}$ at 10 mA and the reverse current (I_r) at − 5 V. V_f increasesd only slightly over a range

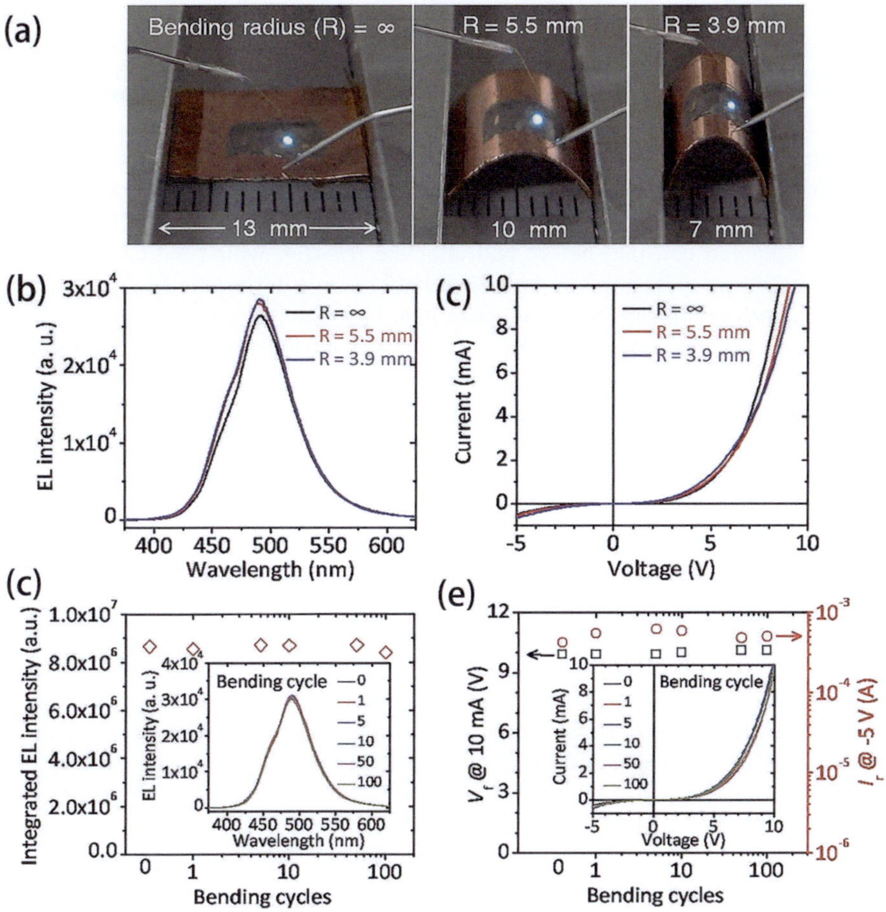

Fig. 6.5 **a** Photographs of luminescence with bending radii of ∞, 5.5 and 3.9 mm, respectively. **b** EL spectra and **c** I-V characteristic curves as a function of bending radius. **d** Plot of integrated emission intensity as a function of bending period. The inset showed the corresponding EL spectra. **e** Forward bias voltage (V_f) at 10 mA and reverse current (I_r) at 5 V are plotted according to the bending period. The inset showed the corresponding IV characteristic curves [33]

of ≈ 2%, while I_r remained nearly constant at ≈ 5 × 10^{-4} A. Fabricated on graphene thin films, all these properties of nanostructured LEDs proved their reliable operation in flexible form. Thus the development of unconventional inorganic LED optoelectronic devices in metastable, flexible or stretchable form can take full advantage of the excellent electrical and optical properties of inorganic semiconductor nanostructures and inorganic semiconductors, as well as expanding the range of applications of LED devices in the market, providing a technological route to flexible lighting and flexible displays.

6.1.3 Flexible HEMTs

Hiroki et al. [34] fabricated AlGaN/GaN HEMTs on h-BN/sapphire substrates and successfully transferred them from the host substrate to a copper plate using h-BN as a release layer. Figure 6.6 showed the typical I-V characteristics and transfer characteristics of the AlGaN/GaN HEMT before release from the substrate and after transfer to the copper plate. The source resistance estimated by the end-resistance method remained constant before and after the transfer, which indicated that the contact resistance and access resistance did not change during the transfer. For the I-V characteristics in Fig. 6.6a, the drain bias V_{ds} was swept from 0 to 20 V, and the gate bias V_{gs} was changed from + 2 to − 4 V in steps of 1 V. For the transfer characteristics in Fig. 6.6b, V_{gs} was swept from + 2 to − 5 V, with a constant V_{ds} of 6 V. Good cutoff and saturation characteristics were obtained both before the release and after the transfer. Before release, at large positive V_{gs}, a significant decrease in drain current I_d with increasing V_{ds}, i.e., negative differential resistance, was observed in the saturation region. At $V_{gs} = 2$ V, I_d decreased by 30% as V_{ds} increased from the saturation voltage to 20 V. The negative differential resistance of the HEMT was also observed in the saturation region. In contrast, for the shifted HEMT, the decrease in I_d was only 8%. The presence of negative resistance was usually attributed to the self-heating effect. The transfer from sapphire (j = 40 W/m K) to copper plate (j = 390 W/m K) improved the heat dissipation efficiency and suppressed the decrease in drain current under high bias operation. At $V_{ds} = 6$ V, suppressing the negative resistance increased the maximum transconductance from 100 mS/mm to 120 mS/mm in Fig. 6.6b. On the other hand, the threshold voltage (3.5 V) hardly changed before and after the transmission.

Figure 6.7 showed the IR temperature images of the sample during pre-release and post-transfer operations. Figure 6.7a, c showed the HEMT images before release without gate bias at 0 and 20 V_{ds}. Figure 6.7b, d showed the corresponding images after the transfer. The white dashed line indicated the outline of the sample. For the

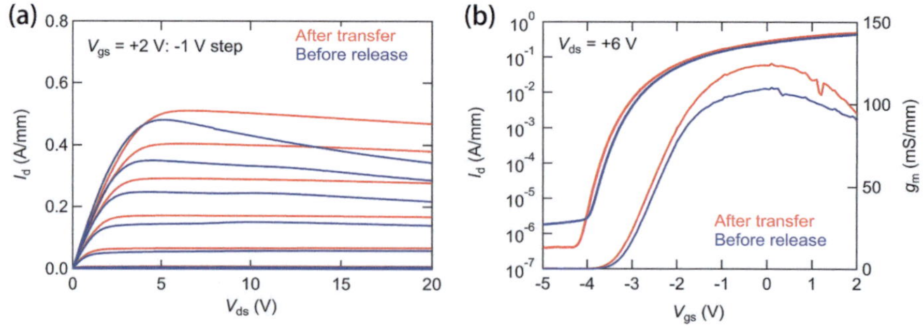

Fig. 6.6 **a** I_d-V_{ds} characteristics of the AlGaN/GaN HEMT before release from sapphire (blue line) and after transfer to a copper plate (red line). **b** A magnified view of the I_d-V_{ds} curve in the saturation region [34]

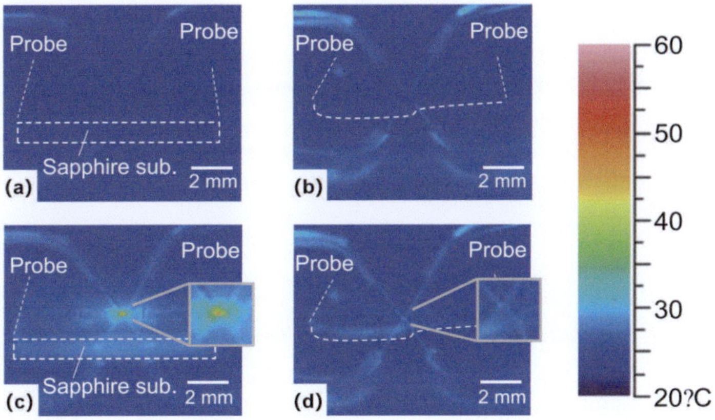

Fig. 6.7 Sample temperature maps taken with the infrared camera Neo Thermo 700. (**a**) and (**b**) showed the off state of the unreleased and transferred devices, respectively. (**c**) and (**d**) showed the on state of the unreleased and transferred devices at $V_d = 20$ V, respectively. A magnified view of the operating region was also shown in (**c**) and (**d**) [34]

HEMT, the power consumption during operation was about 0.8 W (40 mA) and 1 W (50 mA) before release and after transfer, respectively. In Fig. 6.7c, a hotspot with a temperature of about 50 °C was observed (temperature rise of 27 °C). In contrast, the temperature of the shifted HEMT was as low as 30 °C (temperature rise 7 °C), despite operating at higher power. This suggested that suppression of the negative differential resistance as well as improved DC characteristics, comed from suppressing the self-heating effect. Note that although quantitative estimates were difficult to make, the channel temperature may be even higher than that measured by thermography. Temperature increases of even up to 180 °C, with a power consumption of 0.6 W, had been reported for AlGaN/GaN HEMTs on sapphire [35]. In sharp contrast, this study demonstrated that the Joule heat generated during HEMT operation can be reduced by transferring the device to a copper plate with high thermal conductivity.

6.2 Technical Routes for Nitride Flexible Stripping and Transfer Methods

Due to the amorphous surface atomic arrangement of flexible amorphous substrates, which typically have lower melting points relative to single crystal substrates, they cannot be stabilized at the growth temperatures required for high quality III-group nitrides (typically greater than 1000 °C) and therefore cannot be used for direct epitaxial growth of single crystal thin films. Therefore, to realize the fabrication of III-group nitride film-based optoelectronic devices on flexible substrates, it is needed to try to delaminate the epitaxial nitride film from a single-crystal substrate and further transfer it to a flexible substrate, thus realizing nitride-based flexible optoelectronic devices, which can provide the basis for wearable and foldable electronic devices based on III nitrides. However, stripping single-crystal III-group nitride films from epitaxial substrates is very difficult because of the strong sp^3-type covalent bonds between the substrate and the epitaxial layer. To overcome this problem, a large number of stripping methods have been investigated and practiced, including heat release by laser radiation, labeling-based printing, chemical etching, and mechanical stripping.

6.2.1 Laser Radiation Heat Release Peel Transfer Film

Epitaxial stripping combined with wafer bonding can be used as a direct method of nitride-substrate integration. Taking advantage of the transparency of the sapphire substrate and the AlN overlayer, the process can be carried out at higher fluxes without noticeable decomposition. Kelly et al. [36] realized that the transparency of sapphire could be used in conjunction with the thermally decomposable nature of GaN to separate the GaN film from the sapphire substrate, resulting in the first successful process of delamiateing a GaN film from sapphire. The process took advantage of the thermally activated decomposition of GaN, which began to occur above about 800 °C and leaded to nitrogen exudation. This decomposition can be induced with high spatial resolution by heating the material with short laser pulses (less than 10 ns). The rapid generation of heat permited high localized temperatures to be reached before heat was conducted away from the illuminated area. In this way, surface patterning was achieved at an illumination intensity of about 0.2 J/cm^2 or more at a wavelength of 355 nm, above the absorption threshold of GaN. And in order to realize the peeling of the film, it was realized to activate at the internal interface instead of the surface. This required that the interface can be illuminated with a certain wavelength of light, which was then absorbed there. The interface can be illuminated with 355 nm light through the sapphire substrate, which was absorbed by the GaN with an extinction length close to 100 nm. The decomposition of the GaN produced nitrogen at the interface, which caused the two interface sides to expand and separate. To delaminate the interface, a membrane material with a low light absorption threshold can be inserted. $In_xGa_{1-x}N$ films fulfilled this requirement in GaN thin film systems. Here, the 355 nm third harmonic of a Q-switched Nd:YAG laser with a nominal pulse duration of 6 ns was used. The beam diameter was 7 mm, and depending on the angle of incidence of the sample, a pulse of 100–200 mJ was required. Both sides of the sapphire substrate were polished to illuminate the interface after growth. Stripping was performed with a single pulse. After disintegration, excess gallium was removed by briefly holding it in concentrated hydrochloric acid. An electron micrograph of GaN/sapphire after thermal radiation was shown in Fig. 6.8. The GaN film had been completely removed in the area of the fully heated interface, leaving a clean substrate surface. The trench widened away from the interface, leaving sloping sidewalls on the rest of the stage. This wedge-shaped form of the removed material was indicative of the explosive nature of the separation and therefore required further investigation to determine the

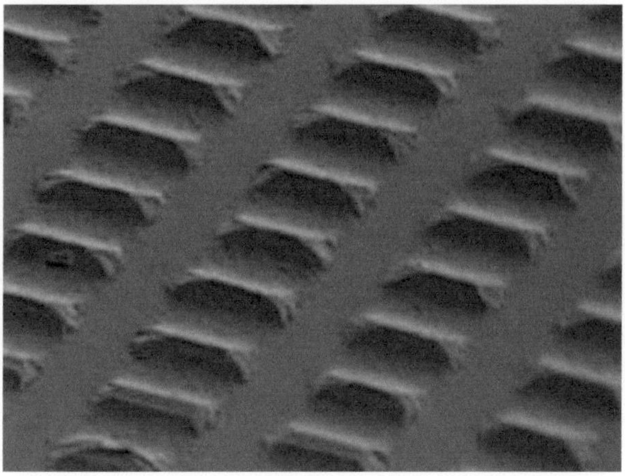

Fig. 6.8 Electron micrograph of structured GaN film showing exposed sapphire substrate lines and sloping sidewalls [36]

extent to which this form can be controlled. As shown in Fig. 6.8, when the spacing of the strips was sufficiently small compared to the film thickness, adjacent grooves converged and formed a roof or pyramidal peak on the table surface. Above this peak, the entire film had been removed. To protect peeling, the front side of the film was secured to the support material by an adhesive. A hard fusible adhesive can be used to attach the film to a glass microscope slide. A laser pulse passed through the sapphire, which then caused the substrate to detach, thus leaving most of the film on the support plate. The film can then be released by dissolving the adhesive in acetone. This process was most successful for relatively thick films (greater than 5 μm), producing complete films up to 1–2 mm in size. For larger film areas, optimization of laser intensity and other process parameters as well as film handling was required.

Wong et al. [17] further demonstrated the efficacy of the above technique at shorter wavelengths (248 nm) and the fact that the GaN crystal quality did not degrade during the separation and transfer process. A GaN film with a thickness of 3 μm epitaxially grown on a single-sided polished sapphire substrate was used as the starting material. Boron-doped p-type Si wafers were bonded to the surface of the GaN film with epoxy resin to form a sapphire/GaN/epoxy/Si structure. The backside of the sapphire substrate was polished with diamond paper before wafer bonding. All laser processing of the sapphire/gallium nitride/epoxy/silicon structures was performed in air using a Lambda Physics Lextra 200 KrF pulsed excimer laser (38 ns pulse width) with the incident beam passing directly through the sapphire substrate. The attenuation of a KrF excimer laser beam penetrating through a 0.5 mm thick sapphire substrate was about 20–30% at 248 nm. The GaN film was successfully delaminated and transferred from the sapphire to the Si substrate after a single laser pulse at an injection greater than or equal to 400 mJ/cm^2. After laser processing, the film was separated from its sapphire substrate by heating the sample on a hot plate to a temperature higher than the melting point of Ga ($T_m = 30$ °C)0.3 × 4 mm GaN film was successfully transferred to 10 × 10 mm Si substrate. The cross-sectional SEM image in Fig. 6.9a showed the GaN/epoxy/ Si structure after separation from the sapphire substrate. The efficacy of this laser processing technique was shown by the analysis of the GaN film by X-ray wiggle curve. Figure 6.9b showed the X-ray wiggle curve of the GaN 0002 reflection before lifting off the GaN deposited on sapphire. It showed fairly high quality GaN, such as a half-peak width of about 0.1°. Figure 6.9c showed the rocking curve of the same film after stripping using a laser fluence of 600 mJ/cm^2, and the GaN (0002) half-peak width was unchanged from the value measured before stripping. The width of the rocking curve increased if the film suffered thermal or mechanical damage during the separation process resulting in microcracks or warpage. Thus there was no damage to the film quality before or after the stripping process for this nitride film.

6.2.2 Transferring Films by the Transfer Method

The transfer method directly addresses the problem of adapting multiple material classes in many situations, while still the best-performing material known for adapting to challenging and unusual environments in specific applications. Figure 6.10 shows three different transfer categories: additive transfer, subtractive transfer, and deterministic assembly. These three options are significant due to their natural compatibility with high-performance single-crystal semiconductor materials (e.g., Si, GaAs, GaN, InP, etc.) with micro- or nanostructured forms [37, 38] Transferring such materials with a soft, elastic compression mold allows them to be assembled to any type of substrate with high throughput and precise alignment accuracy. When repeated, this process will provide high-resolution, large-area assembly techniques with 2D- or 3D layouts and heterogeneous levels of material integration, all of which are beyond the capabilities of any other method.

In all three scenarios of Fig. 6.10, we refer to the first case (Fig. 6.10a) as additive transfer, which is an efficient way for many types of organic and inorganic materials to be manipulated in real time [39], where the transfer occurs between the entire ink layer of those surfaces that have been deposited on the impression and the receiving substrate, or a selected portion of it. Inking of the impression can be achieved by a variety of methods, ranging from solution casting or physical vapor deposition. Layers printed by additional techniques may provide different functions, such as forming integrated device components or serving as etch masks for subsequent processing [40, 41]. The second method, subtractive transfer, as shown in Fig. 6.10b, utilizes a stamp to selectively recover areas of the overlay film. This type of printing can be used to pattern the active layer directly or to open a window in the etch mask, thus allowing subsequent processing of the lower layers of the target material. As in the case of additively printed transfers, chemical correction of the receiver, heat, or a thin intermediate polymer bonding promoter layer between the impression and the ink can be effective in aiding the transfer. Unlike additive printing, this mode requires cohesive rupture of the ink layer, so other aspects need to be considered in the engineering design. In some cases, subtractive transfer represents the inking step of the add-on transfer process. Such methods or more conventional processing

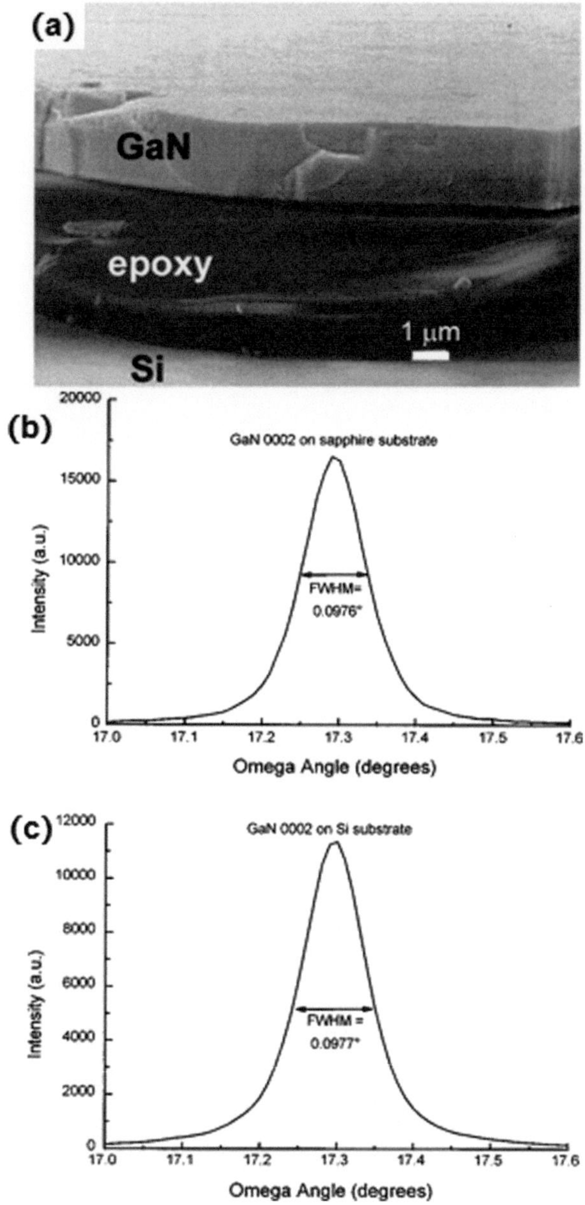

Fig. 6.9 **a** SEM of a cross-section of an isolated GaN membrane bonded to a silicon substrate. X-ray wiggle curves of the GaN membrane (0002): **b** GaN on sapphire, **c** after transferring from sapphire to a silicon substrate [17]

or growth techniques can form defined structures on the donor substrate for use in deterministic assembly techniques, as shown in Fig. 6.10c. This strategy is valuable because it greatly expands the possibilities for fabricating materials by separating the growth and processing of the ink from the impression and the receiving substrate.

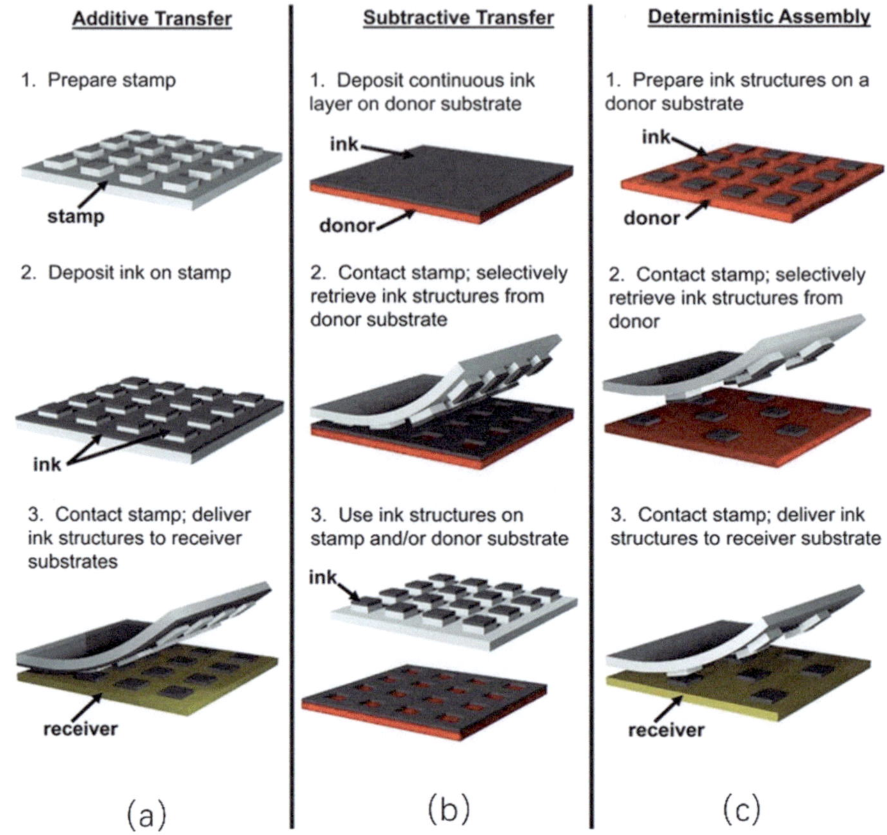

Fig. 6.10 Schematic representation of the three basic modes of transfer printing. **a** Additional transfer utilizes a stamp "embossed" with the target material using processes such as physical vapor deposition, solution casting/assembly, or physical transfer. Bringing this inked impression into contact with the target substrate and then stripping it affects the transfer of the material. **b** Subtractive transfer begins with a continuous film material deposited on a donor substrate. An impression in contact with the film selectively removes material from the contact area, leaving a patterned film on the donor and transferring the material over the impression. The stamp inked in this manner can then be used for additional transfers, as shown in (**a**). Alternatively, the patterned material on the donor itself may be used for further device processing. **c** Deterministic assembly involves contact of the stamp with a donor substrate supporting a prefabricated micro- or nanostructure. Stripping the stamp results in the removal of the selected structure from the donor substrate. Printing on the receiving substrate completes the process

These methods typically rely on a dynamic, reversible approach to regulating adhesion to the impression. The main advantage of the processes outlined in Fig. 6.10 is that they can rapidly transfer material in sparse or dense layouts over larger substrate areas, particularly through multiple stamp inking and printing cycles.

6.2.3 Chemical Etching to Peel Transfer Films

Chemical etching, as a simpler and more effective method of delamination thin films from the substrate, is based on the corrosion resistance of nitrides and utilizes a chemical etching solution to separate the thin film from the substrate. Typically, chemical etching requires the addition of an easily corroded insertion layer between the substrate and the epitaxial nitride film, which is then corroded off as a sacrificial layer in the subsequent stripping process to achieve the purpose of detaching the epitaxial layer from the substrate.

Rogers et al. [42] grew GaN on ZnO/c-Al_2O_3 using low-temperature or low-pressure MOCVD using nitrogen as a carrier and dimethylhydrazine as an N source, as shown in Fig. 6.11a. GaN and ZnO shared the same stacking order, in addition to identical fibronectin structures and smaller lattice mismatches (1.8%). Therefore, stacking mismatch boundaries and antidomain boundaries should be less likely in GaN grown on ZnO. In addition, the thermal expansion coefficient mismatch of c-axial GaN/ZnO was greatly reduced compared to GaN/c-Al_2O_3. Moreover, the known thermal strain was compressive for GaN/ZnO and tensile for GaN/c-Al_2O_3. This meaned that the possibility of rupture of GaN grown on ZnO can be reduced. Another attractive feature of ZnO was that it can be chemically etched in most acids and bases. Since GaN was resistant to most of these same etchants, ZnO can be selectively removed in order to delaminate GaN-based devices from the substrate. After mechanical delineation of the sub-millimeter stage with a diamond scribe, the GaN/ZnO/c-Al_2O_3 samples were immersed in a dilute hydrochloric acid solution to perform the stripping. Figure 6.11b showed the optical microscope photographs of the chemical stripping process over time sequence. After a few minutes, the ZnO was completely dissolved by the hydrochloric acid solution and the GaN film floats to the surface of the etchant.

Goto et al. [43] proposed to prepare a GaN thick layer on c-face sapphire with a CrN buffer layer by HVPE and to delaminate the epitaxial GaN thick layer from the sapphire substrate by chemical stripping. Although CrN had a rock salt structure, the bond length and thermal expansion coefficient of CrN were between those of GaN (0001) and c-face sapphire. Therefore, the use of CrN as a buffer material for epitaxial growth of GaN films can be chosen to realize high-quality epitaxy of nitrides. In addition, CrN can be easily chemically etched while the nitride remains stable. Prior to the growth of GaN using the HVPE method, a buffer layer of CrN was formed by sputter deposition of a Cr film on c-plane sapphire under NH_3 atmosphere followed by nitriding. The GaN layer was then grown using a two-step growth method (low-temperature followed by high-temperature) to obtain epitaxial thin films with flat surfaces, as shown in Fig. 6.12a. Chemical stripping of thick GaN layers was executed by selectively etching the CrN buffer layer using a solution of cerium diammonium nitrate (IV) and perchloric acid at 50 °C. To lift off a sample with an area of $5 \times 5\ mm^2$, approximately 3 h were required. The GaN layer was completely isolated from the sapphire substrate by etching the CrN buffer layer. Concentric cracks in the GaN layer were occasionally observed during the etching process from

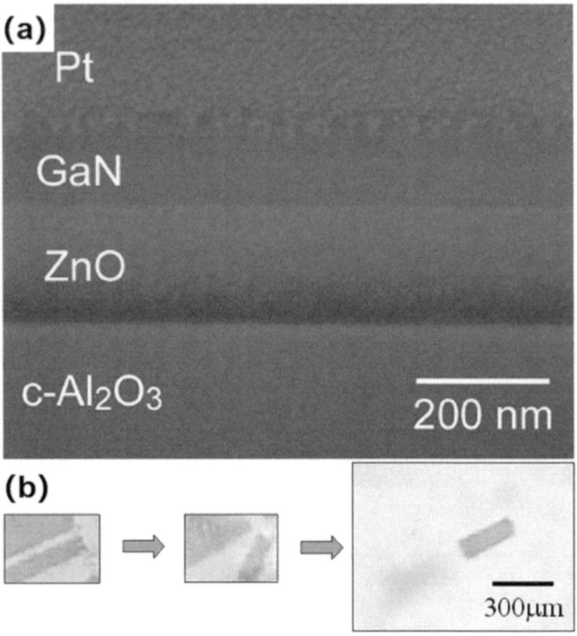

Fig. 6.11 **a** SEM image of Pt/GaN/ZnO/c-Al_2O_3 cross-section, and **b** optical microscope image of chemical exfoliation [42]

the edge to the center of the sample, whereas the thickness of the GaN layer was only 20 μm. By looking at both surfaces through a planar scanning electron microscope shown in Figs. 6.12b, c, it can be seen that both the front surface and the back surface of the GaN layer were smooth and free of pits or cracks after the stripping was achieved. In order to evaluate the effect of the chemical stripping process on the crystal quality of the GaN layer, Fig. 6.12d illustrated the (0002) X-ray wiggle curves of the GaN thick layer before and after stripping. The FWHM values of the (0002) ω-scan of X-ray diffraction from the sample were 241 arcsec before and 230 arcsec after chemical stripping, respectively. Thus, it illustrates that chemical etching has only a small effect on the crystal quality of the GaN layer.

6.3 Realization of Flexible Stripping and Transfer of Quasi van der Waals Epitaxial Nitrides

In addition to being used as a buffer layer to assist the growth of nitride on it, another major use of 2D materials is that they can be used as an inserted weak bonding layer to realize the mechanical stripping of devices epitaxially grown on their substrates. Conventional stripping methods, such as chemical etching and laser stripping as described above, the former takes a long time, the substrate has a certain degree of selectivity, the latter leads to device thermal damage, and process conditions are difficult to control. If device stripping is realized through the weak interaction

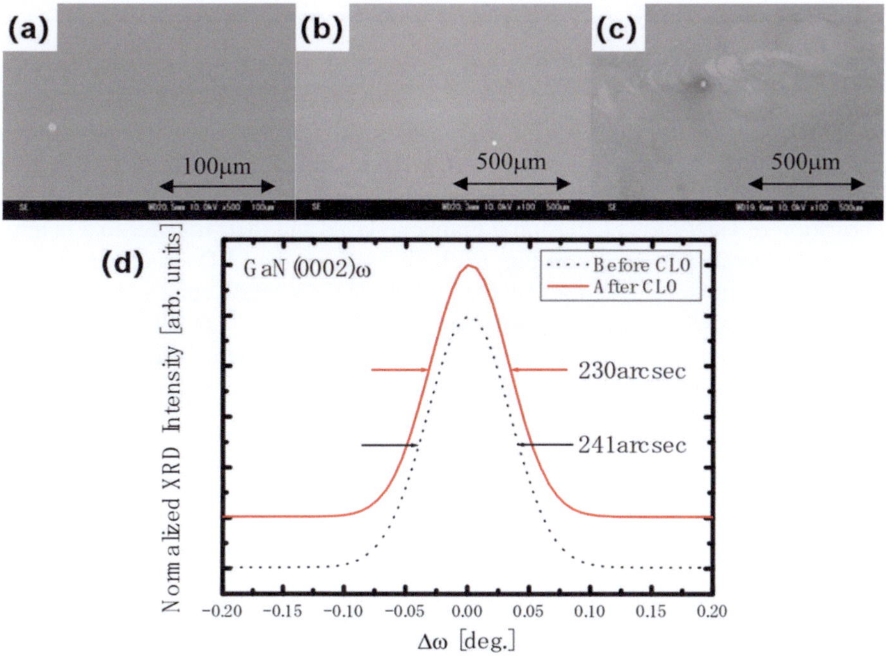

Fig. 6.12 **a** SEM image of GaN epitaxial layer grown on CrN/Al_2O_3; SEM image of a stand-alone GaN thick film exfoliated by the chemical stripping process **b** growth side, **c** backside (substrate side). **d** X-ray wiggle curves of the samples, with dashed lines indicating the wiggle curves of the GaN samples before chemical stripping (0002). **e** X-ray wiggle curve of the GaN sample (0002) after chemical stripping [43]

force between 2D material layers, there are many advantages, including: (1) no laser damage, resulting in a smooth separation surface, (2) no time-consuming chemical treatment is required, and (3) reduced equipment costs. Several elements also need to be met in order to achieve mechanical stripping of devices:

(1) 2D material: the thickness is suitable, 2–5 nm best, too thick is not conducive to the growth of nitride, too thin can not be fully shielded from the role of the substrate and is not conducive to the subsequent peeling, integrity is good to ensure that the device of large-area peeling; At the same time, the surface of the 2D material is small undulation, there is a certain number of dangling bonds to meet the high quality of the epitaxial layer of the growth;
(2) High-temperature film layer: the growth temperature can not be too high, in order to prevent the 2D materials in the growth process of heat diffusion or chemical reactions; Thickness can not be too thin, to ensure that the film has a certain degree of self-supporting after peeling down the film will not be broken or produce a large number of cracks, so that it is unable to further processing.

6.3.1 Graphene as an Insertion Layer to Assist Nitride Flexible Exfoliation

In 2010, Chung et al. [15] prepared high-quality nitride films on graphene-covered substrates for the first time, and achieved device exfoliation after further growth of GaN-based blue LEDs, as shown schematically in Fig. 6.13. A schematic of the fabrication of GaN-based LEDs on a graphene layer and their transfer to a heterogeneous substrate was shown in Fig. 6.13. To fabricate the LED structures, a 2 μm thick Si-doped n-GaN layer, three-cycle $In_xGa_{1-x}N$/GaN MQWs and a 350 nm thick Mg-doped p-GaN layer were deposited on the planar surface as well as a GaN film grown on the graphene layer. For EL spectra, a metallic graphene layer under n-GaN and a nickel/gold (Ni/Au) bilayer deposited on the top surface of p-GaN were used as n-type and p-type contacts, respectively. After fabricating the devices, the thin-film LEDs on the graphene layer were simply removed from the original substrate using mechanical force and then transferred to heterogeneous substrates of metal, glass, and plastic. All of the transferred LED devices emitted a very strong blue light that was visible to the naked eye under normal indoor lighting. As shown in the optical microscope image in Fig. 6.14a, the luminescence was very uniform over a 300 μm × 300 μm area, which may be due to the uniform distribution and injection of the current through the metal bottom electrode consisting of a graphene layer. Each substrate in Fig. 6.14a has unique benefits for LED applications. The metal substrate provided good thermal and electrical conductivity for high-power LEDs, while the glass or plastic substrate allowed inorganic LEDs to be made into large-area, full-spectrum LED displays in flexible or stretchable form. In addition, GaN-based epitaxial films on graphene layers can be easily used as functional components in photovoltaic devices. The EL properties of LEDs transferred onto plastic were further investigated by measuring the power-dependent EL spectra. The EL spectra and the corresponding EL images of the LEDs at various applied current levels were represented in Fig. 6.14b. The LED luminescence images and the EL spectra showed good luminescence characteristics with increasing emission intensity with the applied current levels of 5.1, 6.4, and 8.1 mA, respectively. Thus transferring materials and devices fabricated on graphene to heterogeneous substrates can provide advantages in the integration and design of electronic and optoelectronic devices.

Kim et al. [20] achieved the release as well as transfer of GaN thin films on SiC substrates as shown in Fig. 6.15. Ni stress source evaporator was deposited in a vacuum of 1×10^{-5} Torr. The GaN film was then completely released using a thermally released tape treatment layer. The remaining graphene/SiC substrate was then immersed in an $FeCl_3$-based solution to completely remove the Ni residue for the next GaN regrowth. The released GaN film was transferred to a silicon wafer coated with 90 nm SiO_2 by down-pressing the film layer, then the thermal tape was removed by annealing at a release temperature above 90 °C, followed by removal of the residual Ni in a solution of $FeCl_3$. By repeating the above steps for several

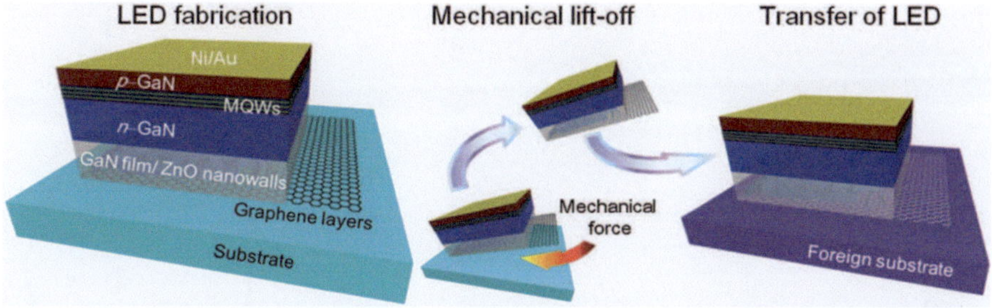

Fig. 6.13 Schematic diagram of the fabrication and transfer process of thin film LEDs grown on graphene layer substrates [15]

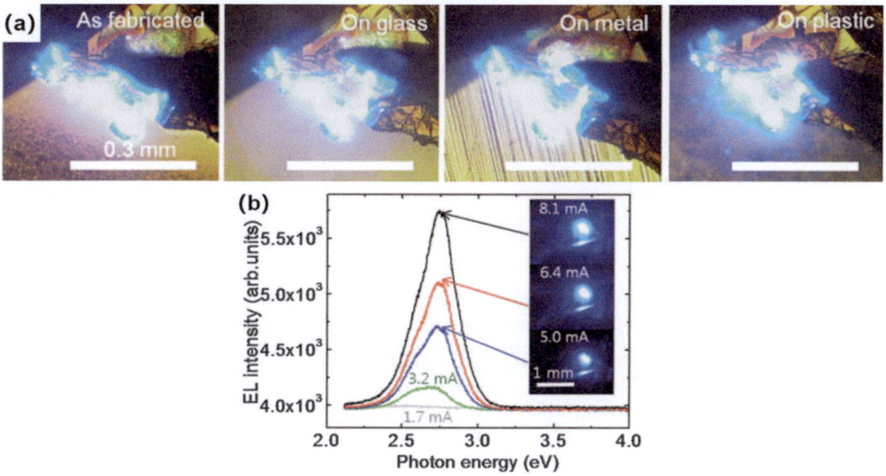

Fig. 6.14 **a** Optical images of the luminescence of the assembled LEDs on the original substrate and the transferred LEDs transferred to heterogeneous substrates (metal, glass, and plastic, respectively). **b** Room-temperature EL spectra of the LEDs transferred to the plastic substrate. Optical microscopy images showed luminescence at different applied current levels from 1.7 to 8.0 mA [15]

transfers, the reuse of the original substrate was achieved. As shown in the cross-sectional SEM image in Fig. 6.16a, the 2.5-mm-thick GaN film was stripped from the substrate by a 2-μm-thick Ni stress source, and the air gap can be clearly seen at the interface. The structure of the released surface of the GaN film imaged by AFM in Fig. 6.16b showed similar steps to those observed on the surface of the pristine graphitized SiC substrate, This implied that the entire GaN film epitaxially growth on the graphitized SiC substrate was accurately removed from the graphene-GaN interface, allowing for the GaN film to be directly bonded to the (100) SiC substrate with the 90 nm SiO_2 layer (100) Si substrate. As shown in the SEM image in Fig. 6.16c, a unique structure of GaN can be realized on the insulator, which can be used for hybrid integration of GaN and (001) Si. In order to accurately determine

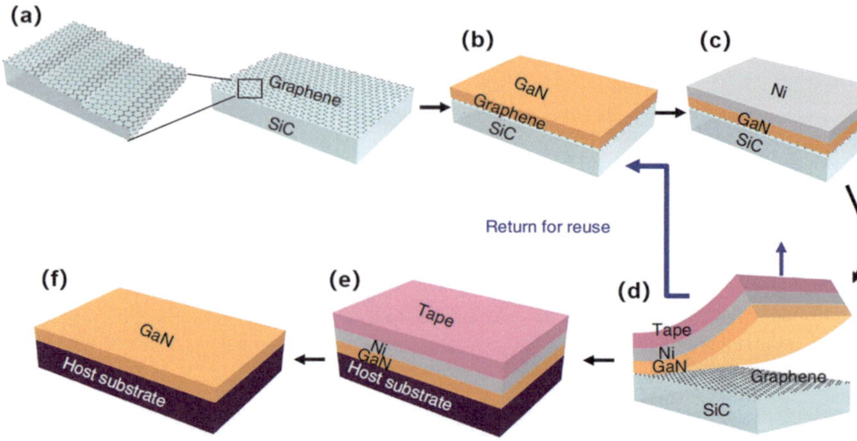

Fig. 6.15 Schematic representation of the method for growing/transferring single-crystal films on graphene. **a** Graphitization of SiC substrate to form epitaxial graphene. **b** Epitaxial growth of GaN on graphene. **c** Deposition of stressor layer (Ni). **d** Release of GaN from substrate with treatment tape. **e** Transfer of released GaN/Ni/tape to heterogeneous substrate. **f** Removal of tape and Ni metal by thermal release and wet etching, leaving a GaN film on the heterogeneous substrate [20]

the released interface, they analyzed the Raman spectra measured from the released surface of the GaN film and the remaining substrate surface as shown in Fig. 6.16d. Raman peaks corresponding to graphene were clearly observed from the surface of the released SiC substrate, but only GaN peaks were observed from the released membrane without any peaks corresponding to graphene in Fig. 6.16d. This clearly demonstrated that the GaN film was released from the graphene surface and not with graphene.

6.3.2 BN as an Insertion Layer to Assist Nitride Flexible Stripping

In 2012, Y. Kobayashi et al. [16] tried a mechanical transfer using a release layer, in which h-BN was used as a release layer for vertical-type InGaN/GaN LED structures. It enabled to grow AlGaN/GaN heterostructures as well as InGaN-based MQWs and LED structures on a h-BN layer with atomic planes and to release these structures from the main body to the sapphire substrate and transfer them to the other substrates. Figure 6.17 depicted the material design of the LED and the release and transfer process. A 3 nm thick h-BN release layer [44] was grown on sapphire by MOCVD, and a 300 nm thick Si-doped AlGaN was grown on top of it as a buffer layer. The grown 10-cycle MQWs structure of InGaN/GaN was turned upside down and attached to the indium bonding sheet by a bonding sheet (indium sheet) placed on a heterogeneous substrate in Fig. 6.17b and heated in order to heat the structure of the

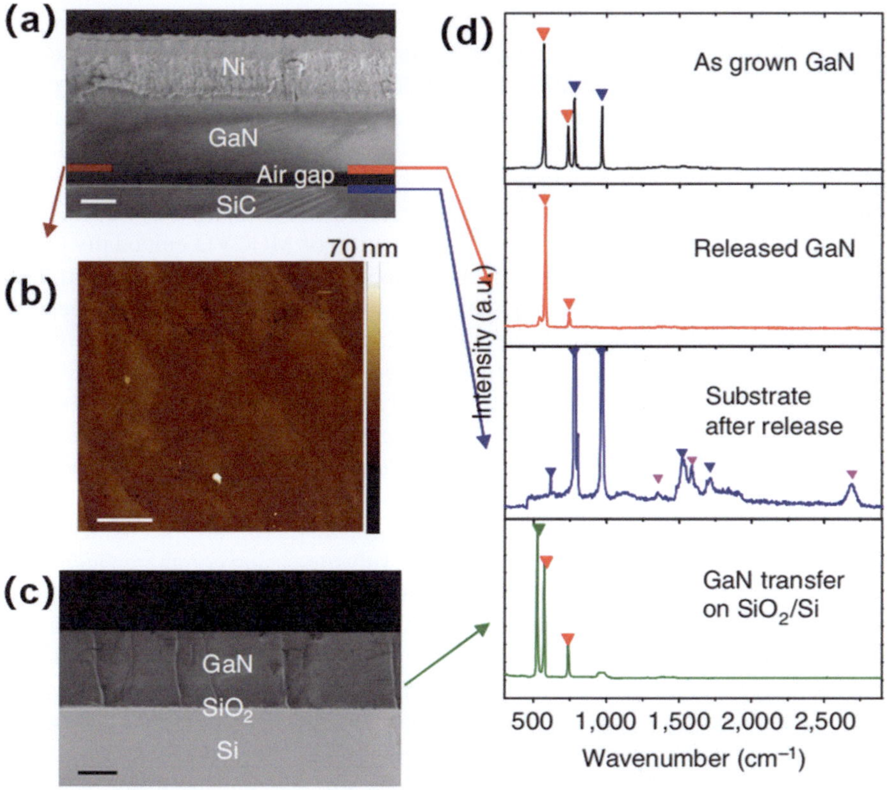

Fig. 6.16 GaN transfer from epitaxial graphene and reusability analysis of graphene/ SiC substrate. **a** Scanning electron microscope image of GaN on graphene/ SiC substrate after deposition of Ni stress source (scale bar, 1 μm). **b** AFM image taken on the release surface of GaN film (scale bar, 3 μm). **c** SEM image of GaN membrane directly transferred onto SiO_2/Si substrate (scale bar, 1 μm). **d** Raman spectra obtained from the grown GaN film on the graphene/ SiC substrate, the GaN film released on the tape, the remaining substrate after the release of GaN, and the transfer of the GaN film to the SiO_2/Si substrate [20]

MQWs on the indium sheet attached to the heterogeneous substrate to temperatures sufficient to heat-seal the indium onto the sapphire and the MQWs, a process which took just a few seconds using a few seconds in Fig. 6.17c. Finally, the MQWs structure was released from the host sapphire substrate by mechanical force, which transferd the MQWs structure to the heterogeneous substrate in Fig. 6.17d. Due to the vdW forces of the h-BN layered structure, the separation occured within the h-BN release layer, and therefore mechanical forces easily separated the MQWs structure from the sapphire substrate, thus allowing the transfer of MQWs structures and other types of nitride devices to various substrates, such as silicon, polycrystalline metals, glass, and transparent plastics, and other flexible substrates. A photograph of a crack-free MQWs structure transferred onto an indium sheet/heterogeneous substrate was shown in Fig. 6.18a. According to AFM, the surface rms roughness was only 0.95 nm in Fig. 6.18b. 2θ/ω-scan XRD results in Fig. 6.18c indicated that the peak positions (peak intensities) were almost unchanged before and after the transfer, which suggested that the transfer process did not damage the quality of the material. The EL intensity was also improved due to the reflection of the indium sheet on the backside in Fig. 6.18d. It was well illustrated that the BN material as an insertion layer provided a way to release various nitride semiconductor devices and transfer them to large-area, flexible substrates.

Ayari et al. [45] achieved wafer-scale exfoliation of MOCVD epitaxially grown InGaN/GaN MQWs structures from a 2-inch sapphire substrate utilizing a 5 nm-thick h-BN layer as a low-viscosity sacrificial layer. Figure 6.19a, b showed a schematic of the structure after growth and a photograph of a wafer of this structure grown on sapphire covered with an h-BN layer, respectively. SEM images, as shown in Figs. 6.19c, d, confirmed that the surface quality of the samples with the h-BN layer was similar to that of the samples with the conventional GaN template. In both cases, the AFM measurements of the p-GaN surfaces demonstrated a comparable roughness (about 1.6 nm), which was a typical value for such surfaces. The weak vdW bonds between the h-BN atomic layers were easily broken, thus permitting mechanical stripping of the MQWs structure from the sapphire substrate using commercial adhesive tape. Figure 6.19e, f showed photographs of two MQWs structures peeled using layers of acrylic conductive adhesive on aluminum and copper foils, respectively. The entire MQWs structure was successfully peeled from the 2-inch substrate with excellent reproducibility. After peeling, these layers exhibit many crack-free surfaces that are several square millimeters in size. This was larger than the typical device size of commercially available LEDs (300 μm × 300 μm). The post-separation surface morphology of the exfoliated MQWs structures was investigated using SEM and AFM. Figure 6.19g showed a SEM image of the n-type buffer surface in the released MQWs structure, which exhibited a smooth surface morphology. According to the AFM image, a surface roughness of 1.14 nm was obtained over a region of 5 μm × 5 μm. This smooth surface suggested that the exfoliation occured by breaking the bonds between the 2D layered BN layers rather than through the bonds at the interfaces of the sp^3-bonded type III nitride layers, since breaking the sp^3-bonded type III nitride layers would lead to an increase in the surface roughness. A cathodoluminescence (CL) of 454 nm existed before stripping, with a small change to 458 nm after stripping. These results indicated that ultrathin h-BN with high crystalline quality can be used as an effective sacrificial layer and can maintain the original photovoltaic properties compared to the conventional two-step growth method, while also reducing the variations in thickness and temperature required for growing GaN buffer layers in the conventional two-step growth process. The results supported the use of h-BN as a low-viscosity sacrificial layer for GaN-based device structures and demonstrated the feasibility of large-area stripping and transfer to any template, which was of great importance for industrial-scale production.

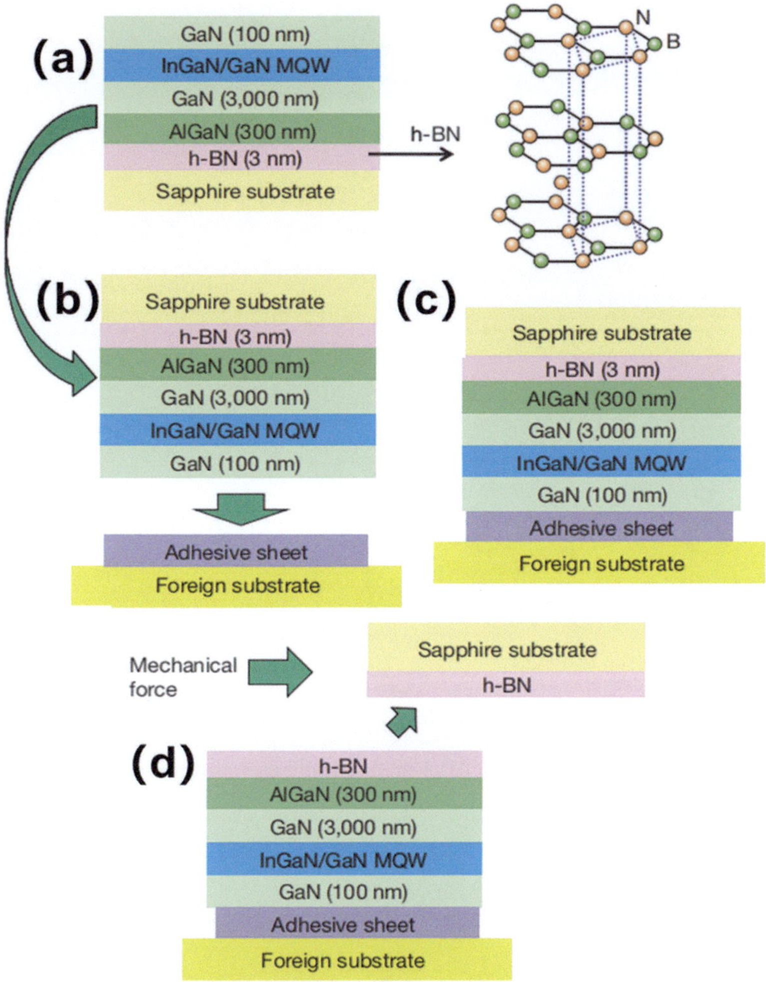

Fig. 6.17 **a** Growth of a single-crystal h-BN release layer on a sapphire substrate, followed by the growth of a fibrillar zincite AlGaN layer and MQWs structure. **b** The MQWs structure is turned upside down and placed on a heterogeneous substrate by bonding a sheet (indium sheet). **c** The MQWs structure attached to the sheet is heated to a temperature sufficient to heat-seal the indium, and then the MQW structure is released from the host sapphire substrate and transferred to the heterogeneous substrate by mechanical force (**d**) [16]

6.4 Summary

Flexible optoelectronic devices have attracted widespread attention in scientific research and industry due to their wearable and foldable special application scenarios. The question of how to achieve stable preparation of III-group nitride-based flexible optoelectronic devices has become a major research hotspot. However, due to the

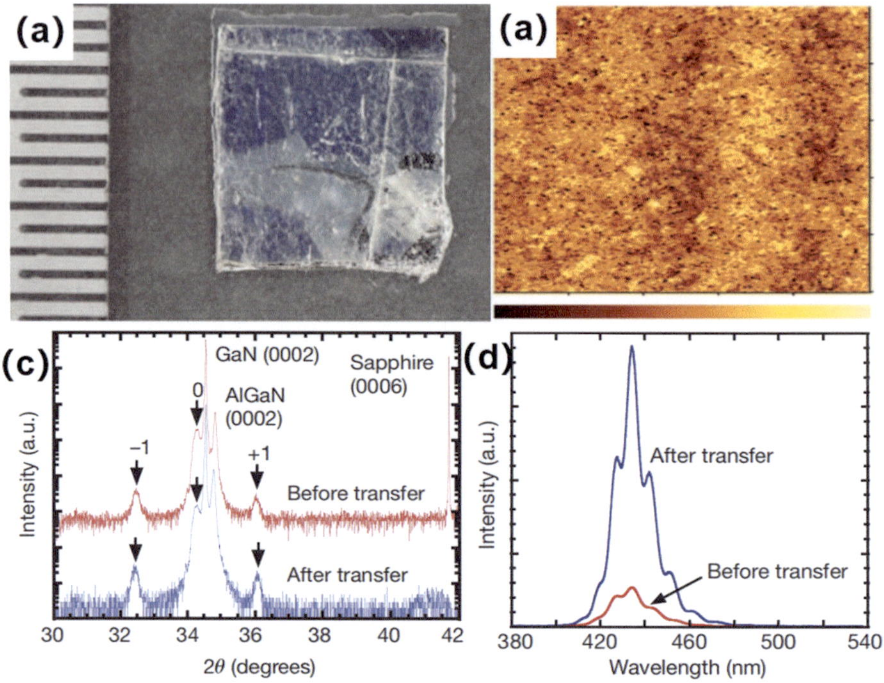

Fig. 6.18 **a** Photograph of the transferred MQWs structure, **b** AFM image of the surface of the transferred structure, **c** X-ray diffraction (2θ/ω-scan) of the InGaN/GaN MQWs before and after the transfer, and **d** PL spectra of the MQWs before and after the transfer, at room temperature [16]

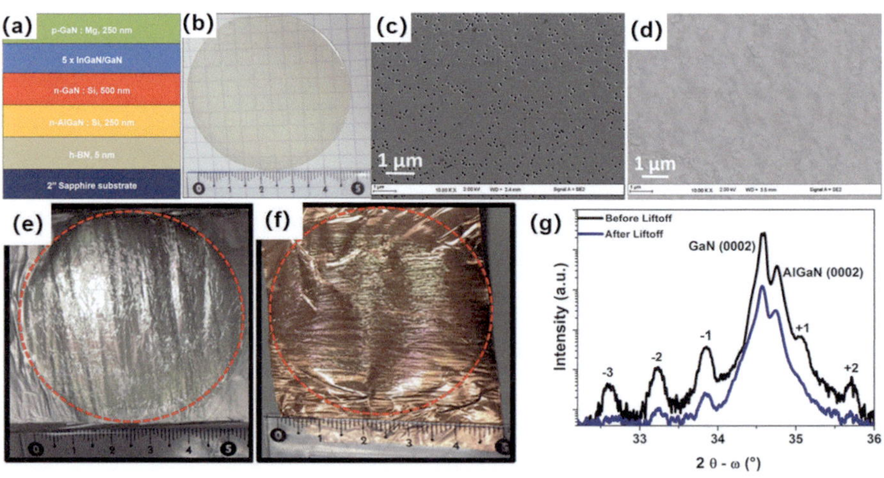

Fig. 6.19 **a** Schematic of the grown structure and **b** photograph of the transparent wafer after epitaxial growth. SEM images of the surface of the MQWs structure **c** on a conventional GaN template and **d** on h-BN. Photographs of the MQWs structure after exfoliation using **e** an aluminum and **f** a copper foil. **g** X-ray diffraction using the MQW structure before and after transferring [45]

strong covalent bond interaction between the substrate and the epitaxial layer, it is very difficult to peel off the single-crystal III-group nitride thin film. This chapter first introduces the relatively primitive methods to achieve the peeling and transfer of epitaxial nitrides from the single-crystal substrate through laser radiation heat release, chemical etching, and mechanical peeling. It also introduces the technical method of transferring functional thin films to heterogeneous flexible substrates by taking advantage of the weak vdW interaction between 2D material layers. The 2D materials not only act as a buffer layer but also provide a release layer for the mechanical peeling of the nitride epitaxial layer. This method provides an excellent technical route for achieving large-scale, large-size, low-cost flexible III-group nitride devices. It then focuses on the use of qvdWE methods to achieve the peeling and transfer of various types of devices, its advantages in optoelectronic applications compared to traditional methods, further illustrating the advanced nature of this technical route in the flexible transfer process.

References

1. Akasaki I (2007) Key inventions in the history of nitride-based blue LED and LD. J Cryst Growth 300:2–10
2. Nakamura S, Senoh M, Nagahama SI, Naruhito I, Yamada T, Matsushita T, Kiyoku H, Sugimoto Y (1996) InGaN multi-quantum-well-structure laser diodes with cleaved mirror cavity facets. Jpn J Appl Phys 35:217–220
3. Xu GY, Salvador A, Kim W, Fan Z, Lu C, Tang H, Morkoç H, Smith G, Estes M, Goldenberg B, Yang W, Krishnankutty S (1997) High speed, low noise ultraviolet photodetectors based on GaN p-i-n and AlGaN(p)-GaN(i)-GaN(n) structures. Appl Phys Lett 71:2154–2156
4. Dahal R, Pantha B, Li J, Lin JY, Jiang HX (2009) InGaN/GaN multiple quantum well solar cells with long operating wavelengths. Appl Phys Lett 94:063505
5. Bernardini F, Fiorentini V, Vanderbilt D (1997) Spontaneous polarization and piezoelectric constants of III-V nitrides. Phys Rev B 56:10024–10027
6. Barker JM, Ferry DK, Koleske DD, Shul RJ (2005) Bulk GaN and AlGaN/GaN heterostructure drift velocity measurements and comparison to theoretical models. J Appl Phys 97:063705
7. Shur MS (1998) GaN based transistors for high power applications. Solid-State Electron 42:2131–2138
8. Yoshida S, Ishii H, Li J, Wang D, Ichikawa M (2003) A high-power AlGaN/GaN heterojunction field-effect transistor. Solid State Electron 47:589–592
9. Nakamura S, Harada Y, Seno M (1991) Novel metalorganic chemical vapor deposition system for GaN growth. Appl Phys Lett 58:2021–2023
10. Powell RC, Lee N-E, Kim Y-W, Greene JE (1993) Heteroepitaxial wurtzite and zinc-blende structure gan grown by reactive-ion molecular-beam epitaxy. Growth kinetics, microstructure, and properties. J Appl Phys 73:189–204
11. Fujito K, Kubo S, Nagaoka H, Mochizuki T, Namita H, Nagao S (2009) Bulk GaN crystals grown by HVPE. J Cryst Growth 311:3011–3014
12. Fujii T, Gao Y, Sharma R, Hu EL, Denbaars SP, Nakamura S (2004) Increase in the extraction efficiency of GaN-based light-emitting diodes via surface roughening. Appl Phys Lett 84:855–857
13. Mo C, Fang W, Pu Y, Liu H, Jiang F (2005) Growth and characterization of InGaN blue LED structure on Si (111) by MOCVD. J Cryst Growth 285:312–317
14. Härle V, Hahn B, Lugauer HJ, Bader S, Brüderl G, Baur J, Baur J, Eisert D, Strauss U, Zehnder U, Lell A, Hiller N (2000) GaN-based LEDs and Lasers on SiC. Phys Status Solidi A 180:5–13

15. Chung K, Lee CH, Yi GC (2010) Transferable GaN layers grown on ZnO-coated graphene layers for optoelectronic devices. Science 330:655–657
16. Kobayashi Y, Kumakura K, Akasaka T, Makimoto T (2012) Layered boron nitride as a release layer for mechanical transfer of gan-based devices. Nature 484:223–227
17. Wong WS, Sands T, Cheung NW (1998) Damage-free separation of GaN thin films from sapphire substrates. Appl Phys Lett 72:599–601
18. Carlson A, Bowen AM, Huang Y, Nuzzo RC, Rogers JA (2012) Transfer printing techniques for materials assembly and micro/nanodevice fabrication. Adv Mater 24:5284–5318
19. Tamboli AC, Haberer ED, Sharma R, Lee KH, Nakamura S, Hu E (2007) Room-temperature continuous-wave lasing in GaN/InGaN microdisks. Nat Photonics 1:61–64
20. Kim J, Bayram C, Park H, Cheng CW, Dimitrakopoulos C, Ott JA, Reuter KB, Bedell SW, Sadana DK (2014) Principle of direct van der Waals epitaxy of single- crystalline films on epitaxial graphene. Nat Commun 5:4836
21. Bour DP, Nickel NM, Walle CGVD, Kneissl MS, Johnson NM (2000) Polycrystalline nitride semiconductor light-emitting diodes fabricated on quartz substrates. Appl Phys Lett 76:2182–2184
22. Geim AK, Grigorieva IV (2013) van der Waals heterostructures. Nature 499:419–425
23. Utama MIB, Zhang Q, Zhang J, Yuan Y, Belarre FJ, Arbiol J, Xiong Q (2013) Recent developments and future directions in the growth of nanostructures by van der Waals epitaxy. Nanoscale 5:3570–3588
24. Fan Z, Razavi H, Do J-W, Moriwaki A, Ergen O, Chueh Y-L, Leu PW, Ho JC, Takahashi T, Reichertz LA, Neale S, Yu K, Wu M, Ager JW, Javey A (2009) Three dimensional nanopillar array photovoltaics on low cost and flexible substrate. Nat Mater 8:648–653
25. Rogers JA, Someya T, Huang Y (2010) Materials and Mechanics for Stretchable Electronics. Science 327:1603–1607
26. Kim T-I, McCall JG, Jung YH, Huang X, Siuda ER, Li YH, Song JZ, Song YM, Pao HA, Kim R-H, Lu CF, Lee SD, Song I-S, Shin G, Al-Hasani R, Kim S, Tan MP, Huang YG, Omenetto FG, Rogers JA, Bruchas MR (2013) Injectable, cellular-scale optoelectronics with applications for wireless optogenetics. Science 340:211–216
27. Harris KD, Elias AL, Chung HJ (2016) Flexible electronics under strain: a review of mechanical characterization and durability enhancement strategies. J Mater Sci 51:2771–2805
28. Gustafsson G, Cao Y, Treacy GM, Klavetter F, Colaneri N, Heeger AJ (1992) Flexible light-emitting-diodes made from soluble conducting polymers. Nature 357:477–479
29. Forrest SR (2004) The path to ubiquitous and low-cost organic electronic appliances on plastic. Nature 428:911–918
30. Yablonovitch E, Gmitter T, Harbison JP, Bhat R (1987) Extreme selectivity in the lift-off of epitaxial GaAs films. Appl Phys Lett 51:2222–2224
31. Ponce FA, Bour DP (1997) Nitride-based semiconductors for blue and green light-emitting devices. Nature 386:351–359
32. Liu Y, Xu Y, Cao B, Li Z, Zhao E, Yang S, Wang C, Wang J, Xu K (2019) Transferable GaN films on graphene/SiC by van der Waals epitaxy for flexible devices phys. Status Solidi A 216:1801027
33. Lee CH, Kim YJ, Hong YJ, Jeon SR, Bae S, Hong BH, Yi GC (2011) Flexible inorganic nanostructure light-emitting diodes fabricated on graphene films. Adv Mater 23:4614–4619
34. Hiroki M, Kumakura K, Kobayashi Y, Akasaka T, Makimoto T, Yamamoto H (2014) Suppression of self-heating effect in AlGaN/GaN high electron mobility transistors by substrate-transfer technology using h-BN. Appl Phys Lett 105:193509
35. Kuball M, Hayes JM, Uren MJ, Martin T, Birbeck JCH, Balmer RS, Hughes BT (2002) Measurement of temperature in active high-power AlGaN/GaN HFETs using Raman spectroscopy. IEEE Elect Dev Lett 23:7–9
36. Kelly MK, Ambacher O, Dimitrov R, Handschuh R, Stutzmann M (2001) Optical process for liftoff of group III-Nitride films. Phys Status Solidi A 159:R3–R4
37. Meitl M, Zhu Z, Kumar V, Lee K, Feng X, Huang YY, Adesida I, Nuzzo RG, Rogers JA (2006) Transfer printing by kinetic control of adhesion to an elastomeric stamp. Nat Mater 5:33–38

38. Sun Y, Kim S, Adesida I, Rogers JA (2005) Bendable gaas metal-semiconductor field-effect transistors formed with printed GaAs wire arrays on plastic substrates. Appl Phys Lett 87:083501
39. Loo YL, Willett RL, Baldwin KW, Rogers JA (2002) Additive, nanoscale patterning of metal films with a stamp and a surface chemistry mediated transfer process: applications in plastic electronics. Appl Phys Lett 81:562–564
40. Kim KH, Bong KW, Lee HH (2007) Alternative to pentacene patterning for organic thin film transistor. Appl Phys Lett 90:093505
41. Seo SM, Kim JH, Lee HH (2006) Dry formation of polymer hole injection layer for top emitting organic light emitting diodes. Appl Phys Lett 89:253515
42. Rogers DJ, Hosseini TF, Ougazzaden A, Gautier S, Divay L, Lusson A, Durand O, Wyczisk F, Garry G, Monteiro T, Correira MR, Peres M, Neves A, McGrouther D, Chapman JN, Razeghi M (2007) Use of ZnO thin films as sacrificial templates for metal organic vapor phase epitaxy and chemical lift-off of GaN. Appl Phys Lett 91:071120
43. Goto H, Lee SW, Lee HJ, Lee HJ, Yao T (2008) Chemical lift-off of gan epitaxial films grown on c-sapphire substrates with crn buffer layers. Phys Status Solidi C 5:1659–1661
44. Kobayashi Y, Tsai CL, Akasaka T (2010) Optical band gap of h-BN epitaxial film grown on c-plane sapphire substrate. Phys Status Solidi C 7:1906–1908
45. Ayari T, Sundaram S, Li X, Gmili YE, Voss PL, Salvestrini JP, Ougazzaden A (2016) Wafer-scale controlled exfoliation of metal organic vapor phase epitaxy grown InGaN/GaN multi-quantum well structures using low-tack two-dimensional layered h-BN. Appl Phys Lett 108:171106

Chapter 7
Quasi van der Waals Epitaxy Nitride Device for Heat Dissipation

The rapid increase in power density of electronic devices has made efficient heat dissipation be a key issue in the development of information, communication, and energy storage technologies [1, 2]. The development of next-generation integrated circuits, 3D integration, and ultra-fast high-power density communication devices has made the thermal management extremely demanding, and efficient heat dissipation has become an important research content for the performance and reliability of modern electronic, optoelectronic, and photonic devices and systems. The expanded application of nitride high-power devices also faces similar problems, mainly manifested in optoelectronic and microelectronic device applications such as LEDs and HEMTs. For example, as the power of LEDs continues to increase, the thermal problems they face are becoming more and more serious. How to solve the heat generation limitation on the device applications has become a hot research topic in the international academic and industrial circles. Therefore, to solve the heat dissipation problem in LEDs, it is necessary to study the material properties of high-power LEDs, device heating characteristics, and the heat dissipation problem of the entire system during device application. If the heat dissipation problem cannot be solved, it will cause many serious problems, including:

(I) The light output of the device will continue to decrease as the junction temperature rises. The light output of the LED and the junction temperature satisfy the following relationship [3]:

$$\phi(T_2) = \phi(T_1)e^{-k\Delta T_j} \tag{7.1}$$

where $\phi(T_2)$ is the luminous flux when the junction temperature is T_2, $\phi(T_1)$ is the luminous flux when the junction temperature is T_1, k is the temperature coefficient, ΔT_j is the temperature difference between T_2 and T_1. As the junction temperature rises, the light output of the LED decreases faster.

T. Wei et al., *Quasi van der Waals Epitaxy of Nitride Semiconductor and Its Applications*, Wide Bandgap Semiconductors,
https://doi.org/10.1007/978-981-95-3469-2_7

(II) The wavelength of the light emitted by the LED will shift as the temperature rises. The light emitted by the LED satisfies the relationship:

$$\lambda_{peak} = \frac{hc}{E_g} \tag{7.2}$$

where h is Planck's constant, c is the speed of light, and E_g is the bandgap width. The bandgap width changes with the temperature, causing a shift of wavelength. Especially for blue LED, when the efficiency drops and the wavelength shifts, it will cause changes in color temperature and color rendering index (CRI), which is fatal damage for high-quality lighting.

(III) The lifespan of the chip will shorten as the temperature rises. The half-life of the LED changes exponentially with the junction temperature [4]. The so-called half-life refers to the time it takes for the light flux to decay to half of its initial value, which is generally used to measure the lifespan of the LED. It can be express as follows:

$$\tau = BN_D^{-1} \cdot j^{-1} e^{\frac{E_g}{KT}} \tag{7.3}$$

where B and K are a coefficient, N_D is the dislocation density, and j is the current density. When the junction temperature is high, the half-life of the chips decreases faster.

(IV) The stability of the device decreases as the temperature rises. The expansion coefficients of the internal materials of the LED are different. When the temperature changes significantly, thermal stress will occur inside, which will cause the chip to deform, reduce the stability of the device structure, and thereby reduce the reliability of the device.

7.1 Rule of Two-Dimensional Materials Assisting Heat Dissipation

Currently, improvements in device heat dissipation performance are mainly focused on packaging technology, but there are many difficulties in the miniaturization and increased costs. Traditional thermal interface materials filled with heat-conducting particles require a filler volume fraction f up to ~ 50% to achieve a thermal conductivity K of composite materials in the range of 1–5 W/mK at room temperature. It is found that 2D materials, such as graphene and h-BN, due to their excellent thermal conductivity, are often used as media to improve device heat dissipation performance. 2D materials and sapphire complement each other in terms of conductivity, thermal conductivity, and flexibility, which can improve the problems of poor thermal conductivity and insulation of sapphire. Nitrides can be directly grown on 2D material substrates, and the excellent thermal conductivity of 2D materials can be used to improve the heat dissipation capacity of the device.

As early as 2012, the design of high-power GaN transistors had been achieved, using graphene-based lateral heat dissipation devices [5]. Currently, there are two main methods to enhance the heat dissipation performance of nitride devices using 2D materials: The first is to grow the nitride device structure first, and then transfer the 2D material to the chip during the device preparation. A common method is to use 2D materials to replace traditional metal electrodes, achieving both conductivity and heat dissipation. The second is to directly grow nitride-based device structures on 2D material substrates, where the 2D materials mainly serve to assist in heat dissipation. Here, we will focus on the second point, growing nitrides on excellent thermal conductivity 2D materials.

New 2D materials generally form various heterostructures through vdW forces or covalent bonds. Many literature reports have calculated and verified the thermal conductivity of intrinsic materials such as graphene, h-BN, carbon nanotubes, and BN nanotubes. For armchair graphene and BN nanoribbons with a width of 2.5 nm, as well as carbon nanotubes and BN nanotubes, the room temperature thermal conductivity when their length is infinitely large is 1316 W m^{-1} K^{-1}, 526 W m^{-1} K^{-1}, 901 W m^{-1} K^{-1}, and 369 W m^{-1} K^{-1}, respectively. In the temperature range of 100–1200 K, due to the intensification of phonon scattering, the thermal conductivity decreases with the increase of the temperature. At low temperature, the thermal conductivity of BN nanostructures is higher than the corresponding carbon structures. However, as the temperature increases, the thermal conductivity of BN nanostructures will turn lower compared to carbon nanostructures. When the nanotube is compressed, due to the influence on the non-harmonic vibration of the lattice, the overall axial thermal conductivity will decrease with the increase in compressive strain. However, unlike BN nanotubes, the low-frequency phonons of carbon nanotubes will be excited under small strains, leading to an increase in thermal conductivity.

The energy quanta in materials include photons, phonons, and electrons. Among them, phonons are the normal mode energy quanta of lattice vibrations, obeying the Bose–Einstein

$$f^{\circ}_{BE} = \frac{1}{\exp\left(\frac{E}{k_B T}\right) - 1} \tag{7.4}$$

where, E represents energy, k_B is the Boltzmann constant, and T represents absolute temperature.

Phonons are the most important heat carriers in semiconductors, insulators and carbon materials. Since the group velocity of optical branch phonons is very small, the contribution to thermal conductivity mainly comes from acoustic phonons. As the material with the strongest thermal conductivity discovered so far, the in-plane thermal conductivity of single-layer graphene can be as high as 5000 W m^{-1} K^{-1} [6], and the thermal conductivity coefficient of graphene is also closely related to its number of layers [7]. There are three types of phonon vibration modes in graphene: the longitudinal acoustic (LA) mode vibrating along the wave propagation direction, the transverse acoustic (TA) mode vibrating laterally, and the out-of-plane acoustic (ZA) mode [8].

In comparison, h-BN is composed of B atoms and N atoms stacked in a honeycomb structure of single atomic layers. Because its structure is very similar to graphene and it is transparent, it is also known as white graphene. Experimental reports on the thermal conductivity of h-BN are still relatively scarce. On the one hand, this is due to the difficulty in preparing high-quality h-BN, and on the other hand, it is due to the very weak sensitivity of its Raman signal. Jo et al. [9] measured the thermal conductivity of 11-layer h-BN at room temperature using the thermal bridge method, which was about 360 W m^{-1} K^{-1}, close to the thermal conductivity of bulk materials. Wang et al. [10] also measured the thermal conductivity of dry transfer bilayer h-BN at room temperature using the thermal bridge method, which was about 484 W m^{-1} K^{-1}, while the theoretical value of h-BN reached 2000 W m^{-1} K^{-1} [11]. MoS_2, as an important member of the 2D material family, has a unique sandwich structure, and its bandgap energy depends on the thickness and number of layers. Therefore, MoS_2 may also act as a very potential electronic material. Using the Raman method, Yan et al. [12] and Sahoo et al. [13] measured the room temperature thermal conductivity of single-layer and 11-layer MoS_2 to be 34.5 ± 4 W m^{-1} K^{-1} and 52 W m^{-1} K^{-1}, respectively.

In the study of the thermal properties of low-dimensional nanomaterials, in addition to the attention to their own thermal conductivity, the interface has a significant impact on the thermal conduction at the nanoscale. These low-dimensional materials, due to their atomic-level thickness and ultra-high specific surface area, will introduce more interface surface area compared to traditional materials, and are more sensitive to the binding conditions of the interface. Different 2D materials [14], the thermal transport behavior of phonons at these interfaces will have an important impact on the overall thermal performance of the device, which is also an important content that must be studied for heterojunction heat dissipation research.

For high-performance electronic devices, high thermal conductivity and low interface thermal resistance can effectively dissipate the Joule heat inside the device and reduce the temperature at the hot spot. Due to the difference in materials and structures on both sides of the interface, phonons will be severely scattered when crossing the interface, and the local temperature on both sides of the interface will jump to produce a temperature difference ΔT, from which the interface thermal resistance (ITR) can be defined:

$$R_i = \frac{\Delta T}{Q} \tag{7.5}$$

where Q is the heat flow density passing through the interfaces, and the reciprocal of the interface thermal resistance is generally called the interface thermal conductivity.

7.2 Device Heat Dissipation of Different Two-Dimensional Materials

As devices continue to miniaturize and integrate, the power density inside the device will increase sharply, generating a large amount of heat per unit area, and even forming a "hot spot" with a large heat flow and high temperature locally. The local temperature rise caused by high heat brings a series of serious consequences, severely limiting the stability and reliability of device operation. The heat dissipation problem of electronic devices is currently an important bottleneck restricting the development of the semiconductor industry, and thermal control at the nanoscale has become a very critical link in device design [15–20].

Semiconductor devices generate a lot of Joule heat when operating at high currents, and temperature changes affect the movement of electrons and holes in the device, further inhibiting device performance [21]. For example, when a GaN-based LED operates at high current, the number of non-radiative electron holes increases with the rise of junction temperature, resulting in a decrease in the optical power or luminous efficiency of the device. With the increase in power and injection current, the failure of LED devices, especially thermal failure, becomes increasingly serious, affecting the luminous characteristics and causing serious consequences such as light decay and color drift. With the promotion of applications, thermal management analysis is also attracting more and more attention. 2D materials (such as graphene, h-BN, etc.) have excellent thermal conductivity far higher than Si (145 W/mK), copper (400 W/mK) and traditional sapphire substrates (30 W/mK) [22]. Therefore, 2D materials can be introduced into nitride devices to improve their heat dissipation performance.

7.2.1 Few-Layer Graphene

Self-heating is a severe problem for high-power GaN electronic and optoelectronic devices. Yan et al. [5] first showed that thermal management of GaN transistors can be substantially improved via introduction of alternative heat-escaping channels implemented with few-layer graphene (FLG). The graphene-graphite quilts were formed on top of AlGaN/GaN transistors on SiC substrates. It was required to transfer FLG films on top of AlGaN/GaN devices placing them precisely at the drain electrode location. The channel region between the gate and the drain generated most of heat. They used PMMA as the supporting membrane for transfer to the desired location. Transfer process of FLG films to exact locations on AlGaN/GaN field-effect transistors (HFETs) device structure was demonstrated in Fig. 7.1. By introducing the graphene–graphite quilt with the heat sink at $D = 10\ \mu m$, one could achieve a drastic 68 °C reduction in the hotspot temperature. The proposed local heat spreading with FLG at nanometer scale represented a transformative change in thermal management.

Recently, bi-layer graphene interlayer was directly grown on a 4-inch PSS substrate by LPCVD [23]. First, the 500 nm–thick Cu thin film was grown onto PSS

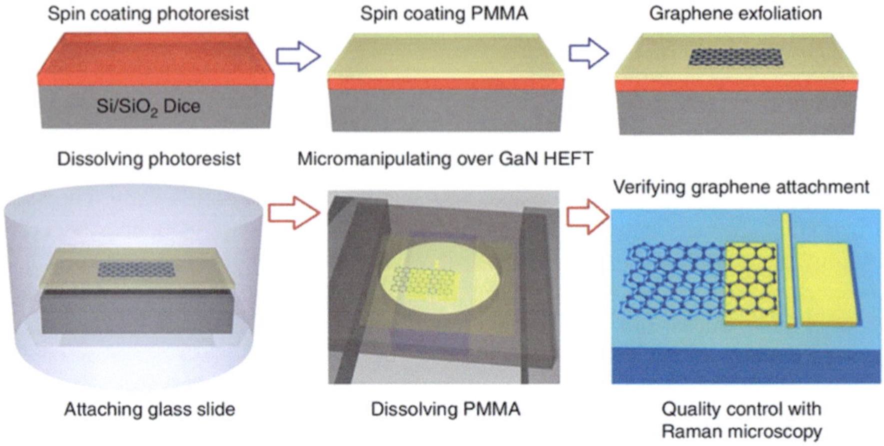

Fig. 7.1 Transfer of FLG films to exact locations on AlGaA/GaN HFET device structure [5]

using a thermal evaporation system. Then, the sample was loaded into a low pressure chemical vapor deposition LPCVD system to grow graphene. The residual Cu thin film was removed using an acid solution of $NH_4S_2O_8$. Thus epi-ready graphene/PSS was used to grow AlN NL and subsequent GaN and InGaN/GaN MQWs. Figure 7.2 showed the photos of the graphene interlayer on a 4-inch PSS substrate before and after removing residual copper thin film. In Fig. 7.2c, the characteristic peaks at 1353, 1586 and 2704 cm^{-1} were assigned to the D-band, the G-band, and the 2D-band, respectively. The peak intensity ratio of 2D-band/G-band was 0.96 and FWHM of the 2D-band was 48 cm^{-1}, indicating high quality of bi-layer graphene interlayer. Here, more compressive strain of GaN template was released after inserting the graphene interlayer for the LED structure.

They presented effects of graphene on improving device performances of wavelength stability and heat dissipation. The thermal measurement was operated in a vacuum chamber at 0.1 mTorr to suppress the upward transport of the heat to the air. Thus, the path of heat transport remained downward, beginning from InGaN MQWs through the graphene interface/PSS substrate to the Cu sink. The photos of the sample inside the vacuum chamber and operated at 60 mA of driving current were shown in Fig. 7.3a. The arrangement of the LED chip and Si diode mounted by means of a silver (Ag) paste on a 1×1 cm^2 Cu foil as a heat sink was also shown in Fig. 7.3a. The thermal conduction model of graphene-LED on a Cu sink was created, as shown in Fig. 7.3b, assuming that the heat generation in the LED chip transports to the Cu sink. The relationship between the LED chip temperature and injection current for LEDs without and with graphene was shown in Fig. 7.3c. The chip temperature in injection current of 100 mA increased to 61.1 °C and 36.7 °C for reference LED and LED with graphene, respectively. Here, the thermal heat conduction can be expressed as follows [24]:

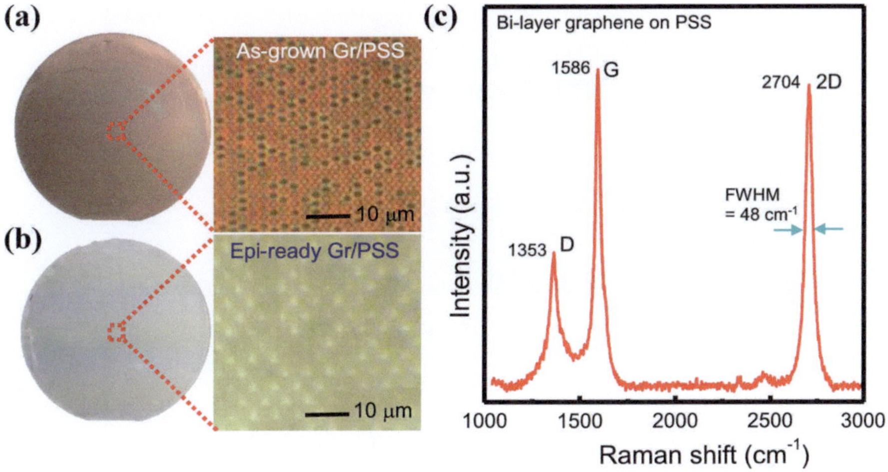

Fig. 7.2 The photos of **a** as-grown bi-layer graphene/PSS substrate and **b** epi-ready graphene interlayer/PSS substrate after removing residual copper thin film. **c** The Raman spectrum of graphene interlayer/PSS substrate [23]

$$(1 - \eta_{WPE}) \cdot W = \frac{A \cdot (T_{LED} - T_S)}{\sum_i (L_i/k_i)} \tag{7.6}$$

where η_{WPE} was the wall-plug efficiency, W was the LED power consumption, A was the effective heat dissipation area between GaN template layer and PSS substrate, T_{LED} was measured LED chip temperature, T_S was Cu sink temperature with an assumption of room temperature, L_i was the thickness of each model components, and k_i was the thermal conductivity of each model components. By balancing the heat generated from the LED chip and thermal conduction to the Cu heat sink, the chip temperature with various injection currents could be well estimated. The estimated effective heat dissipation areas were 5.7×10^{-4} and 1.6×10^{-3} cm^{-2} for LEDs without and with graphene, respectively. Here, the heat conduction path for LED with graphene was combined by both low defect density of AlN NL on the unpatterned region and sidewall of the pattern region. In contrast, the heat conduction path of LED without graphene mainly relied on the sidewall of the pattern. Thus, the heat accumulation phenomenon on the PSS substrate surface for LED could be improved significantly by embedding the graphene interlayer resulting in a low chip temperature.

In addition, the TDs density and compressive strain of InGaN LEDs grown on a graphene interlayer/PSS substrate also decreased significantly. The low compressive strain and piezoelectric field diminished the effect of screen polarized field resulting in a stable EL emitting wavelength and even increasing injection current to 100 mA which was more beneficial for high-current driving InGaN LEDs. The low chip temperature of InGaN LEDs was achieved, which demonstrated that bi-layer

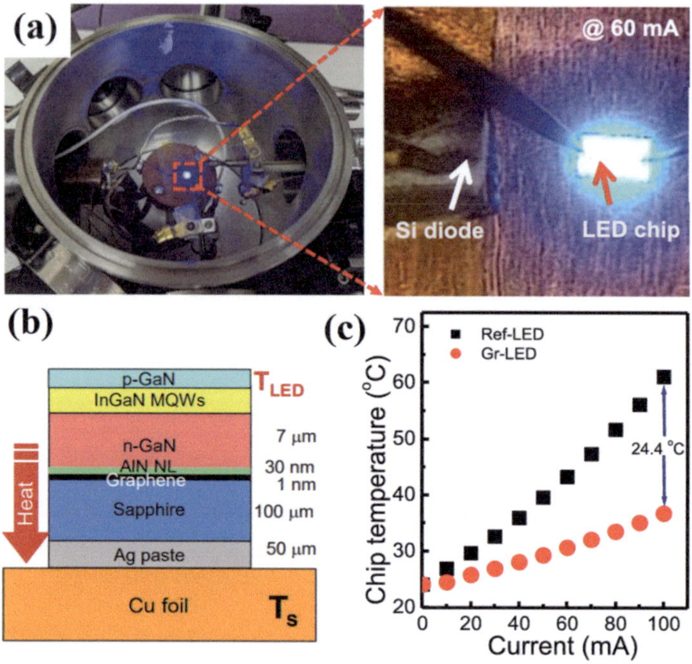

Fig. 7.3 **a** The photos of the sample inside a vacuum chamber and operated at 60 mA of driving current. **b** The thermal conduction model for the calculation of thermal effect in two samples. **c** The relationship between the chip temperature and injection current of Ref-LED and Gr-LED [23]

graphene interlayer/PSS was a promising substrate for high-power semiconductor devices.

7.2.2 Reduced Graphene Oxide

Graphene is composed of carbon atoms with sp^2 bond to form the single-atom layer. Due to its lack of chemical reactivity, it is difficult for nitride materials to directly epitaxially grow on graphene, and thus the integration of graphene and GaN-based devices still poses great challenges. Based on these considerations, Han et al. [25] from the National Jeonbuk University of South Korea chose to directly epitaxially grow GaN on a reduced graphene oxide (rGO) template. Here, rGO served as a buffer layer for the lateral growth of high-quality GaN epitaxy, providing better heat dissipation due to its high thermal conductivity. They used a combination of photolithography and spray methods to produce low-cost, scalable, and processable rGO patterns on sapphire substrates, as shown in Figs. 7.4a–d. First, a circular array of photoresist masks was formed on the sapphire substrate by photolithography, followed by the deposition of rGO on the sapphire substrate using a spray method. The

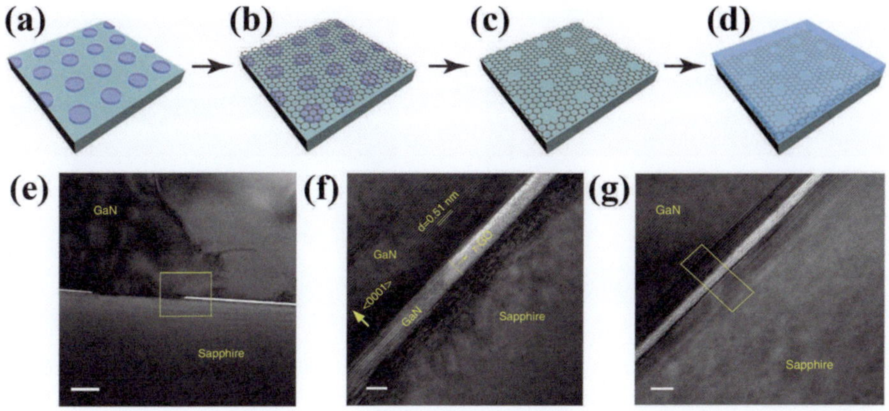

Fig. 7.4 Schematic of the key steps involved in the process of formation of graphene oxide-embedded LED. **a** The formation of circular arrays of photoresist masks on the sapphire substrate by photolithography process. **b** Deposition of graphene oxide on the sapphire substrate by the spray-coating method. **c** Removal of photoresist using acetone. **d** Thermal reduction and direct growth of GaN epilayer structure on the rGO pattern. **e** Cross-sectional bright field image of GaN/rGO/sapphire interface with scale bar of 100 nm. **f**, **g** High-resolution snapshots of GaN/rGO/sapphire interface obtained in the rectangular region marked in **e** with scale bar of 5 nm [25]

formation of the circular photoresist array and the deposition of rGO dispersion on the sapphire substrate were completed by photolithography and a spray gun system. Subsequently, acetone was used to remove the photoresist, forming a hole pattern on the entire rGO-covered substrate. Then, in-situ thermal annealing was performed at 1100 °C for 10 min in an H_2 environment to finally obtain rGO, followed by the growth of GaN-based LED structure using MOCVD.

With the rGO pattern, the GaN growth began selectively at the exposed sapphire regions. When the growth time increased, the GaN mounds also increased in size and coalescence occured owing to the epitaxial lateral overgrowth. After a growth time of 90 min, a mirror-like GaN film was obtained. XRD results also signified that the crystal quality of the GaN template grown on rGO pattern was comparatively better compared with the one grown on a sapphire substrate. The HRTEM images also provided the evidence that the rGO was embedded between the sapphire and GaN templates in Fig. 7.4e–g. The lattice spacing between the adjacent planes was asured to be 0.51 m in Fig. 7.4f, corresponding to the -spacing of (0001) plane of a wurtzite GaN crystal [26]. This result ccompanied by the observed regular diffraction patterns further erified the single-crystalline nature of the GaN grown on the rGO pattern. These results indicated the feasibility of using rGO for high quality of GaN epilayer.

Furthermore, InGaN/GaN MQWs blue LED chips with and without embedding the rGO pattern were fabricated. The infrared imaging technique was used to monitor the temperature profile on the surface of the LED chips. The infrared intensity images for the temperature distribution on the chip surface were demonstrated at 100 mA in Fig. 7.5. The peak temperature of the chip surface rose up to 58 and 53.2 °C,

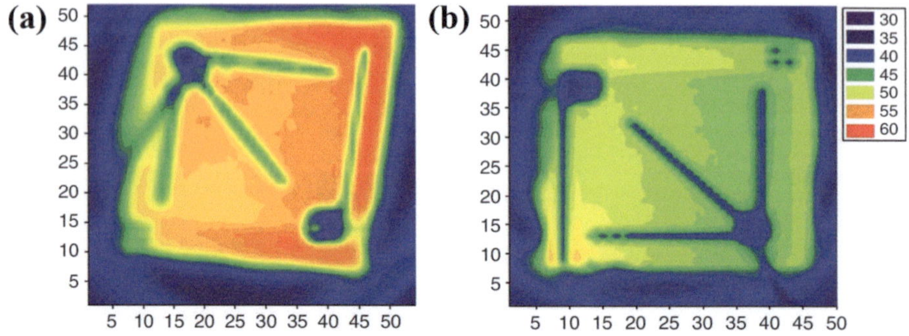

Fig. 7.5 Infrared thermal imaging camera photographs of the chip surfaces under 100 mA current injection. **a** Temperature distribution on the surface of conventional LED chip. **b** Temperature distribution on the surface of rGO-embedded chip [25]

respectively, for the conventional and rGO-embedded LEDs, measured within the same temperature range of 30–60 °C. In addition, the average temperature distribution on the chip surface was decreased for the rGO-embedded LED (47.06 °C) compared with its conventional counterpart (51.4 °C). This result signified that with rGO at the GaN/sapphire interface, the heat generated in the interior of the chip was certainly reduced as compared with the conventional case. Although the absolute value of the temperature reduction was not large, it may greatly extend the lifespan of the LED.

In LED packaging, T_j and R_{th} are the two physical quantities generally considered for the evaluation of heat dissipation. To explore the extent of heat dissipation that the rGO can give, the above two parameters were studied using transient thermal analysis [27]. Figure 7.6a showed the transient cooling curve of the LED package under a driving current of 150 mA, where each curve represented the time dependence of the chip junction temperature rise under constant injection current. Figure 7.6b elucidated the measured T_j as a function of injection current for the three LED packages. It should be noted that the T_j increased with the injection current in all cases, a behaviour common to standard LED packages. For all the driving currents studied, the T_j of the rGO-embedded LED was lower than its conventional counterpart. For instance, at 150 mA, the measured T_j of the conventional and rGO-embedded LEDs were found to be 32.3 and 24 °C, respectively, indicating a 25.7% reduction in T_j for the rGO-embedded LED.

The differential structure functions which were derived by direct mathematical transformations from the experimental cooling curves were shown in Fig. 7.6c. Considering the chip as a sandwich structure, where each layers had the same area, the local peaks and valleys in the structure function represent the arrival of new materials in the heat flow path. In Fig. 7.6c, the first peak represented the heat flow from the chip to die-attach, and the saturation region denoted the heat flew from die-attach to the heat sink. In the structure function shown in Fig. 7.6c, the characteristics of the two LED packages can be distinguished. Under a current injection of 150 mA, the values of R_{th} for the rGO-embedded LED and the conventional LED

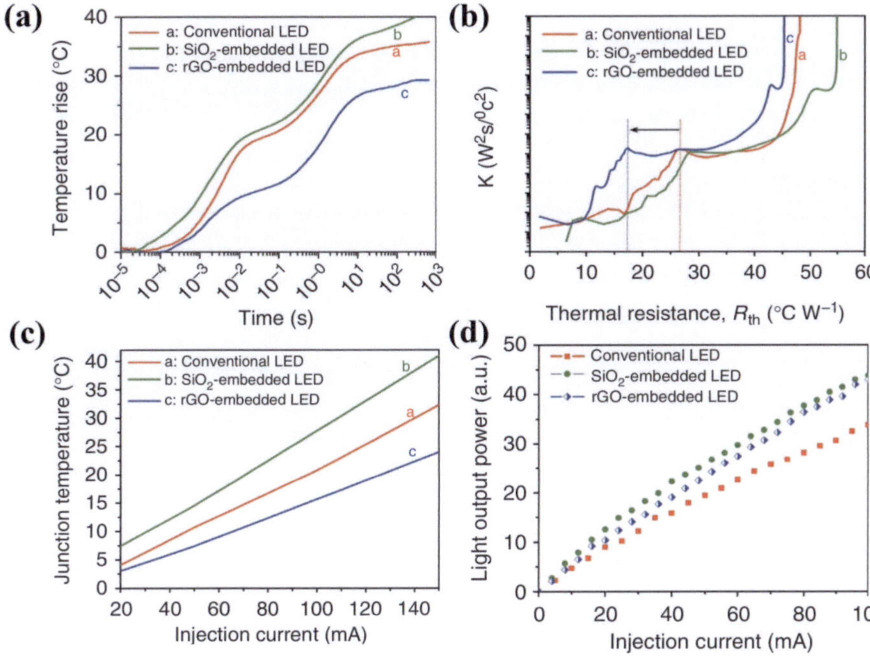

Fig. 7.6 Transient thermal and optical characteristics of different GaN LEDs. **a** Experimental transient cooling curves of the LED packages at an injection current of 150 mA. **b** Variations in junction temperature as a function of injection current. **c** Differential structure functions (K) of conventional, SiO_2- and rGO-embedded LED packages under 150 mA current injections. **d** Light-output power of the LED chips as a function of injection current [25]

were 17.3 and 26.2 °C W^{-1}, respectively. It was worth noting that the R_{th} of the rGO-embedded LED chip was about 8.9 °C W^{-1} lower than that of the conventional counterparts. Under the same experimental conditions, measurements were made on lots of LED chips for each sample, further verifying the reliability of the data. These results confirmed the observations from infrared imaging and junction temperature results, the rGO layer in the LED chip significantly improved heat dissipation.

The electrical and optical characteristics of the rGO-embedded LED were compared with the conventional LED. The embedded rGO had no influence on the current–voltage characteristics of the device. Figure 7.6d showed the variations in the light output power as a function of injection current. It was observed that the light output power of the rGO-embedded LED was slightly enhanced over that of conventional LED, which may be likely due to the improved IQE of the devices. Additional experiments conducted to gain insights into the feasibility for high-power LEDs showed about 33% enhancement in the light output power at an injection current of 350 mA for the rGO-embedded LED. Interestingly, the observed enhancement in the thermal and optical characteristics of the rGO embedded LED was analogous to the results reported for vertical LEDs on graphite or Cu substrates [28]. Due to the heat propagation along the in-plane direction of graphene being much larger than its perpendicular direction, the embedded rGO was likely to ease out the heat from the chip by acting as a heat spreader. Therefore, the present findings promise that graphene-based materials will soon find their way into high-power LEDs for thermal management.

In order to exclude the effect of improving crystalline quality of GaN for heat dissipation, a similar LED epilayer structure was also grown on a low thermal conductivity amorphous SiO_2 buffer layer with patterns analogous to that of the rGO patterns. The device design can help to understand the impact of thermal conductivity of the buffer layer on the heat spreading. All the factors studied above were taken into consideration. Both the GaN thickness and the etch pit density were found to be comparable to that of the values observed for the sample grown with rGO buffer layer. However, from the transient thermal measurements, very high R_{th} (28.4 °C W^{-1}) and T_j (40.9 °C) were indicated for the device grown on SiO_2 buffer layer in Fig. 7.6. Here, higher T_j and R_{th} associated with the device grown with SiO_2 buffer layer could be attributed to additional thermal boundary resistance (TBR) added to the GaN/sapphire interface negating the benefits offered by the improved crystal quality. Although the lateral overgrowth of GaN on SiO_2 micropatterns may result in high crystal quality, the poor thermal conductivity of the SiO_2 restricted the heat flow across the interface, as evidenced from the results.

Furthermore, the same group have also demonstrated the fabrication of a prototype LED structure which combines an rGO heat spreading layer with PSS, to reduce the junction temperature. Despite being similar in electrical and optical characteristics to conventional PSS LEDs, the rGO-embedded LED offered a better thermal stability and a lower TBR as evidenced by the low Tj and surface temperature. Using the forward voltage method, the junction temperature Tj of the rGO embedded LED was found to be reduced by about 17 °C from the ~ 62 °C of the LED grown without the rGO buffer layer. This further proved the application prospect of rGO in industrialization.

7.2.3 *Vertical Graphene*

The multilayer graphene 2D structure is based on the weak vdW interaction between the layers. Experiments show that graphene has excellent thermal conductivity in the horizontal direction (within the plane), but poor thermal conductivity in the vertical direction (outside the plane) due to the weak vdW interaction between the layers. Considering the thermal conductivity anisotropy of graphene, that is, the heat transfer ability is strong in the plane direction (ab direction), but weak in the vertical direction [6, 29]. Therefore, for multilayer graphene, heat is mainly transferred through the interface layer that directly or closely contacts the heat source. Due to the limited thermal conductivity in the vertical direction, the contribution of the graphene layer far from the heat source to heat dissipation is small. Compared with the common lateral multiple layer graphene, vertically oriented graphene (VG)

[30] has a 3D structure and unique vertical direction, which can show good thermal transmission performance in both vertical and horizontal directions. Thus, it may effectively promote the heat dissipation performance between the epitaxial layer and the substrate of the device, thereby improving the device performance. Using the traditional Raman method [31–33], the thermal conductivity of the suspended vertical graphene nanowall buffer layer has been measured to be 680 Wm^{-1} K^{-1}, which is much higher than the heat dissipation effect of GaN (130 Wm^{-1} K^{-1}), SiC (200 Wm^{-1} K^{-1}) and copper (400 Wm^{-1} K^{-1}) materials in LED devices [34].

Usually, horizontally oriented graphene can be obtained by the CVD method. Zhongfan Liu's group from Peking University used the vertical electric field generated by the DC-PECVD system to regulate graphene growth and graphene tended to grow in the vertical direction. The grown graphene is almost perpendicular to the substrate, with a large number of edge defects, which are mainly caused by the electric field during the plasma generation process [35–37]. The sharp edges, non-stacked morphology, and larger surface-to-volume ratio of VG can provide a sufficient number of edge defects as the nucleation sites for nitrides. Then, the MOCVD method is used to grow AlN structure on the vertical graphene/sapphire substrate. In contrast to the common horizontal graphene layer, vertical graphene nanowalls have a unique 3D structure that enable to make use of their excellent vertical thermal conductivity to enhance the heat dissipation in LED applications.

Figure 7.7a schematically depicted the design using VG nanowalls to enhance the heat dissipation of AlN epilayer. First, VG nanowalls were grown by PECVD at LT ($\approx$ 580 °C) on a c-plane sapphire substrate. Then, a 2.0-μm thick AlN film was grown on the VG-sapphire substrate by a one-step HT MOCVD process. Finally, for the obtained AlN-VG/sapphire structures, the VG nanowalls exhibited good heat dissipation under the illumination. Figure 7.7b–e showed the representative SEM morphology, Raman spectrum, AFM and TEM images of the as-grown VG nanowalls [38]. From the SEM image in Fig. 7.7b, it can be clearly seen that the VG nanowalls partially covered the surface of the sapphire substrate (the bright area represented the VG nanowalls, and the dark area represented the exposed sapphire substrate). The AFM image in Fig. 7.7d showed the distributions of the VG nanosheets with a density of about 81 μm^{-2}. The average height of the VG nanosheets was $\approx$20.5 $\pm$ 5.0 nm according to the height curve inserted in Fig. 7.7d. The Raman spectrum of the VG nanowalls in Fig. 7.7c showed the characteristic Raman peaks of 1357 cm^{-1} (D band), 1592 cm^{-1} (G band) and 2713 cm^{-1} (2D band) [39], indicating relatively high crystallinity of the graphene grown at LT. In these nanostructures, individual nanosheets usually had lateral dimensions of $\approx 0.12 \pm 0.02$ μm and a thickness of only several nanometers, consisting of 8–10 layers. The TEM in Fig. 7.7e showed the curved edges of a single nanowall, with the defects such as dislocations concentrated in the nanoside wall area.

Therefore, the density of AlN nucleation may be effectively controlled, by tuning the density of the VG nanowalls, that was, the lowest density of AlN nucleation occured on bare sapphire and higher densities of AlN nucleation occured on sapphire substrates with higher densities of VG nanowalls as shown in Fig. 7.8a–f. Here, the

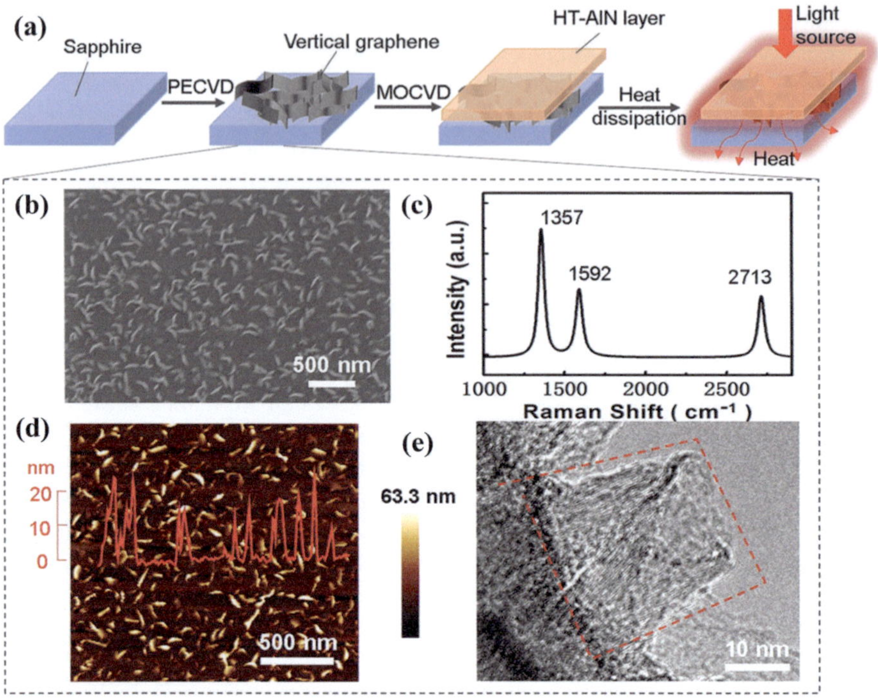

Fig. 7.7 Enhancement of the heat dissipation in an AlN film on sapphire substrate buffered with VG nanowalls. **a** Schematic illustration of heat dissipation enhancement in the AlN-VG-sapphire heterostructures. **b** SEM characterization of the VG nanowalls. **c** Raman spectroscopy showing three typical peaks at 1357 cm^{-1} (D band), 1592 cm^{-1} (G band), and 2713 cm^{-1} (2D band). **d** AFM characterization. The inset shows a height profile of the AFM image, with an average height ≈ 20 nm. **e** A representative cross-sectional TEM image of a vertical graphene nanosheet with ≈ 22 nm in height [38]

defects position may act as the nucleation sites for AlN growth. By simply using one-step MOCVD growth instead of a traditional two-step MOCVD process, a smooth AlN film can be obtained easily, as displayed in the SEM image in Fig. 7.8h. The AFM image in Fig. 7.8i and the inset height profile showed that the surface of the as-grown AlN film was atomically flat, with a measured RMS roughness of 1.8 Å over an area of 1.5 μm × 1.5 μm. However, it should be noted that further increasing the VG coverage (high-density) degraded the crystal quality of the AlN film due to the generation of high density dislocations. Therefore, the coverage (density) of VG nanowalls should be optimized for the LED applications.

Near-ultraviolet LEDs are mainly used in resin curing, ink printing, anti-counterfeiting identification, etc., while deep-ultraviolet (DUV) LEDs have important application value in sterilization, biochemical detection, medical health, covert communication, etc. Especially in the field of sterilization, DUV LEDs mainly use high-energy UV light to irradiate microorganisms (bacteria, viruses, spores, and

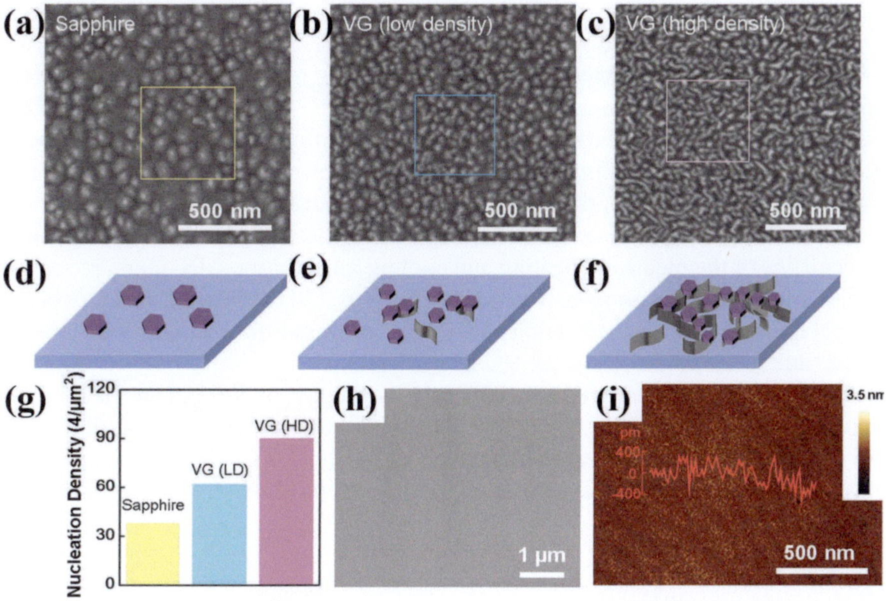

Fig. 7.8 Effects of the VG nanowall buffer layer on the nucleation and quality of the AlN film grown on the sapphire substrate. **a–c** At the initial nucleation stage, SEM images of the AlN on the bare sapphire substrate (**a**), the substrate with a low density of VG nanowalls (**b**), and the substrate with a high density of VG nanowalls (**c**). **d–f** The corresponding schematic illustrations of the initial AlN nucleation conditions on the different substrates. **g** The statistics for the number of AlN nucleation sites in a 500 nm × 500 nm area as plotted from (**a–c**). **h** SEM and **i** AFM images of a flat AlN film grown on a low density VG nanowall substrate. The inset in **i** showed the height profile of the AFM image, confirming its flatness at the atomic scale [38]

other pathogens) and destroy nucleic acid structures, thereby achieving the purpose of microbial inactivation. Compared with traditional sterilization techniques (such as chlorination, ozonation, ultraviolet mercury lamps, etc.), DUV LEDs have the advantages of high sterilization efficiency, strong applicability (broad spectrum), no chemical pollutants, simple operation, etc., and can be widely used in air, water, and surface sterilization. With the rapid development of AlN-based UV LEDs, they will gradually replace traditional mercury lamps that cause serious pollution. However, the wall-plug efficiency of DUV LEDs is lower, usually below 10%, and the heat dissipation of high-power UV LED lamps becomes a major development bottleneck in its application field.

To overcome aforementioned problems, the research compared the heat dissipation capabilities of AlN films on sapphire substrate with VG and bare substrate. The sample structures were shown in Fig. 7.9a, c, which corresponded to the schematic diagrams of AlN-sapphire structure and AlN-VG-sapphire structure, respectively. Initially, the same temperature for the two samples were maintained. Using a power of 850 mW/cm^2 and a wavelength of 1450 nm laser irradiation, the temperature of the two samples increased rapidly when the laser was turned on and then slowly increased after 1 min, as shown in Fig. 7.9e. The surface temperatures of the AlN-sapphire sample and the AlN-VG-sapphire sample presented similar trends versus the irradiation time. However, the surface temperature of AlN-VG-sapphire was the lower of the two. Particularly, after laser exposure for 5 min, the peak temperature detected for the AlN film buffered with VG nanowalls was 36.7 °C, which was approximately 1.4 °C lower than that of the film without VG nanowalls. After turning the laser off, the temperatures of both samples decreased gradually. A histogram of the statistics of the measured heating and cooling rates for the AlN-sapphire sample and AlN-VG-sapphire sample were shown in Fig. 7.9f, indicating more effective thermal transport during the heating and cooling processes for the AlN-VG-sapphire sample due to the existence of the VG nanowalls.

Furthermore, by quantitatively evaluating the impact of VG nanowalls between sapphire and AlN on heat dissipation, a 2D temperature heat distribution simulation was conducted based on finite element analysis in Fig. 7.9b, d. Two structures were constructed according to the actual sample configuration. The AlN-sapphire structure

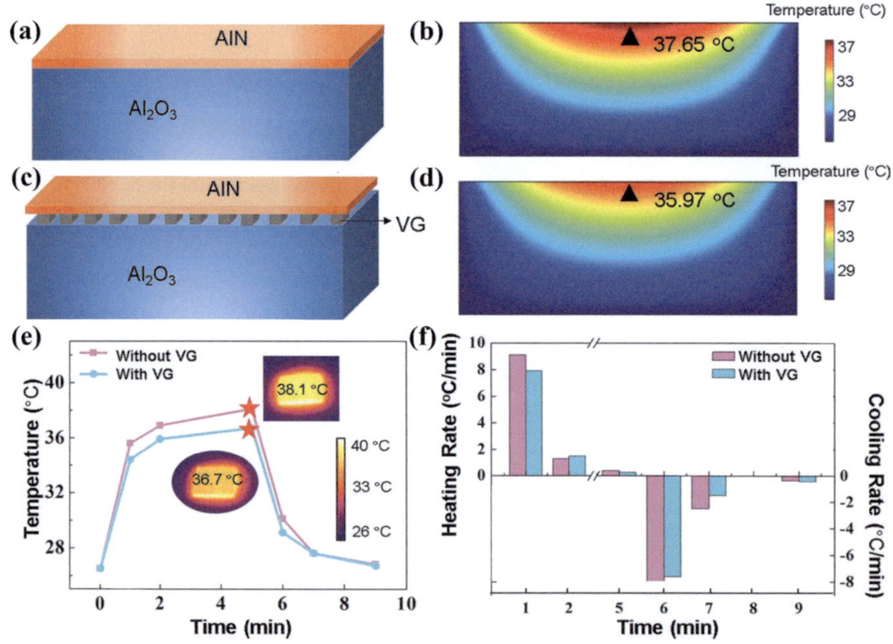

Fig. 7.9 **a** Heat dissipation in an AlN film on sapphire substrate with and without VG nanowalls. Schematic illustrations of an AlN-sapphire structure **a** and an AlN-VG-sapphire structure (**c**). Simulated 2D distributions of the heat mapping of an AlN-sapphire cross-section **b** and an AlN-VG-sapphire cross-section (**d**). **e** The measured temperatures of AlN-sapphire and AlN-VG-sapphire are plotted as a function of the time as detected by an infrared thermal imaging camera. The insets show the temperature distributions under laser irradiation for 5 min. **f** The calculations for the heating rate and cooling rate versus time ($\Delta T/\Delta t$) for the AlN- sapphire and AlN-VG-sapphire samples [38]

consisted of a 1 cm (length) × 300 um (height) sapphire layer and 1 cm (length) × 2 um (height) AlN layer. For the simulated AlN-VG-sapphire structure, in addition to the above structure, there were regular intervals of vertical graphene (100 nm apart) between the sapphire layer and the AlN layer. A Gaussian distributed heat source with a central heat flow of 10 W and a laser spot radius of 2 mm were used to simulate the laser source. The results showed that after 5 min of laser irradiation, the peak temperature of the sample with vertical graphene was 35.97 °C, while the peak temperature of the AlN layer on bare sapphire was 37.65 °C (1.68 °C higher), which was consistent with the experimental results. Therefore, the temperature difference can effectively lower the junction temperature of the LEDs, thereby improving the light output power and enhancing the stability and life of the devices [40].

The improvement of junction temperature of the LEDs had a positive effect on LED device properties. InGaN-based UV-LED structures were fabricated on the as-grown AlN/VG-nanowall/sapphire substrate and AlN/sapphire substrate, schematically shown in the inset of Fig. 7.10a, b, respectively. Figure 7.10a, b showed the EL characteristics of the fabricated LEDs with VG nanowalls and without VG nanowalls. The EL peak wavelength of the LED using VG nanowalls as a buffer layer showed a 1.4 nm shift as the applied current increases from 40 to 500 mA, obviously reducing the wavelength shift compared to that of the LED not using a VG buffer layer (9.5 nm). The different current–voltage (I–V) characterization curves in Fig. 7.10c indicated a turn-on voltage of approximately 3.2 V for the 390 nm UV-LED device, confirming the better quality of the LED device with the VG nanowalls as a buffer layer. Figure 7.10d showed the light output power curves of the related two LED devices as a function of the injection current. For the LED device without a VG buffer layer, the saturated LOP was 137 mW at an injection current of 450 mA. However, for the LED device with a VG buffer layer, the LOP reached 200 mW at 450 mA and still increased with increasing injection current (i.e., 230 mW at 600 mA). The EQE was increased $\approx$ 33.33% due to the presence of VG nanowalls under 350 mA current. Moreover, the smaller EL peak wavelength shift and higher saturation injection current could be explained in terms of the better heat dissipation and strain relieving effects of the VG nanowalls.

7.2.4 Multilayer h-BN

Currently, the luminous efficiency of DUV LEDs is very low compared to blue LEDs. To meet application requirements, multi-chip integrated packaging is often used to obtain high optical power DUV LED modules. However, in the pursuit of high optical power density, the heat flow density per unit area is greater, and the accumulation of heat causes the junction temperature of the DUV LED to rise, thereby affecting the luminous efficiency and reliability of the DUV LED. On the one hand, this is due to the increase in carrier energy caused by the rise in temperature in the active area of the chip, which increases the probability of current leakage and reduces the IQE of the DUV LED. On the other hand, the rise in junction temperature will cause the defects

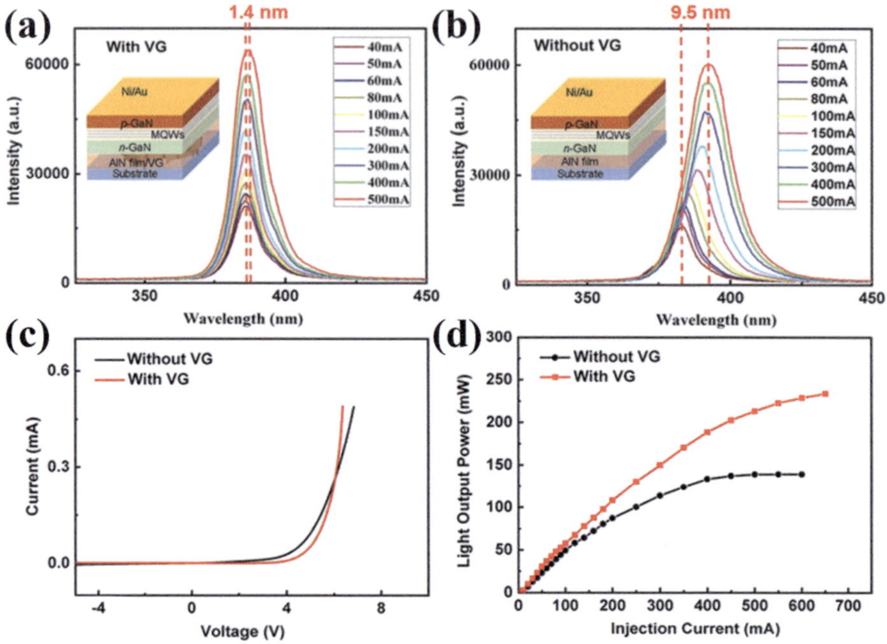

Fig. 7.10 Application of the as-grown VG-nanowall/sapphire in UV-LEDs. EL spectra of the UV-LEDs with **a** VG nanowalls and without **b** VG nanowalls measured by varying the injection current from 40 to 500 mA. The insets in **a** and **b** are schematic diagrams of the InGaN-based UV-LED structure for the AlN/VG-nanowall/sapphire and AlN/sapphire substrates. **c** *I–V* characterization curves of the UV-LEDs fabricated with VG nanowalls and without VG nanowalls. **d** Light output power of the UV-LEDs fabricated with VG nanowalls and without VG nanowalls as a function of the injection current [38]

and impurities in the epitaxial material to form deep energy levels, increasing the probability of non-radiative recombination and reducing the EQE of the DUV LED. Therefore, in order to reduce the adverse effects of the rise in junction temperature, it is necessary to improve the heat dissipation performance of the deep ultraviolet LED package and maintain the long-term performance of the device. To reduce the junction temperature of DUV LEDs, high thermal conductivity h-BN materials can be introduced as insertion layers during the growth.

Li et al. [41] from Peking University have accomplished the direct CVD growth of h-BN thin films on functional quartz and sapphire glass, by using an optimized LPCVD route. The as-obtained h-BN exhibited excellent optical transparency, ultra-flat surface, good thermal conductivity, etc. They also studied the thermal conductivity of h-BN, drew the temperature distribution of the sample with an infrared camera, and formed a hotspot in the center of the sample with a focused infrared laser beam (1450 nm). The experimental setup was shown in Fig. 7.11a, with a fixed laser power of 850 mW/cm^3 and a current density of 1.71 A. For bare glass samples, due to the limited heat dissipation capacity of pure glass (blue line in Fig. 7.11b), the temperature at the hotspot was 50 °C, and the temperature at the edge of the sample was only 33 °C. Obviously, the temperature peak of the thick h-BN/quartz glass was reduced to 48 °C due to the enhanced heat diffusion effect (red line in Fig. 7.11b). Overall, the temperature change of the entire h-BN coated quartz glass was reduced to 10 °C in Fig. 7.11c, which was much smaller than the temperature change of uncoated h-BN quartz glass (17 °C, Fig. 7.11d). Therefore, directly applying an h-BN coating on the glass can significantly improve the thermal performance of insulating glass.

The directly grown, transparent, thermal conductive h-BN/ sapphire was specifically served as an ideal substrate for the vdW epitaxial growth of AlN, a fundamental material for developing high-efficiency AlGaN-based DUV-LEDs. The AFM image in Fig. 7.12a showed a smooth surface of the epitaxial AlN film featured by well-defined steps and terraces, reflecting the stepflow growth mode of AlN materials on h-BN/sapphire. The RMS roughness in the $5 \times 5\ \mu m^2$ area was only $\approx$0.16 nm, corresponding to an atomic level fluctuation. Furthermore, the cross-sectional HRTEM image in Fig. 7.12b revealed the relative high crystal quality of the AlN layers, as well as the interface of the h-BN buffer layer. In Fig. 7.12c, the FWHM values for the XRCs of AlN films grown on h-BN/sapphire were found to be 127 and 1099 arcsec for (0002) and (10–12) reflections, respectively, which was comparable with those of directly grown AlN on sapphire by a similar MOCVD route. Furthermore, by

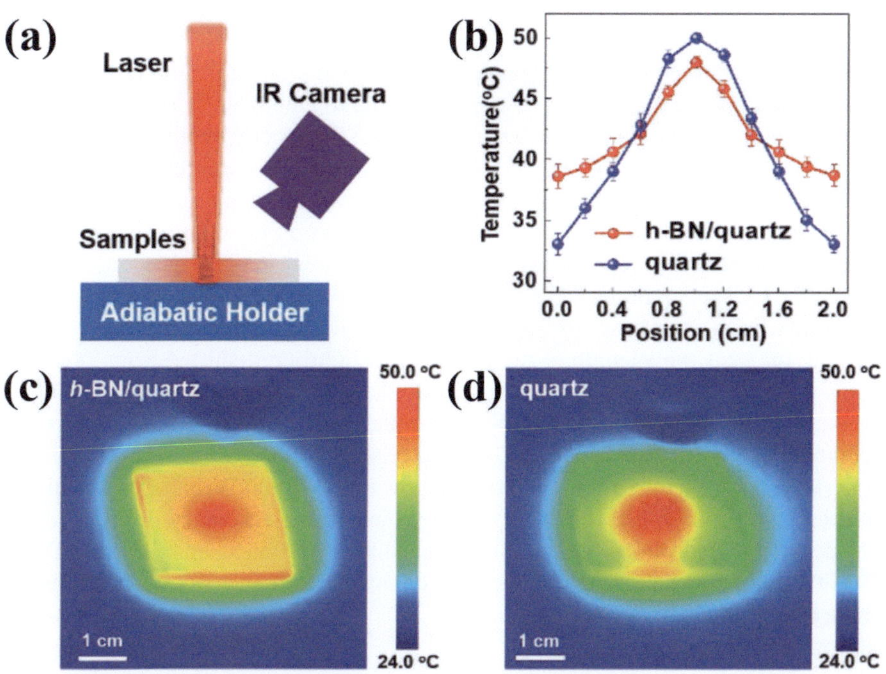

Fig. 7.11 **a** Schematic illustration of the temperature distribution map when applying a hot spot by a focused laser beam. **b** Temperature plots and temperature distribution images of **c** h-BN coated quartz and **d** pristine quartz, respectively [41]

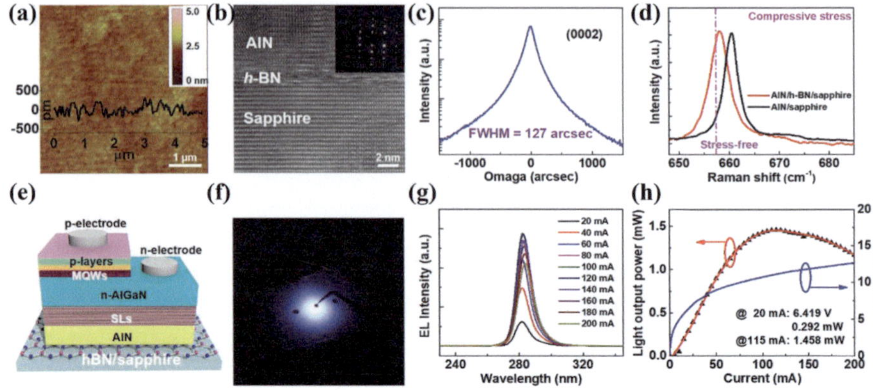

Fig. 7.12 Applications of as-grown h-BN in DUV-LEDs. **a** AFM image of the AlN epilayer grown on h-BN/sapphire, with the corresponding line profile shown as insets. **b** Cross-sectional HRTEM image of the AlN/h-BN/sapphire (inset: SAED pattern of epitaxial AlN). **c** X-ray rocking curve from AlN epilayer on-axis (0002). **d** Raman spectra of the E_2 (high) peaks of AlN/h-BN/sapphire (red line) and AlN/sapphire (black line), respectively. **e** Schematic diagram of the AlGaN-based DUV-LED structure fabricated on h-BN/sapphire. **f** Photograph of the EL from the DUV-LED wafer. **g** EL spectra of the DUV-LEDs measured by varying the injection current from 20 to 200 mA with an interval of 20 mA. **h** Light output power (red line) and the I-V characteristic curve (blue line) of the DUV-LEDs plotted as a function of injection current [41]

employing h-BN/sapphire as the epitaxial substrate, the direct heteroepitaxial interface changed into vdW-epitaxial interface, leading to significantly reduced strain in the epitaxial AlN. The E_2 (high) Raman peak of AlN directly synthesized on h-BN was centered at 658 cm^{-1} in Fig. 7.12d, corresponding to a compressive stress of $\approx$ 0.095 GPa, much less than that of AlN directly grown on sapphire.

The AlGaN-based DUV-LEDs structure was accordingly fabricated on the as-grown AlN/h-BN/sapphire template, schematically shown in Fig. 7.12e. Figure 7.12g showed the EL spectra of the DUV-LED structure when the injection current changes to 200 mA, with an interval of 20 mA, showing DUV emission characteristics. In addition, the peak wavelength of the EL spectra slightly red-shifted from 281.4 nm (20 mA) to 283.8 nm (200 mA), which was due to the thermal-induced bandgap shrinkage of the AlGaN-based MQWs under high current injection. At an injection current of 115 mA, the I-V and I-L characteristics of the realized DUV-LEDs in Fig. 7.12h showed a maximum light output power of 1.458 mW. Notably, the highly transparent h-BN film may not affect the light extraction efficiency of DUV-LEDs, considering its absorption edge at 200 nm. In summary, it can be imagined that DUV-LEDs fabricated on h-BN can exhibit ideal device performance, which is promising for the development of stress-free, durable LEDs in the next generation of solid-state lighting.

Since h-BN displays a high visible-light transmittance in addition to a high thermal conductivity, it is expected to find effective use as a heat-transfer medium for LEDs without much affecting their optical and electrical characteristics. Chen et al. [42]

reported the use of nanocomposites with h-BN nanosheets and demonstrated potential of thermal-management applications for LEDs and electronic devices. In addition, Choi et al. [43] transferred multiple-layer h-BN sheet with an average thickness of 11 nm onto the back of a green LED. As a reference, an LED without the hBN (Ref-LED) was also prepared. Differential scanning calorimetry (DSC) was used to analyze the degree of the heat flow through the LED chips. Figure 7.13a showed a schematic of the transferring of h-BN to the back of the LED wafer. Among various thicknesses of h-BN, the heat dissipation characteristic of LEDs was improved at a condition of 11 nm or more. Here, heat generated inside LEDs largely escapes through the substrate and the ambient air. Figure 7.13b showed an optical microscope image of the fabricated h-BN LED wafer, with the relatively dark area indicative of where the h-BN was transferred. After fabricating the h-BN LED wafer, a single chip was extracted from the wafer through a scribing process, which was schematically shown in Fig. 7.13c.

The role of the h-BN in contributing to heat transfer and dissipation of the h-BN LED became more apparent when analyzing the change in T_{max} over time after switching off the LEDs. Figure 7.13d showed the T_{max} distribution over time after stopping the injection of current into the LED chips. The inset showed an expanded plot to compare in more detail the results for the two LEDs. For the h-BN LED, the time required for the T_{max}, measured at the edge n-type electrode, to decrease to 37.5 °C was 35 s, whereas the Ref-LED required 265 s. In the case of the Ref-LED, the heat generated inside the LED flowed directly to the glass and finally

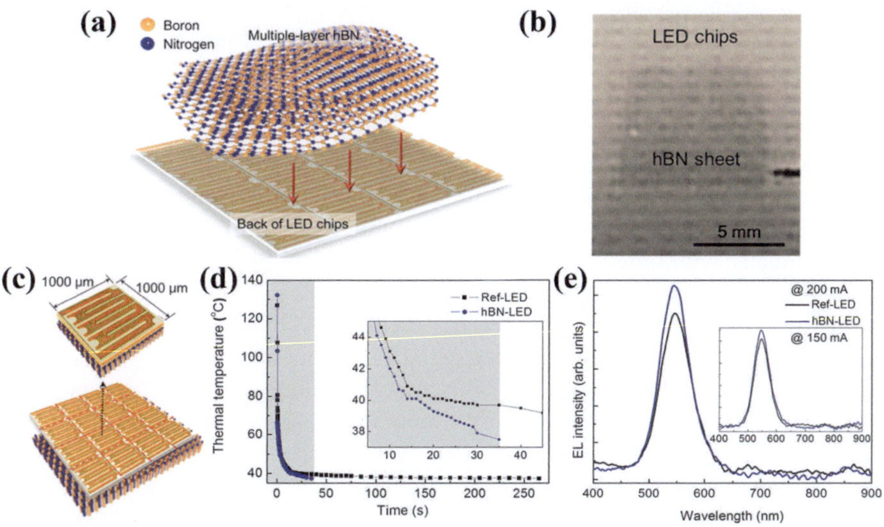

Fig. 7.13 **a** Schematic diagram of the transfer of multiple-layer h-BN onto the back of a LED wafer. **b** Optical image of LEDs with the h-BN sheet. **c** Schematic of a single LED with attached h-BN, made by scribing the wafer containing the h-BN sheet. **d** T_{max} as a function of time after stopping the injection of current into the LED chips. **e** EL spectra of the Ref-LED and h-BN LED [43]

dissipated. On the other hand, the heat that flowed to the h-BN layer for the h-BN LED diffused horizontally via the high thermal conductivity medium and was emitted through the relatively large area of the glass. That was, the effective volume of the glass contributing to heat emission increased, thereby increasing the heat-dissipation efficiency. As a result, the heat dissipation in the h-BN LED was relatively rapid compared to that of the Ref-LED. Once again, h-BN with a high thermal conductivity was found to be a very effective medium for the heat transfer and dissipation of the InGaN/GaN-QW green LEDs. Figure 7.13e showed EL spectra for the LEDs, measured at an injection current of 200 mA. The EL intensity of the h-BN LED showed 1.09 times higher than that of the Ref-LED. As shown in the inset, the EL spectra measured at an injection current of 150 mA showed the same tendency. This result was attributed to the reduction in thermal effect by introducing h-BN as a heat-transfer medium.

Figure 7.14a showed the schematic structure of a single h-BN LED (left) and a package consisting of the h-BN LED and two Cu wires attached to a piece of glass (right). In typical LED packages, metal and/or ceramic were commonly used as heat-dissipation materials [44]. Here, the LED package structure without a heat-sink material was used to analyze more deeply how h-BN affected the heat-transfer and dissipation characteristics of the chips. Temperatures at various positions on the LED chips were measured as a function of time using an infrared thermographic (IRTG) camera. Figure 7.14b, c showed IRTG images of Ref-LED and h-BN LED, respectively, measured at 180 s after first applying a current of 120 mA. The color of the IRTG images of the LED chips was converted to temperature by using the analysis program. According to these IRTG images, the highest temperature of each LED was at the edge Cr/Au electrode shining brightly fabricated on an n-type GaN cladding layer. The lowest temperature was measured at the top area between the finger-type electrodes formed on the p-GaN cladding layer. Figure 7.14d showed T_{min} as a function of time after injecting a current of 120 mA to the LED chips. Throughout the duration of the experiment, the T_{min} of the h-BN LED was lower than that of the Ref-LED. The T_{min} of the h-BN LED initially increased at a rate with 7.63 °C/s and reached a value of 53.4 °C in 7 s, after which the temperature increased more gradually with time. Similarly, the T_{min} of the Ref-LED initially increased at a rate of 8.33 °C/s and reached a value of 58.3 °C in 7 s, and then increased more gradually, at a rate of 0.13 °C/s, to reach 70 °C in 98 s. Figure 7.14e–g showed the T_{max} as a function of time after injecting the 120 mA current, with the T_{max} measured at the edge n-type electrode of each LED chip. Overall, the T_{max} versus time curve for the h-BN LED was found to be very similar to that for the Ref-LED. But a close inspection showed that the time required to reach a T_{max} of 136.1 °C was 6 s for the h-BN LED and 11 s for the Ref-LED. These differences in the temperature distribution over time between the h-BN LED and the Ref-LED were obviously related to the role of h-BN as a heat-transfer medium. The heat that flowed through the sapphire diffused faster in the horizontal direction along the multiple-layer h-BN than in the perpendicular direction, resulting in an increase in the probability that heat escaped to the edge of the LED chip. Thus, the process of forming h-BN and applying it to an LED is relatively simple.

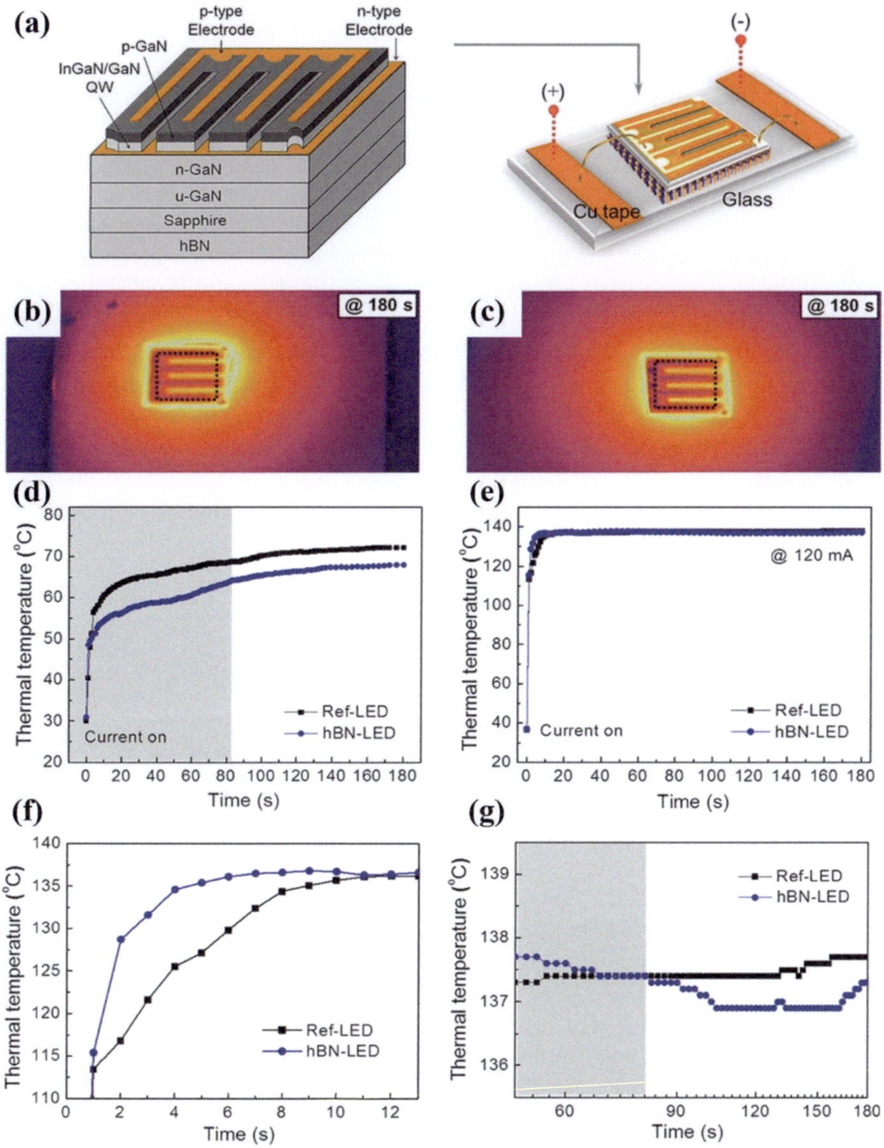

Fig. 7.14 **a** Schematic structure of an InGaN/GaN-QW green LED. IRTG images of the **b** Ref-LED and **c** hBN-LED chips measured at 180 s after the onset of an injection current of 120 mA. **d** T_{min} and **e** T_{max} evaluated from the IRTG images measured as a function of time. Expanded views of the plots of the T_{max} **f** up to 13 s and **g** from 50 to 180 s after the onset of the injection current [43]

7.3 Summary

In the early applications of LEDs, due to the low power, the heat generated is very small, and the heat dissipation problem does not need to be considered. With the increasing application of high-power nitride optoelectronic devices, strict requirements have been put forward for the thermal reliability of the device. For example, DUV LEDs have very low luminous efficiency, and most of the electrical energy is converted into the heat. As the integration density of multi-chips increases, more heat will be generated per unit area, and heat accumulation will cause the junction temperature of DUV LEDs to rise, thereby affecting the working life of DUV LEDs. Therefore, different packaging structure optimization and effective thermal management methods are tried to enhance the heat dissipation performance of LEDs, improving the reliability of LEDs. This chapter starts from the perspective of material growth, considers the heat dissipation effects during the application when growing the device structure, and uses high thermal conductivity 2D materials as the insertion layer during nitride growth. Thus, an effective heat dissipation layer between the nitride and the epitaxial substrate may be formed. This is conducive to the design and manufacture of high-efficiency, high-reliability, high-power LEDs, and promotes the development of high-power LED technology.

References

1. Kang JS, Li M, Wu H, Nguyen H, Koki T, Hu YJ (2021) Integration of boron arsenide cooling substrate into gallium nitride devices. Nat Electron 4:416–423
2. Schubert EF, Kim JK, Luo H, Xi JQ (2006) Solid-state lighting-A benevolent technology. Rep Prog Phys 69:3069–3099
3. Arik MP, Weaver J (2002) Thermal challenges in the future generation solid state lighting applications: light emitting diodes. In: Proceedings of IEEE intersociety conference on thermal phenomena. San Diego, USA, pp 113–120
4. Tsao JY (2004) Solid state lighting-lamps-chips, and materials for tomorrow. IEEE Circuits Devices Mag 20:28–37
5. Yan Z, Liu G, Khan JM, Balandin AA (2013) Graphene quilts for thermal management of high-power GaN transistors. Nat Comm 3:827
6. Balandin AA, Ghosh S, Bao WZ, Calizo I, Teweldebrhan D, Miao F, Lau CN (2008) Superior thermal conductivity of single-layer graphene. Nano Lett 8:902–907
7. Balandin AA (2011) Thermal properties of graphene and nanostructured carbon materials. Nat Mater 10:569–581
8. Seol JH, Jo I, Moore AL, Lindsay L, Aitken ZH, Pettes MT, Li XS, Yao Z, Huang R, Broido D, Mingo N, Ruoff RS, Shi L (2010) Two-dimensional phonon transport in supported graphene. Science 328:213–216
9. Jo I, Pettes MT, Kim J, Watanabe K, Taniguchi T, Yao Z, Shi L (2013) Thermal conductivity and phonon transport in suspended few-layer hexagonal boron nitride. Nano Lett 13:550–554
10. Wang CR, Guo J, Dong L, Aiyiti A, Xu XF, Li BW (2016) Superior thermal conductivity in suspended bilayer hexagonal boron nitride. Sci Rep 6:25334
11. Zhou HQ, Zhu JX, Liu Z, Yan Z, Fan XJ, Lin J, Wang G, Yan QY, Yu T, Ajayan PM, Tour JM (2014) High thermal conductivity of suspended few-layer hexagonal boron nitride sheets. Nano Res 7:1232–1240

12. Yan RS, Simpson JR, Bertolazzi S, Brivio J, Watson M, Wu XF, Kis A, Luo TF, Walker ARH, Xing HG (2014) Thermal conductivity of monolayer molybdenum disulfide obtained from temperature-dependent Raman spectroscopy. ACS Nano 8:986–993
13. Sahoo S, Gaur APS, Ahmadi M, Guinel MJF, Katiyar RS (2013) Temperature-dependent Raman studies and thermal conductivity of few-layer MoS_2. J Phys Chem C 117:9042–9047
14. Liang DD, Wei TB, Wang JX, Li JM (2020) Quasi van der Waals epitaxy nitride materials and devices on two dimension materials. Nano Energy 69:104463
15. Pop E, Varshney V, Roy AK (2012) Thermal properties of graphene: Fundamentals and applications. MRS Bull 37:1273–1281
16. Pop E (2010) Energy dissipation and transport in nanoscale devices. Nano Res 3:147–169
17. Freitag M, Steiner M, Martin Y, Perebeinos V, Chen ZH, Tsang JC, Avouris P (2009) Energy dissipation in graphene field-effect transistors. Nano Lett 9:1883–1888
18. Cahill DG, Ford WK, Goodson KE, Mahan GD, Majumdar A, Maris HJ, Merlin R, Phillpot SR (2003) Nanoscale thermal transport. J Appl Phys 93:793–818
19. Saaskilahti K, Oksanen J, Tulkki J, Volz S (2014) Role of anharmonic phonon scattering in the spectrally decomposed thermal conductance at planar interfaces. Phys Rev B 90:134312
20. Xu ZP (2016) Heat transport in low-dimensional materials: a review and perspective. Theor Appl Mech Lett 6:113–121
21. Wei TB, Wu K, Chen Y, Yu J, Yan Q, Zhang YY, Duan R, Wang J, Zeng Y, Li JM (2012) Improving light output of vertical-stand-type InGaN light-emitting diodes grown on a free-standing GaN substrate with self-assembled conical arrays. IEEE Electron Device Lett 33:857–859
22. Florescu DI, Asnin VM, Pollak FH, Jones AM, Ramer JC, Schurman MJ, Ferguson I (2000) Thermal conductivity of fully and partially coalesced lateral epitaxial overgrown GaN/sapphire (0001) by scanning thermal microscopy. Appl Phys Lett 77:1464–1466
23. Ke W, Chiang C, Lin Y, Liao Y, Cheng W, Chang K, Lin J (2023) Improved wavelength stability and heat dissipation of InGaN-based light-emitting diodes using a graphene interlayer on patterned sapphire substrate. Appl Phys Lett 122:203501
24. Incropera FP, Dewitt DP, Bergman TL, Lavine AS (2007) Fundamentals of heat and mass transfer. Wiley, Hoboken, NJ
25. Han N, Cuong TV, Han M, Ryu BD, Chandramohan S, Park JB, Kang JH, Park YJ, Ko KB, Kim HY, Kim HK, Ryu JH, Katharria YS, Choi CJ, Hong CH (2013) Improved heat dissipation in gallium nitride light-emitting diodes with embedded graphene oxide pattern. Nat Comm 4:1452
26. Goldberger J (2003) Single-crystal gallium nitride nanotubes. Nature 422:599–602
27. Tsai MY, Chen CH, Kang CS (2012) Thermal measurements and analyses of low-cost high-power LED packages and their modules. Microelectron Reliab 52:845–854
28. Lee HK, Lee DH, Chung KS, Yu JS (2011) Optical, spectral and thermal characteristics of GaN-based vertical light emitting diodes on graphite substrate. Semicond Sci Technol 26:115011
29. Li HY, Ying H, Chen XP, Nika DL, Cocemasov AI, Cai WW, Balandin AA, Chen SS (2014) Thermal conductivity of twisted bilayer graphene. Nanoscale 6:13402–13408
30. Bo Z, Yang Y, Chen JH, Yu KH, Yan JH, Cen KF (2013) Plasma-enhanced chemical vapor deposition synthesis of vertically oriented graphene nanosheets. Nanoscale 5:5180–5204
31. Zhao J, Shaygan M, Eckert J, Meyyappan M, Rummeli MH (2014) A growth mechanism for free-standing vertical graphene. Nano Lett 14:3064–3071
32. Chen SS, Moore AL, Cai WW, Suk JW, An JH, Mishra C, Amos C, Magnuson CW, Kang JY, Shi L, Ruoff RS (2011) Raman measurements of thermal transport in suspended monolayer graphene of variable sizes in vacuum and gaseous environments. ACS Nano 5:321–328
33. Faugeras C, Faugeras B, Orlita M, Potemski M, Nair RR, Geim AK (2010) Thermal conductivity of graphene in corbino membrane geometry. ACS Nano 4:1889–1892
34. Tsai PY, Huang HK, Sung CM, Kan MC, Wang YH (2013) InGaN/GaN vertical light-emitting diodes with diamondlike carbon/titanium heat-spreading layers. IEEE Electron Device Lett 34:1029–1031

35. Ci HN, Ren HY, Qi Y, Chen XD, Chen ZL, Zhang JC, Zhang YF, Liu ZF (2018) 6-inch uniform vertically-oriented graphene on soda-lime glass for photothermal applications. Nano Res 11:3106–3115
36. Wu YH (2002) Effects of localized electric field on the growth of carbon nanowalls. Nano Lett 2:355–359
37. Wu YH, Yang BJ, Zong BY, Sun H, Shen ZX, Feng YP (2004) Carbon nanowalls and related materials. J Mater Chem 14:469–477
38. Ci HN, Chang HL, Wang RY, Wei TB, Wang YY, Chen ZL, Sun YW, Dou ZP, Liu ZQ, Li JM, Gao P, Liu ZF (2019) Enhancement of heat dissipation in ultraviolet light-emitting diodes by a vertically oriented graphene nanowall buffer layer. Adv Mater 31:1901624
39. Ni ZH, Fan HM, Feng YP, Shen ZX, Yang BJ, Wu YH (2006) Raman spectroscopic investigation of carbon nanowalls. J Chem Phys 124:204703
40. Lee SH, Guan XY, Jeon SK, Yu JS (2016) Temperature-dependent optical, spectral, and thermal characteristics of InGaN/GaN near-ultraviolet light-emitting diodes. Phys Status Solidi A 213:46–51
41. Li QC, Wu QQ, Gao J, Wei TB, Sun JY, Hong H, Dou ZP, Zhang ZP, Rümmeli MH, Gao P, Yan J C, Wang JX, Li JM, Zhang YF, Liu ZF (2018) Direct growth of 5 in. uniform hexagonal boron nitride on glass for high-performance deep-ultraviolet light-emitting diodes. Adv Mater Interfaces 5:1800662
42. Chen J, Huang XY, Sun B, Wang YX, Zhu YK, Jiang PK (2017) Vertically aligned and interconnected boron nitride nanosheets for advanced flexible nanocomposite thermal interface materials. ACS Appl Mater Interfaces 9:30909–30917
43. Choi I, Lee K, Lee CR, Lee JS, Kim SM, Jeong KU, Kim JS (2019) Application of hexagonal boron nitride to a heat-transfer medium of an InGaN/GaN quantum-well green LED. ACS Appl Mater Interfaces 11:18876–18884
44. Yu X, Xie B, Shang B, Chen Q, Luo X (2016) A cylindrical tuber encapsulant geometry for enhancing optical performance of chip-on-board packaging light-emitting diodes. IEEE Photonics J 8:1600709

Chapter 8
Two-Dimensional III-Nitride Materials Growth and Applications

Bulk III-nitride semiconductor materials, such as InN, GaN, AlN and their alloys, have proved to be ideal materials for photoelectric devices, power electronic devices, surface acoustic wave devices, and many other related technologies because of their tunable direct band gaps, high breakdown voltage, and polarization characteristics. Compared with the bulk structure, the 2D group-III nitride semiconductor material has a thin atomic structure and a high activity surface. The combination of these two advantages can make the material exhibit more excellent and interesting properties [1, 2]. Therefore, it is of great significance to study 2D III-nitride materials, which are critical to the developments of flexible electronic and photoelectric devices, single-photon emission devices, spin-based devices, and many other similar advanced technological components.

III-Nitride 2D materials include GaN, AlN, InN, BN etc. Among the 2D III-nitride materials, 2D BN is the most stable system and is relatively easy to develop. However, the preparation of other 2D nitrides is difficult and less studied. Compared with AlN, which has a similar bandgap, h-BN epitaxial layers with p-type conductivity can be obtained. This can enhance p-type conductivity and reduce contact resistance, which will greatly improve free hole injection and quantum efficiency, reduce operating voltage and heat generation, and extend the working life of the devices [3, 4]. h-BN is also the best substrate and gate dielectric material for graphene and other 2D atomic crystals. Heterostructures based on h-BN and graphene can not only maintain the extremely high carrier mobility of intrinsic graphene, and regulate the band structure of graphene, but also are expected to present novel physical properties at the heterojunction interface, which is of great significance for the realization of vdW heterostructure electronic applications [5].

Other III-nitride compounds such as GaN, AlN, InN, etc., are important materials for solid-state lighting. The excellent characteristics of these materials at the 3D scale have aroused the curiosity of researchers. High carrier mobility, high electron drift speed, effective light detection and light emission, etc., endow them with great development potential in high-speed, high-frequency electronic devices and

T. Wei et al., *Quasi van der Waals Epitaxy of Nitride Semiconductor and Its Applications*, Wide Bandgap Semiconductors,
https://doi.org/10.1007/978-981-95-3469-2_8

high-efficiency optoelectronic devices, including solar cells, LD and LED. Generally speaking, as the dimension of the material decreases, the material usually exhibits a variety of physical or chemical properties, which are significantly different from the characteristics of 3D semiconductor materials. Due to the quantum confinement effect, the bandgap of 2D nitride materials will tune with the change of its layers. Its structure is unique, the electronic band structure is different from the others and the extreme quantum confinement effect can control its luminescent properties, which will have a profound impact on various optoelectronic devices. When they are reduced from 3 to 2D, they can make electronic devices operate at ultra-high voltage, frequency or temperature, and can work under the entire spectrum including UV light, visible light and infrared light. If the light emission of single-layer 2D GaN reaches the DUV range, it can achieve the sterilization and water purification applications. Polarized light emission at room temperature can be achieved through uniaxial in-plane strain, which is an ideal material for energy-saving display applications [6, 7].

Since 2005, theoretical research had predicted that GaN and AlN can form 2D stable ML structures and show electronic and optical properties different from bulk materials [8]. There are also studies based on the thermodynamics, dynamics and thermal stability of ML allotropes of several III-V compounds such as GaN and AlN [9], aiming to reveal through calculations whether the mechanical, electrical and optical properties of 2D-GaN and AlN crystals are different from 3D-GaN and AlN crystals, and to find suitable substrates for growing ultra-thin 2D-GaN and AlN [10–12]. The study of these 2D structures is becoming more and more important.

At present, conventional group-III nitride semiconductor materials do not have layered structures like graphene. They have inherent strong covalent bonds in 3D space, which makes it very difficult to mechanically peel them layer by layer to convert them to a 2D morphology, and greatly hinders 2D anisotropic growth [13]. In principle, only single-layer BN can be easily obtained using typical 2D material preparation methods. There have been some reports on the growth of 2D nitrides, hoping to improve the special performance of III-nitrides by changing dimensions. In this part, the recent progress in the study of 2D III-nitride semiconductor materials with special emphasis on the properties, growth methods, and potential applications of GaN and AlN are presented. Some theoretical calculation results of the 2D nitride structure and the fabrication methods of 2D GaN and AlN which appeared in recent years are introduced.

8.1 Properties of 2D III-Nitride Materials

8.1.1 Structure of 2D III-Nitride Materials

As group-III nitride semiconductor materials change from bulk to just one or several layers, they can be expected to exhibit different properties than bulk. ML 2D III-nitride materials exhibit sp^2 or $sp^1 + sp^2$ hybridizations instead of sp^3 hybridizations found in bulk materials, which significantly affect several structural parameters, such as bond lengths and binding energies. Furthermore, the emergence of numerous characteristics such as strong two-dimensional electron gas, expanded lattice constants, bandgap blue shift, strong electron–hole interaction, and strong polarized light emission, has stimulated scientists' interest in exploring the properties of III -nitrides at the 2D limit.

Initially, pioneering research in the field of III-group 2D nitrides mainly focused on whether 2D GaN and AlN compounds of ML or ultra-thin layer could form a stable structure like graphene. In 2006, Freeman et al. [8] first theoretically predicted that ultra-thin films of (0001) plane wurtzite materials (GaN, SiC, and ZnO), when they are only a few atomic layers thick, could possibly transform into a layered graphite structure, and be thermodynamically most stable. Şahin et al. [14] also examined a large number of group III-V compounds to reveal whether they may form 2D honeycomb structure. In recent years, several Group III 2D ML nitride structures predicted by first-principles calculations have been synthesized in laboratories, and research has been conducted on the synthesis of single-walled nanotubes and thick-walled tubular forms of GaN and AlN, ultra-thin layers or nanoplates [15–17], as well as several forms of compounds grown on specific substrates.

Overall, the physical and chemical properties of 2D GaN may show diversity due to its low dimensionality, which can further affect the applications of relevant electronic devices. Qin et al. [18] performed a comprehensive study of the phonon transport properties of ML GaN with planar honeycomb structure and compared the results to bulk GaN. The top and side view of the optimized geometry structure of ML GaN and bulk GaN was presented in Fig. 8.1a, b, respectively. It was clearly seen that the buckled structure in bulk GaN became planar when transformed to the ML GaN, due to the variation of orbital hybridization form. The optimized lattice constant of ML GaN was 3.255 Å, which was larger than that of bulk GaN ($a = 3.190$ Å). However, the Ga-N bond length (1.879 Å) in ML GaN was smaller than that in the nonhorizontal plane of bulk GaN (1.949 Å). The charge density difference for ML GaN and bulk GaN were plotted in Fig. 8.1c, d. Distinctly different from graphene, for ML GaN and bulk GaN, charge transfer between Ga and N atoms occured when the Ga-N bond was formed. Due to the breaking of Ga-N bonds along the c axis (out-of-plane direction) when transforming from bulk GaN to ML GaN, the ability of Ga atoms to lose electrons in ML GaN was weakened compared to that in bulk GaN. Thus, the charge transfer for each Ga-N bond was calculated to be 0.453 e and 0.358 e for ML GaN and bulk GaN, respectively. The more charge transfer in ML GaN meaned the covalent bond was more strongly polarized.

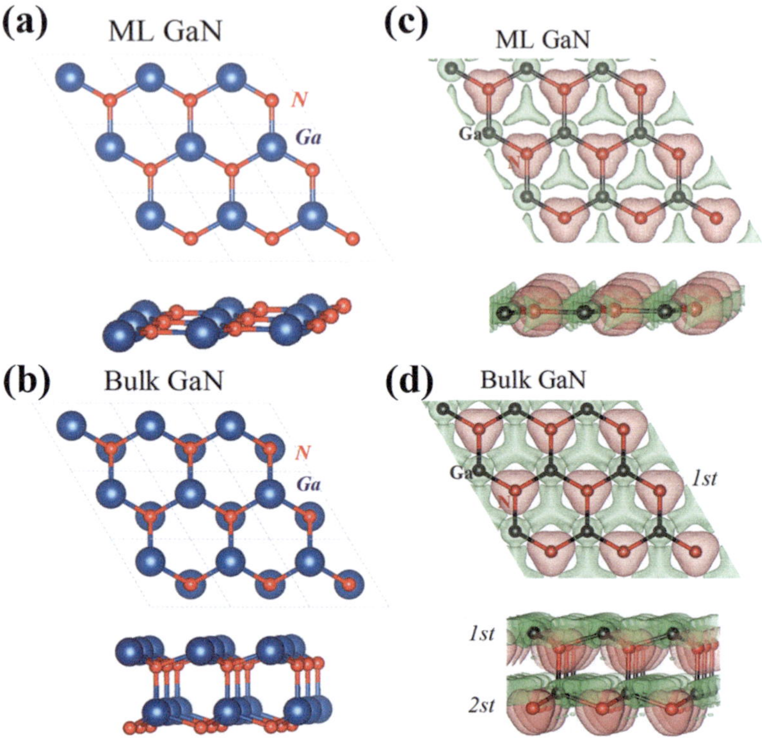

Fig. 8.1 The top and side view of **a** ML GaN and **b** bulk GaN. Top and side view of charge density difference for **c** ML GaN and **d** bulk GaN [18]

By comparing 2D nitrides of planar and buckled structures, Z. Y. Albalushi et al. found that the stability of the buckled structure was improved relative to the plane due to the reduction of binding energy [19]. Surface passivation can affect the stability. When the unsaturated state was not properly passivated, a planar structure would be obtained. When the two layers of nitrides of the planar structure came into contact, a super vdW bonding would occur and a semi-metallic behavior would happen. The structure in the bent form was most stable when the unsaturated state was properly passivated. As can be seen from Fig. 8.2a, compared with the planar structure, the binding energy of the independent buckled structure was more negative. Therefore, regardless of the number of layers, the independent buckled structure was considered as the preferred configuration of the 2D nitride. The change in atomic structure leaded to a significant difference in electronic structure. In 2D GaN, due to quantum constraints, the band gap increased significantly (3.42 eV) relative to the bulk GaN. In addition, the planar structure showed an indirect band gap of 4.12 eV, while the buckled structure still retained a direct band gap of 5.28 eV in Fig. 8.2b. Due to the quantum confinement effect, both of them were larger than the direct Eg of wurtzite bulk GaN (3.42 eV). The buckled structure was the most thermodynamically

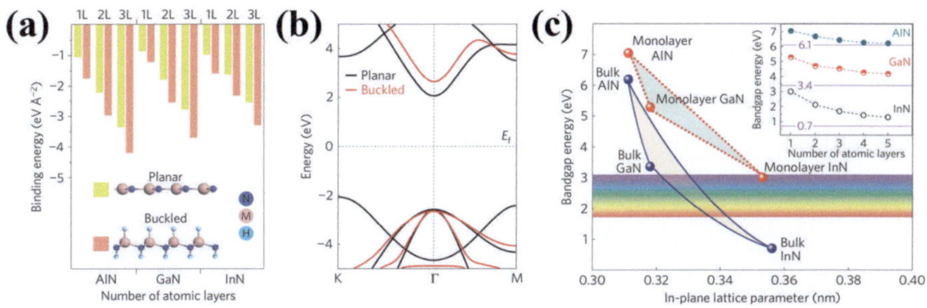

Fig. 8.2 **a** Binding energy calculations of freestanding planar and buckled 2D nitrides (N: nitrogen, M: group-III metal element and H: hydrogen atoms), as a function of layer number (L). **b** Band structure calculations via DFT meta-GGA of freestanding planar and buckled 2D ML GaN. **c** Diagram of bandgap energy versus in-plane lattice parameter for bulk and buckled 2D nitrides. A plot of Eg as a function of number of atomic buckled layers is included as an inset in (**c**) [19]

stable and thus 2D nitrides and their alloys became viable candidates for tunable optoelectronic devices as shown in Fig. 8.2c.

8.1.2 Mechanic Properties of 2D III-Nitride Materials

The mechanistic properties of a system include mechanical strength, Young's modulus, and fracture toughness. The fundamental mechanistic parameters of 2D III-nitride materials are summarized in Table 8.1 [20]. In addition, the defects, doping and absorption also influence the molecular structure and mechanical properties of 2D III-nitride materials. Zhang et al. found that the lattice constants of Si-chain-doped h-AlN showed a decreasing (increasing) trend in the armchair (zigzag) direction with longer Si–Si bonds, and consequently, the electronic and magnetic properties were changed [21]. Similarly, it was also confirmed by DFT calculations that the structure and mechanical properties of 2D III-nitride materials can be tuned by doping Fe chain, C atoms, and C chains, etc. [22] The 2D III-nitride materials get buckled under certain compressive strains. For example, 2D graphene-like GaN and InN were buckled when the compressive strain exceeded a certain threshold value, and consequently, the electronic properties were changed.

J. V. N. Sarma et al. conducted a simulation study on the mechanical properties of ML GaN [23]. They studied the mechanical behavior of a ML of GaN at different temperatures through molecular dynamics simulations. The crack propagation was observed at 700 K and 1300 K, respectively. The transition from brittleness to ductility was confirmed and it was found that the transition was controlled by dislocation migration. Furthermore, Qing Peng et al. also calculated the mechanical properties of an AlN ML [24]. It was found that it was stable under different strain directions and loads. An AlN ML can withstand the relatively large uniaxial strain and small biaxial strain before fracture, and similar to a GaN ML, its second order elastic

Table 8.1 The basic mechanistic parameters of 2D graphene-like structure III-nitride materials

	h-AIn	h-BN	h-GaN	h-InN
Lattice constant [Å]	3.127–3.170	2.504–2.511	3.156–3.24	3.52
Bond length [Å]	1.805–1.83	1.444–1.452	1.822–1.85	2.03
Young's modulus (in plane) [N m^{-1}]	135.7	278.3	109.45	62
Poisson's ratio (in plane)	0.366	0.225	0.43	0.59
Melting point [K]	3500–4000	/	/	/
Thermal conductivity [W mk^{-1}]	264.1	400	16	/

constant (including plane stiffness) increased monotonically with the increase of pressure, while the Poisson's ratio decreased monotonically. The Poisson's ratio of AlN MLs was about 1.6 times that of BN MLs, which was smaller than that of GaN MLs.

The graphene-like 2D III-nitride was found to be stable when the number of layers was less than a certain threshold. Toru Akiyama et al. [25] calculated the cohesive energies of 2D III-nitride materials, having wurtzite and hexagonal graphene-like structures with different numbers of layers, by the following expressions

$$E_{\mathrm{coh}}^{\mathrm{WZ}}(L) = \frac{1}{2}\left(4 - \frac{1}{L}\right)E_{\mathrm{bond}}^{\mathrm{sp}^3-\sigma} \tag{8.1}$$

$$E_{\mathrm{coh}}^{\mathrm{Hex}}(L) = \frac{3}{2}E_{\mathrm{bond}}^{\mathrm{sp}^2-\sigma} + E_{\mathrm{bond}}^{\mathrm{layer}} + \frac{1}{L}\left(E_{\mathrm{bond}}^{pp\pi} - E_{\mathrm{bond}}^{\mathrm{layer}}\right) \tag{8.2}$$

where $E_{\mathrm{bond}}^{\mathrm{sp}^3-\sigma}$ and $E_{\mathrm{bond}}^{\mathrm{sp}^2-\sigma}$ were the σ-bonding energies of sp^3 orbitals in the wurtzite structure and of sp^2 orbitals in the hexagonal graphene-like structure, respectively. $E_{\mathrm{bond}}^{\mathrm{sp}^3-\sigma}$ and $E_{\mathrm{bond}}^{\mathrm{sp}^2-\sigma}$ were the energies of interlayer bonding and π-bonding, respectively, in the hexagonal graphene-like structure. Based on the calculated cohesive energies in Fig. 8.3, it can be concluded that the graphene-like AlN, GaN, and InN were energetically favorable up to 13, 7, and 15 layers, respectively. However, the graphene-like structure was always stable for BN irrespective of the number of layers.

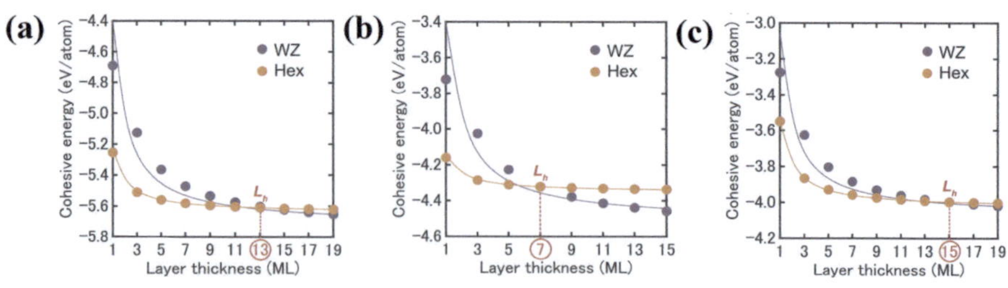

Fig. 8.3 The cohesive energy of graphene-like (Hex) **a** AlN, **b** GaN and **c** InN with different number of layers [25]

8.1.3 Electronic and Optical Properties of 2D III-Nitride Materials

The electronic and optical properties of 2D III-nitride are critical for electronic and optical applications. The electronic properties of 2D III-nitride materials can be tuned by numerous methods, and thereby the scope of the applications of 2D h-III-nitride materials can be broadened. Strain engineering is a powerful method to modulate the properties, particularly the band structures of 2D III-nitride materials. When compressive (tensile) strain is generated within the plane of these systems, the overlap of neighboring atomic orbitals increases (decreases), and affects the band structure of the material. Behzad et al. [26] studied the effects of strain on the band dispersion and dielectric response of a ML as well as a bilayer AlN. It was found that the conduction-band minimum (CBM) and the valence band edge at the Γ point moved downward (upward) under tensile (compressive) strain, while the valence-band edge of the valence-band maximum (VBM) at the K point remained almost unchanged. Hence, the band gap of 2D AlN decreased (increased) under tensile (compressive) strain.

Notably, the band gaps of other 2D III-nitride materials apart from the 2D AlN do not evolve systematically with the change in biaxial strain profiles [27]. However, when these systems are subjected to shear strain, the band gaps of all the materials increase initially and then decrease with increasing shear strain. Besides, the change in the strain profile also determines the band-gap type. For example, it was confirmed that when biaxial strain was generated in the range from − 1.53% to + 10% ($\varepsilon < 0$ indicates compressive strain), 2D h-BN exhibited a direct band gap [28]. 2D GaN showed a direct band gap under the strain ranging from − 10 to − 3.045%. However, with the in-plane biaxial strain variation from − 10 to 10%, the band gap of 2D AlN remained indirect. Although there are controversies on the band gap structure of 2D InN systems, the band gap definitely changes with different strain profiles.

The layer-dependent electronic properties of 2D III-nitride materials have also been studied by many researchers. For example, the stacking mode of the layers in a 2D AlN system was studied to explore other fundamental properties of multilayer 2D AlN systems as shown in Fig. 8.4a–c [29]. The results showed that AA′ stacking (Al over N and N over Al) was the most stable stacking mode with a layer distance of $\approx$ 2.13 Å. It appeared that due to the presence of surface states, the interlayer distance in bilayer h-AlN was slightly larger than that of bulk h-AlN. Hence from ML to bilayer, the indirect character was conserved and the band gap was changed from 2.9 to 3.5 eV. The same architecture can also be obtained in 2D GaN, where the most stable stacking sequence is AA′AA′.... With increasing number of stacking layers, the band gap increases and turns into a direct band-gap type for ten layers of stacking.

The optical properties of 2D III-nitride materials were also studied by DFT calculations [30]. The optical properties can be affected by defects, strain, and number of layers, among other factors. The optical spectra of 1-, 2-, 3-, and 4-layered h-AlN indicated that the prominent absorption took place outside the visible-light regime.

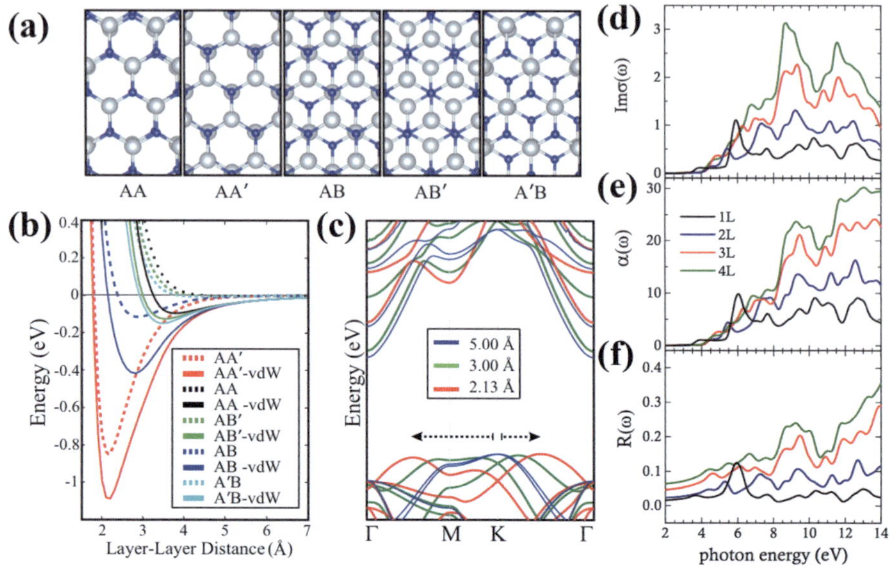

Fig. 8.4 **a** Possible stackings between two layers of h-AlN. **b** Layer-layer interaction energy for different stackings. **c** Interlayer-spacing-dependent band dispersion of AA′ stacked two layers of h-AlN. **d** Frequency-dependent optical conductivity, **e** absorption coefficient, and **f** reflectivity of 1L, 2L, 3L, and 4L h-AlN as a function of $\hbar\omega$ [29, 30]

Within the UV range, absorption intensities increased with the number of layers, approaching the bulk case. Here, the optical conductivity [Im$\sigma(\omega)$], absorption coefficient [$\alpha(\omega)$], and reflectivity spectra [$R(\omega)$] of 1L, 2L, 3L, and 4L *h*-AlN were calculated as a function of photon energy, $\hbar\omega$, and were exhibited in Fig. 8.4d–f. The former two graphs displayed, similarly to the $\varepsilon_2(\omega)$, that the rather substantial absorption was restricted to a range between 4 eV and above 10 eV, predominantly in the region above 8 eV, for all the systems of concern. Evidently, the peak intensities related to 4L *h*-AlN were the strongest, due to the well-distinguished broad peak around 9 eV. As for the frequency-dependent reflectivity function, for all the systems, it was observed that the reflectance over the visible range and slightly above was generally quite small, which pointed to a potentially transparent material. On the other hand, the applied tensile strain gradually redshifted the optical spectra. The many-body effects leaded to a blueshift of the optical spectra, while exciton binding was also observed for 2D *h*-AlN. The possibility of tuning the optoelectronic properties via thickness and/or strain opens doors to novel technological applications of this promising material.

8.2 Preparation of 2D III-Nitride Materials

Although the 2D III-nitride materials possess several unique properties, as introduced in the previous section, the growth of 2D III-nitride materials remains a challenge, especially for the 2D h-AlN and h-GaN systems. This is believed to be driven by the preferred *c*-orientation in the growth and the significant mismatch between the epitaxial layers and their substrates. Moreover, the bonding between the epitaxial layers and substrates also increases the difficulty in growing 2D materials. The fabrication of 2D h-BN is pretty mature and many methods have been adopted to obtain 2D h-BN. Therefore, the h-BN growth will not be discussed in this part. Despite these difficulties in fabricating 2D h-AlN and GaN, several successful attempts have been reported [20].

8.2.1 Preparation Methods of 2D GaN

The epitaxial growth method with the assistant of graphene encapsulation had been put forward to obtain 2D GaN beyond wurtzite structure. In 2016, Zakaria Y. Al Balushi et al. presented the growth of the 2D GaN by a process called migration enhanced encapsulation growth (MEEG). This proposed MEEG process was shown in Fig. 8.5a. In the context of the present study, the issue of 3D island formation had been identified as a consequence of lattice mismatch and surface energy constraints, when undertaking the process of direct wurtzite GaN growth on SiC (0001). Hence, they first deposited a layer of graphene on the surface of the SiC substrate by sublimation epitaxy, and transformed the graphene into quasi-free-standing epitaxial graphene (QFEG) through in situ hydrogenation reactions while reducing the interface energy between the graphene and the substrate, providing the initial interface conditions for the growth of 2D GaN. Subsequently, at 550 °C, it was placed in trimethylgallium for cycling to adsorb gallium atoms to form gallium atom interlayers. Finally, ammonolysis at higher temperatures converted the interlayer gallium into 2D GaN. In short, the entire process was that the bent form of 2D GaN was sandwiched between the top double-layer graphene and the bottom 6H-SiC substrate, and hydrogenation was used to passivate the dangling bonds between the graphene and the C layer in SiC (0001).

As shown in Fig. 8.5b, c, the distribution of the synthesized 2D GaN was observed to be in proximity to the wrinkles of the graphene. This was because, during this process, the wrinkles on the graphene and the point defects in the lattice acted as penetration sites for gallium atoms, causing gallium atoms to accumulate at these locations in Fig. 8.5d and underwent penetration and intercalation through nanoscale tearing in Fig. 8.5e, thereby reacting with the nitrogen atoms generated by ammonolysis to form 2D GaN. Ammonolysis of gallium on SiC (0001) without a graphene capping layer resulted in the formation of 3D doughnut-shaped structures of GaN due to enhanced desorption of gallium from the apex of liquid gallium droplets before

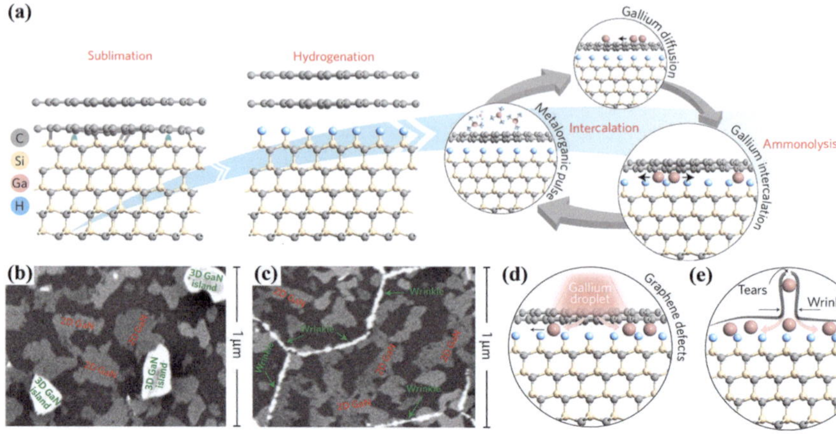

Fig. 8.5 **a** Schematic of the MEEG process. SEM images of 2D GaN in the vicinity of 3D GaN islands (**b**) and graphene wrinkle networks (**c**). **d** Ga penetration through point defects in the graphene lattice. **e** Nanoscale cracks through graphene wrinkle vertices [19]

ammonolysis. Therefore, it is found that graphene plays a key role in stabilizing the direct bandgap (5.0 eV) 2D nitride bent structure (Buckled). The bandgap of the entire structure obtained from the electron energy loss spectrum (EELS) and calculated by the mixed functional first-principles method are 5.53 and 4.89 eV, respectively, which are significantly higher than the 3.42 eV of the GaN thin film. 2D nitrides (or alloying) are viable candidate materials for tunable optoelectronic devices.

Subsequently, Wang et al. [31] used a plasma-enhanced metal–organic chemical vapor deposition (PEMOCVD) method to achieve the epitaxial growth of 2D GaN with a well-controlled lattice structure and band gap on the graphene/Si heterogeneous substrate by controlling the energy and growth temperature of plasma during the reaction. This was accomplished by regulating the plasma energy and temperature during the growth process. As shown in Fig. 8.6a, they initially fabricated graphene on a copper foil using the CVD method. Subsequently, following the dissolution of the copper foil in a $FeCl_3$ solution, the graphene was transferred onto the silicon substrate, thereby yielding the graphene/silicon heterostructure. The graphene/silicon heterostructures were then annealed at 900 ºC for 20 min. In this process hydrogenation broke the C–C bonds and passivated the dangling bonds. Trimethylgallium was utilized as the gallium source to diffuse gallium atoms onto the graphene surface. Figure 8.6b presented the formation process of 2D GaN. The hydrogenation-induced pores on graphene provided the pathways for the diffusion of gallium and nitrogen atoms to the intermediate layer. And then 2D GaN was obtained by cycling the three steps of hydrogenation, gallium atom transfer and ammonolysis in a high-purity nitrogen atmosphere. Finally, 2D GaN was formed as the Ga and N atoms were accelerated into graphene/Si heterostructures as shown in Fig. 8.6b-iii. Here, no 2D GaN layers can be found without the hydrogenation of

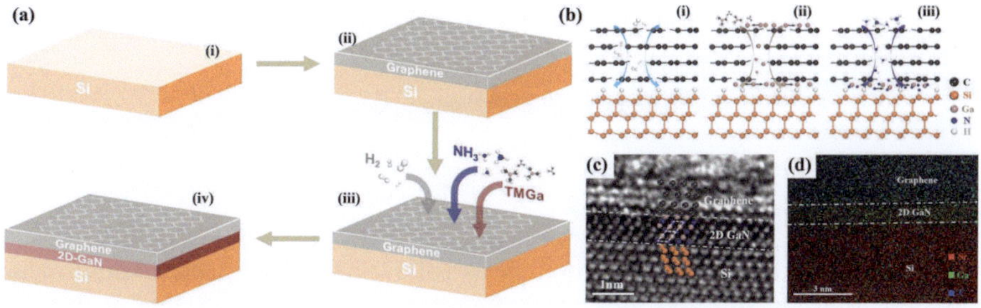

Fig. 8.6 **a** Schematic of the synthesized 2D GaN layer. **b** Schematic diagram of the principle of forming a 2D GaN layer. **c** Cross-sectional TEM image for GaN grown with the assistance of plasma. **d** Elemental EDX mapping Si, Ga, and C [31]

graphene/Si heterostructures before the epitaxial growth. In contrast, 2D GaN was found with the hydrogenation of graphene/Si heterostructures in Fig. 8.6c. Furthermore, EDS mapping in Fig. 8.6d revealed the presence of nitrogen and gallium atoms. The resulting 2D GaN lattice structure was P63MC, which depended on the plasma energy during the reaction process. The 2D GaN with well-controlled lattice structure and bandgap opens up a great potential for the development of DUV LEDs, energy conversion devices, etc.

The aforementioned 2D GaN with a nanometer level through a graphene encapsulation layer still encounters the difficulties in separating its two constituent components. The synthesis of 2D GaN poses higher requirements for exploring its inherent physical and chemical properties. The rigid substrates for growing 2D GaN or other 2D nanomaterials are most commonly used. However, due to the random distribution of surface crystals and the presence of defects, the diffusion is restricted by the boundary of the quality points, and there are shortcomings such as atomic anisotropy. The prepared 2D nanomaterials have high density and random nucleation, uneven composition, and extremely low growth efficiency.

Therefore, some studies have proposed a new idea of growing 2D materials on the surface of liquid metal catalysts [32, 33]. For metals in a molten state, the internal atoms have intense thermal motion and extremely high heat transfer efficiency. When growing 2D nanomaterials on the surface of liquid metal, these advantages make the material grow faster, the grains are spliced more smoothly, and self-limiting layer growth can be achieved. In addition, there are many free electrons with a wide range of movement on the surface of liquid metal. The existence of these free electrons makes the formation of covalent bonds in the chemical adsorption process simpler, thereby making the catalytic activity higher. In 2018, Chen et al. [32] utilized urea as the nitrogen source and employed a CVD apparatus to achieve the direct synthesis of wurtzite structure 2D GaN single crystals through a nitridation reaction on liquid Ga, as shown in Fig. 8.7a. The method to successfully obtain 2D GaN was mainly due to the design of a surface tension-induced stratified structure (Ga/Ga-W), using tungsten foil as the substrate, so that the surface-constrained nitridation reaction occured during the CVD process. The involved reaction equations were as followed,

$$2NH_3(g) \rightarrow^{850\,^{\circ}C} N_2(g) + 3H_2(g) \tag{8.3}$$

$$Ga(g) + NH_3(g) \rightarrow^{>800\,^{\circ}C} GaN(s) + H_2(g) \tag{8.4}$$

$$W(g) + NH_3(g) \rightarrow^{>800\,^{\circ}C} WN(s) + H_2(g) \tag{8.5}$$

The preparation process was roughly as follows: first, the used gallium metal particles, tungsten foil and urea were pretreated to a certain extent, so that the metal gallium particles were decomposed into small droplets in hot ethanol and then frozen into very small gallium spheres. Then, the pretreatment urea was placed in the upstream area (zone1) of the reactor as the precursor, and the gallium spheres were placed on the tungsten foil far away from the central zone2 of the quartz tube furnace. After heating, the gallium droplets would diffuse on the tungsten metal surface to form a Ga/Ga-W structure after reaching a certain temperature. At this point, the quartz tube was moved so that the Ga-W substrate was moved to the central region (zone2). Because tungsten had a stronger nitriding ability than gallium atoms, it can hinder the GaN atoms, thereby hindering the thickening of GaN. The method

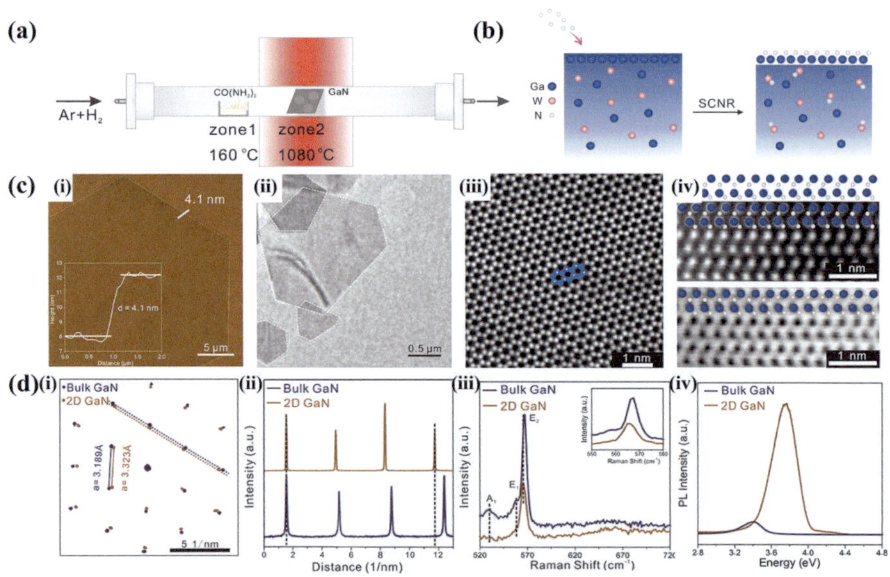

Fig. 8.7 **a** Schematic diagram of the reaction unit. **b** schematic of the SCNR strategy for the growth of 2D GaN single crystals. **c** i: AFM image of the 2D GaN. ii: A low-magnification TEM image of 2D GaN. iii HAADF and iv BF STEM images of the 2D GaN crystal acquired along the [0001] zone axis. **d** i: Characterization of the 2D GaN. ii: Overlay of the SAED spots and intensity profiles extracted from the SAED spots. iii Raman spectra and iv PL spectra for the bulk GaN and 2D GaN with a thickness of 5.2 nm [32]

was called as surface-confined nitridation reaction (SCNR) in Fig. 8.7b. After the SCNR, the 2D GaN was formed on the surface of liquid gallium.

The 2D GaN obtained was 4.1 nm thick and showed an ultrathin hexagonal morphology, as shown in Fig. 8.7c-i, ii. The atomic structure of 2D GaN was confirmed by Fig. 8.7c-iii, iv. The Ga and N atoms were arranged in a hexagonal close packing, in which the Ga atoms located at the tetrahedral sites (point group 3 m). Compared with bulk GaN, the lattice constant of 2D GaN increased. As can be seen from Fig. 8.7d-i, ii, the lattice constant increased by 4.17% along the (0001) plane. In addition, the phonon mode of 2D GaN also changed obviously. As can be seen from the Raman spectrum in Fig. 8.7d-iii, 2D GaN only had a single peak (E_2) at 566.2 cm^{-1} with good symmetry and low intensity, while bulk GaN had three peaks. The symmetry and strength of E_2 determined the high crystallinity of 2D GaN. Furthermore, compared with the bulk GaN, the E_2 of 2D GaN showed an obvious blue shift, indicating that it was in the state of tensile strain. The disappearance of A_1 and E_1 peaks in 2D GaN indicated that the Raman polarization selection rule was maintained. The 2D GaN exhibited obvious PL emission at 330 nm with a photon energy of 3.76 eV, indicating that the 2D GaN showed obvious light emission at short UV wavelengths. Moreover, compared with the bulk GaN (3.40 eV), there was a significant blue shift, showing the quantum confinement effect in the 2D limit.

Li et al. [34] also grew the ultra-thin quasi-2D GaN at room temperature using nitrogen plasma and liquid Ga. Compared with traditional nitridation reactions, the chemical reaction between nitrogen plasma and liquid gallium in a thermodynamically stable excited and ionic state was more prone to generating GaN. Since liquid gallium underwent rapid oxidation in air, after removing the oxide film on the surface of liquid gallium using NaOH and placing it in an N_2 atmosphere, a dense GaN film formed on the surface of the gallium droplet after 10 min of nitrogen plasma treatment. The surface GaN film restricted the nitridation reaction of the inner liquid gallium metal, thereby preventing the film from thickening. The structure of GaN samples was also found to be wurtzite by HRTEM. Even if the nitridation reaction duration was prolonged, the thickness of the GaN film did not increase indefinitely.

8.2.2 Preparation Methods of 2D AlN

Similar to 2D GaN, there are only a few reports about the epitaxial growth of 2D h-AlN systems. In 2013, for the first time, Tsipas et al. [35] fabricated 2D AlN with a graphene like structure using plasma-assisted MBE method. During the respective growth process, the Ag substrate was first cleaned in a vacuum by 1.5 keV Ar ion sputtering for 30 min and then annealed at 500 °C to cure the damages caused by ion sputtering. Various growth temperatures in the range of 136–650 °C and Al deposition rates from 0.06 to 0.22 Ås^{-1} were adopted. The reflection high energy electron diffraction (RHEED) results exhibited extra streaks, confirming that flat AlN layers were grown epitaxially on the Ag(111) substrate, which was also supported by the scanning tunneling microscopy (STM) results, shown in Fig. 8.8a–c. Comparing the RHEED patterns of 2D AlN on a Ag(111) substrate with the bulk AlN on a Si(111) substrate, it was evident that the in-plane lattice constant of 2D AlN was larger than that of the bulk AlN, thereby suggesting that the 2D AlN had been formed. The in-plane lattice constant of the 2D AlN was estimated as 3.14 ± 0.06 Å, which was consistent with the lattice constant acquired from the first principle calculations. Figure 8.8d showed the possible atomic arrays of 2D AlN on the Ag substrate. Three possible configurations compatible with the RHEED patterns of 4 ML-AlN/Ag(111) could visualize, in real space, the registry of AlN and Ag(111) lattices: on top (either Al or N on top of Ag atom), hollow sites and bridge sites. Here, all three aforementioned configurations depicted in Fig. 8.8d can be found within a 13 × 13 coincidence unit cell if a 3.13 A° lattice constant was assumed for a 4 ML-thick AlN as estimated from RHEED data.

In 2015, 2D AlN was also developed on an c-plane sapphire substrate by ammonia MBE [36], and the growth mechanism was similar to the case of 2D AlN grown on a Si(111) substrate by MBE. The 2D AlN was formed by chemical reactions on the surface of the sapphire substrate during the nitridation process. The Al^{3+} ions on the surface of Al_2O_3 were partially reduced to the Al^{2+} oxidation state, which reacted with the absorbed NH_2 to form the 2D AlN. Precision measurements of 2D AlN lattice parameter (3.01 Å) during the nitridation indicated the good coincidence of 2D AlN lattice and modified surface lattice of the Al_2O_3. This value minimized the elastic stresses at the AlN/(0001)Al_2O_3 interface.

However, the experimental synthesis of 2D group AlN is still meeting with tremendous difficulties due to the fact that group III-nitride semiconductors more easily orientate along *c*-axis than any other axis and the lateral mobility for precursor atoms

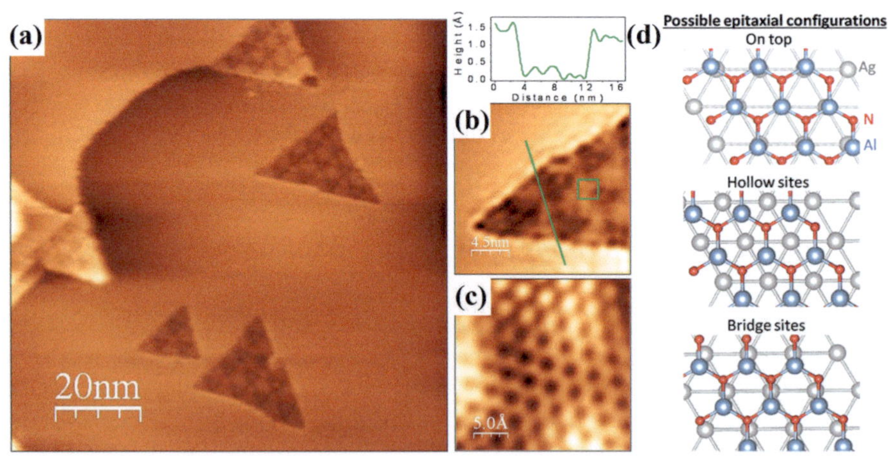

Fig. 8.8 **a** STM topography (100 × 100 nm^2) of 2D triangle-shaped islands epitaxially grown on Ag(111) substrates. **b** Magnification (20.5 × 20.5 nm^2) of a single AlN island and line profile along the green line. **c** Atomic resolution (2.5 × 2.5 nm^2) STM topography inside an island. **d** Top view of the possible configurations (on top, hollow sites and bridge sites) of graphitic AlN on Ag(111) in real space. Gray, blue, and red balls represent Ag, Al, and N atoms, respectively [35]

of AlN, is very low [37], which result in the formation of 3D crystal structure. In 2019, Wenliang Wang et al. [38] had successfully realized the epitaxial growth of 2D AlN layers sandwiched between graphene and Si substrates for the first time by MOCVD. The 2D AlN layers were grown by the processes shown in Fig. 8.9a. To further study the formation mechanism of 2D AlN layers and the effect of hydrogenation on the formation of 2D AlN layers, theoretical calculations with first-principles calculations based on DFT were carried out carefully. Figure 8.9b illustrated the schematic mechanism of formation for 2D AlN layers sandwiched between graphene and Si substrates. Primarily, a three-layer graphene was physically transferred to Si(111) surfaces, bonding with van der Waals force. Subsequently, H_2 was provided to passivate dangling bonds of Si(111) surface and to break the integrated structure of graphene, causing surface defects like wrinkles. Afterward, TMAl and NH_3 were decomposed into Al and N atoms used as precursors on the surface of graphene and accessed to the interlayer through the pathway. Eventually, the activated Al and N atoms were reacted in the interlayer and formed 2D AlN layers.

The region of 2D AlN layers can be found between graphene and Si substrates, consisting of seven layers of atoms with hydrogenation process, as illustrated in Fig. 8.9c, where the atomic layers were defined based on the atoms located on the same horizontal line. Based on the results, the three-layer graphene/Si heterostructures were suitable for the growth of 2D AlN in this situation. Finally, the graphene/2D AlN/Si substrate structure was cleaned with ultrasonic deionized water, and the topmost graphene layers were washed off, exposing the 2D AlN layer in Fig. 8.9d. The removal of graphene can be confirmed by Raman spectra Fig. 8.9e, due to the reducing of D, G, and 2D peaks. Finally, the E_g of the 2D-AlN layer was predicted to be $\approx$ 9.63 eV using HF-LDA theory, and was determined to be 9.20–9.60 eV and 9.26 eV by UPS-VB and UV–visible spectroscopic ellipsometry experiments. This ultra-wide bandgap semiconductor material has broad application prospects in the DUV optoelectronics field and is expected to support the innovation and front-end development of optoelectronic devices.

In addition, Yang et al. [39] also designed a method of physical vapor transport (PVT) to achieve independent control of substrate temperature and source temperature, and prepared a large number of layered 2D AlN under high temperature and high supersaturation conditions. They put the AlN powder into a tungsten crucible and then put it together in a furnace to heat it. The source temperature was increased to 2100 °C, and the substrate temperature was always 50–100 °C higher than the source temperature during the heating process. After holding for a period of time, the substrate temperature was reduced to 100–150 °C lower than the source temperature. After heat preservation and natural cooling, a large number of 2D AlN can appear on the W substrate. The 2D AlN showed the XRD characteristic peaks of wurtzite AlN and belongs to the hexagonal crystal structure. The TEM image of AlN showed the hexagonal lattice of Al and N atoms, with the interspacing of the (100) and (010) planes the same as the theoretical value (0.27 nm). All these indicated that the prepared AlN crystals have a uniform morphology and high crystalline quality.

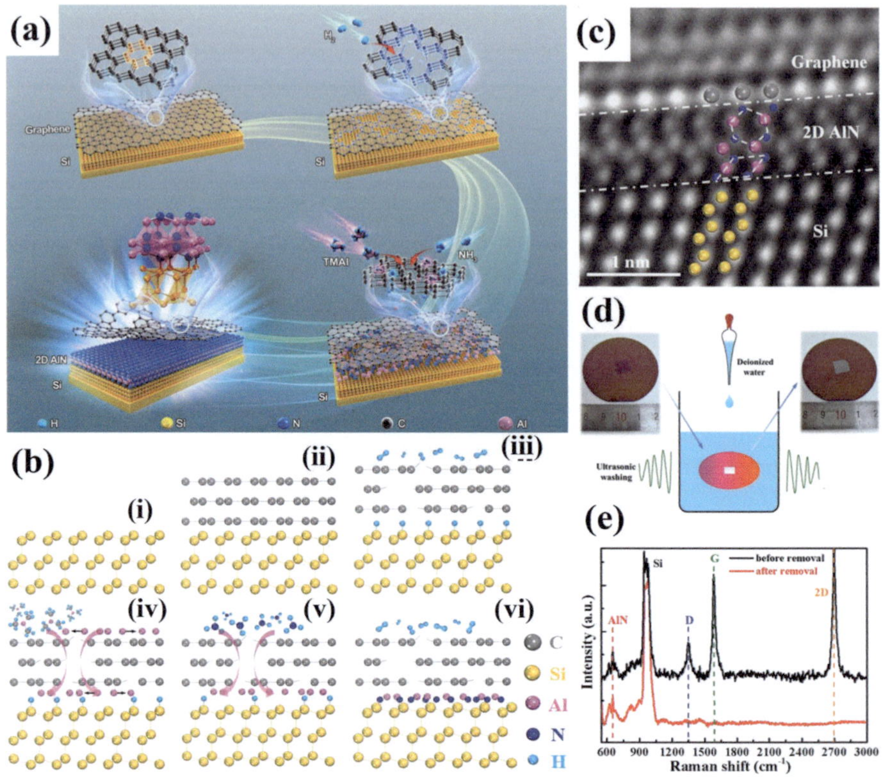

Fig. 8.9 **a** Schematic diagram for the growth of 2D AlN layers sandwiched between graphene and Si substrates. **b** Schematic structures for the formation of 2D AlN layers sandwiched between graphene and Si substrates. **c** Cross-sectional TEM images for AlN layers. **d** Schematic illustration of ultrasonic washing process to remove the graphene from graphene/2D AlN/Si, and a typical photograph for 2D AlN layers before and after the removal of graphene. **e** Raman spectrum for 2D AlN layers before and after the removal of graphene [38]

8.3 Applications of 2D III-Nitride Materials

Besides the advantage to fabricate flexible and wearable devices, the 2D III-nitride materials also have many other interesting characters. The 2D III-nitride materials combine the advantages of traditional 2D materials and wide-bandgap semiconductor materials, which make them suitable for use in the fabrication of electronic and photoelectric devices. Through chemical modification by hydrogen, fluorine atoms, etc., and the application of external electric fields, [40] it is found that the bandgap of ML III-nitride nanoplates can be adjusted, and due to their spontaneous polarization, bandgap sensitivity to the direction and intensity of the electric field, they may bring possibilities for the next generation of optoelectronic devices, such as photodetectors or full-color LEDs. The electronic properties of III-nitride 2D materials range from metal to semiconductor, with the bandgaps spanning the visible light range.

When the bandgap is within the visible light range, it can be used for photoelectric or photocatalytic applications. In addition, the special stacking order of ML honeycomb structure nitrides can be used to design QW structures, reveal the dimensional effects in heterostructures, and explore the potential of composite structures in designing new nanomaterials. If ultra-wide bandgap 2D nitride semiconductor materials can be further prepared, they will have broad application prospects in the DUV optoelectronics field, such as the preparation of high-output power UV light sources, and new generation of lasers, electronic devices and sensors.

However, due to the limitations of the difficult growth processes, it is still hard to prepare 2D III-nitride materials other than the h-BN system, although many of the properties of 2D h-AlN, h-GaN, and h-InN are comparable or even better than the 2D h-BN material. However, many studies on the performance of 2D GaN show that 2D GaN and AlN have more interesting performance compared with bulk nitrides, which can make up for some shortcomings of bulk nitride and make them applied in a wider range of fields. The following is an introduction to some of the devices integrated using 2D GaN and AlN.

8.3.1 Electronic and Photoelectronic Devices

GaN has been intensively studied for many years since they have extensive applications in high-efficiency electronic and optoelectronic equipment such as FET, HEMT, LED, detector, and so on by theory and experiment. Thererfore, 2D GaN will be potential channel materials for next-generation electronic and optoelectronic devices. In 2018, Yunxu Chen et al. constructed 2D GaN FET devices, using mechanically exfoliated graphene films as the contact layer in order to achieve better Ohmic contact [32]. Figure 8.10 showed the device prepared by using the 2D GaN synthesized by this method and its related characterization. The I_{dS}–V_G characteristic curve with $V_{DS} = 1$ V and the gate-dependent I_{dS}–V_{ds} characteristic curves were obtained, as shown in Fig. 8.10d. Finally, the mobility of 160 $cm^2V^{-1}S^{-1}$ and the on/off ratio of about 10^6 were calculated. The fabricated 2D GaN with a high electronic mobility hold substantial promise for future nanoelectronics.

Compared with the bulk GaN materials, various nanoscale effects of the low-dimensional GaN materials can display better photoelectric properties. Qian Li et al. demonstrated for the first time to print the ultrathin quasi-2D GaN films on SiO_2/Si substrates through introducing liquid metal-based synthesis and printing processes at room temperature [34]. They also fabricated lateral-gate GaN FETs using Ag as the metal contacts. The performance of this device exhibited a near-ideal subthreshold swing (*SS*) of 98 mV per decade, and the average field-effect mobility at room temperature was 53.1 cm^2/V s. Electrical characteristics of 30 FETs obtained in the same manner were evaluated, revealing average logarithmic on/off ratios, carrier mobilities, and SS of 5.31 $\pm$ 0.52, 53.1 $\pm$ 4.5 cm^2/V s, and 97.6 $\pm$ 2.42 mV/dec, respectively. The consistency of such test results, indicating high stability, along

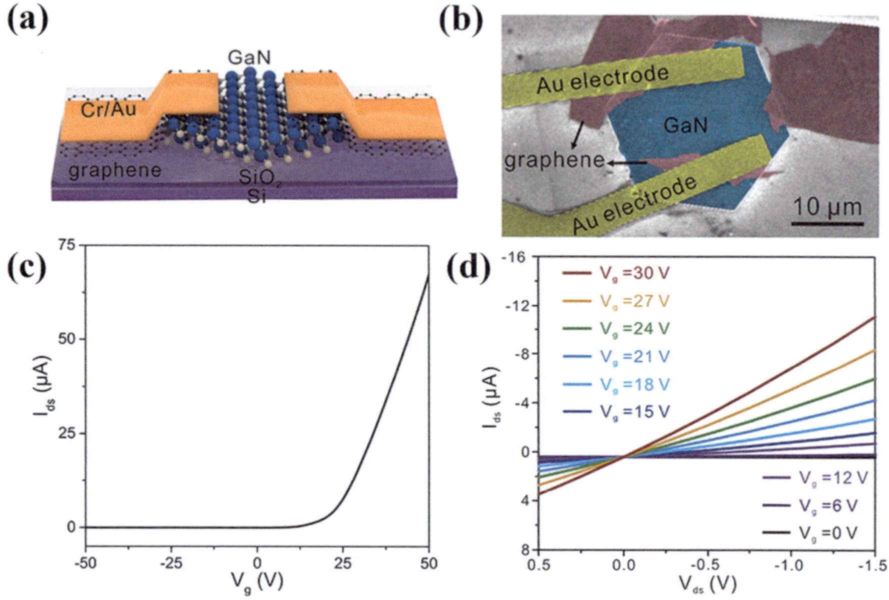

Fig. 8.10 FET performance of 2D GaN single crystal. **a** Schematic of 2D GaN FET. **b** The false-colored SEM image of a 2D GaN crystal FET. **c** I_{ds}–V_g characteristic curve recorded at $V_{ds} = 1$ V. **d** Gate-dependent I_{ds}–V_{ds} characteristic curves. The gate voltages of individual I_{ds}–V_{ds} curve are given [32]

with its large-scale and low-cost production process, rendered it to have extensive application prospects in areas such as integrated circuits.

In the field of photoelectric devices, 2D III-nitride materials can be used to fabricate solar cells and LEDs. A new active region was theoretically proposed, namely, three cycles of quasi-2D GaN layers replacing traditional AlGaN MQWs. Subsequently, this quasi-2D GaN layer was grown into a quasi-2D GaN well structure sample using sub-ML digital alloying (SDA) technology with MBE [41]. Due to the extreme quantum confinement effect, the ground state of electrons and holes in the QWs moved towards their respective high-energy directions, and the corresponding radiative recombination energy ΔE was greater than the bandgap width of GaN bulk material, thus UV or even DUV emission can be achieved.

To illustrate the advantages of the proposed new quasi-2D layer over traditional AlGaN MQWs, the crystal structures and charge density distributions of the new quasi-2D QWs and traditional AlGaN QWs were shown in Fig. 8.11a, b. Two types of QWs were designed for mid-UV (approximately 280 nm) emission. For comparison, both the quasi-2D GaN well and the $Al_{0.43}Ga_{0.57}N$ QW structures used $Al_{0.75}Ga_{0.25}N$ barriers. For the quasi-2D structure, three periods of a single ML quasi-2D GaN layers spaced by 2 ML $Al_{0.75}Ga_{0.25}N$ matrix. However, conventional QWs structure included $Al_{0.43}Ga_{0.57}N$ QW/$Al_{0.75}Ga_{0.25}N$ barriers. The three red arrows marked the positions of the quasi-2D GaN layers. As shown in Fig. 8.11c, in the quasi-2D GaN layer, the charge densities of both the CBM and VBM were greater than the charge density in the nearby $Al_{0.75}Ga_{0.25}N$ barriers. The carriers were mainly distributed in the quasi-2D GaN layer, promoting carrier localization in both the vertical and lateral directions. This localization of carriers in the quasi-2D well increased the overlap of the electron–hole wave functions, improving the luminescence efficiency. In contrast, for the conventional AlGaN QW, the strong polarization field caused the electrons in the CBM and the holes in the VBM to be mostly located on opposite sides of the QW in Fig. 8.11d, resulting in less overlap of the electron–hole wave functions and lower luminescence efficiency. Additionally, since the quasi-2D GaN layer growed on the AlGaN layer, there were fewer mismatch dislocations, better crystal quality, and TE-polarized light emission along the c-axis orientation.

To avoid the difficulties of p-type doping high Al component AlGaN and the absorption of UV light by p-type GaN, the electron beam pumping method were used instead of the conventional electrical injection method. The schematic diagram of the sample structure is shown in Fig. 8.12a. The quasi-2D GaN structure was deposited on the AlN/c-plane sapphire grown by MOCVD. The blue dashed line represented the quasi-2D GaN layer, which was a discontinuous GaN layer with a thickness of 1 ML in the experiment, representing a fractional ML. The sample structure was free of n-contact, p-contact layers, or electron-blocking layers, since the active region was excited by an e-beam impinging on the surface. Figure 8.12b showed STEM-HAADF image of the sample. The periodic quantum structure of the ultrathin brightest stripes was clearly recognized with a period length of about 33–35 nm. An unintentional periodic intensity variation of the $Al_xGa_{1-x}N$ matrix with a period length of about 8–9 nm. Figure 8.12c showed a magnified section as depicted by the white frame in Fig. 8.12b. Three periods of fractional ML GaN layers, which appeared brighter due to their higher atomic number, spaced by 2 MLs of AlGaN matrix were clearly recognized. They corresponded to effective ≈ 2 nm thick AlGaN QWs. No misfit dislocations were found at the GaN/AlGaN interface, confirming a coherent growth of the quasi-2D GaN layers.

With these approaches, the optical output power of the quasi-2D structure in dependence of the e-beam current under pulse-scan and continuous-wave (CW) pumping modes at RT were realized as shown in Fig. 8.12d, e, respectively. The e-beam pumped emission spectrum in the CW pumping mode was shown in the inset of Fig. 8.12e with the emission centered at 285 nm. The full width at half maximum of the emission spectrum is broadened to be about 40 nm. Under pulsed excitation, the optical output power increased linearly with the e-beam current and went up to a maximum of about 113 mW at an e-beam current of 1.1 mA, which showcased its broad prospects and strong competitiveness in the non-line-of-sight communication and medical disinfection fields.

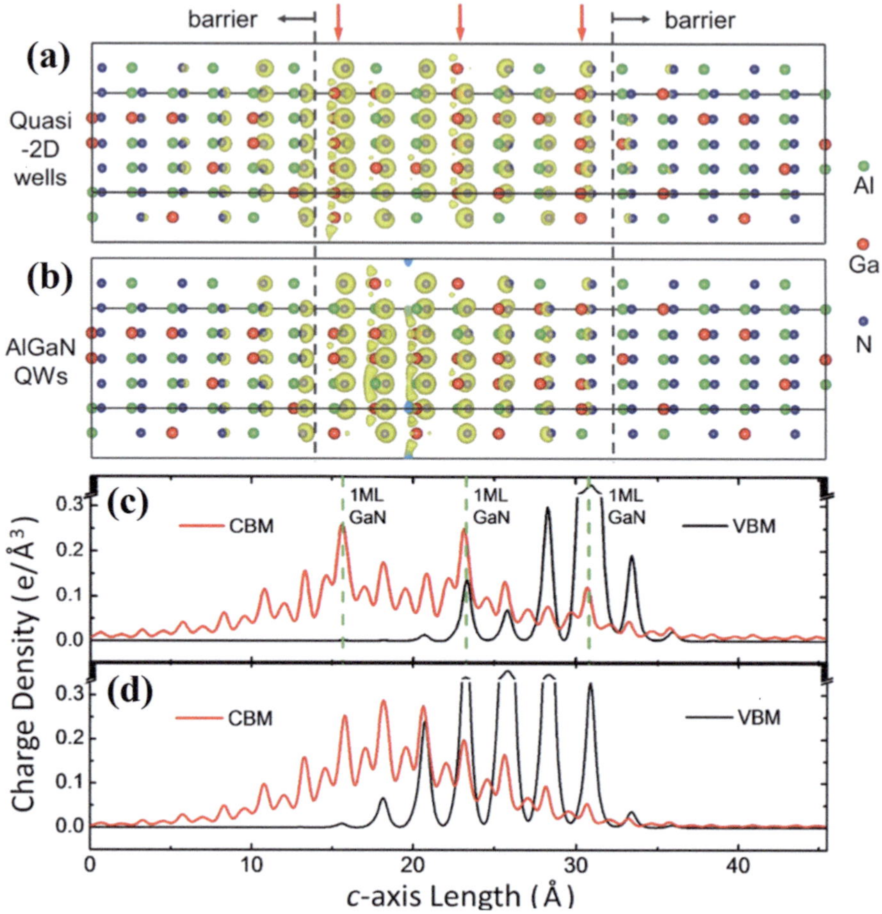

Fig. 8.11 Schematic image of crystal structure combined with the isosurface charge density plot ($\rho = 0.03$ e $Å^{-3}$) of the CBM at the Γ point for (**a**) the quasi-2D wells and (**b**) the $Al_{0.43}Ga_{0.57}N$ QWs. Local charge densities of CBM and VBM for **c** quasi-2D wells and **d** $Al_{0.43}Ga_{0.57}N$ QWs [41]

8.3.2 Detector and Sensor

Gas-sensing devices are becoming increasingly mandatory for homeland security, medical and environmental monitoring, space exploration, automotive applications, and chemical process control. To meet these demands, the sensing properties (mainly including sensitivity, selectivity, stability, and speed) of gas sensors should be greatly improved, and these have awakened enormous interest in the development and selection of optimal sensing materials. Due to their outstanding electronic properties, 2D nitride materials are usually good candidates for gas sensors. For example, H_2 adsorption results in n-type doping of AlN nanostructures, while molecules like CO_2, CO,

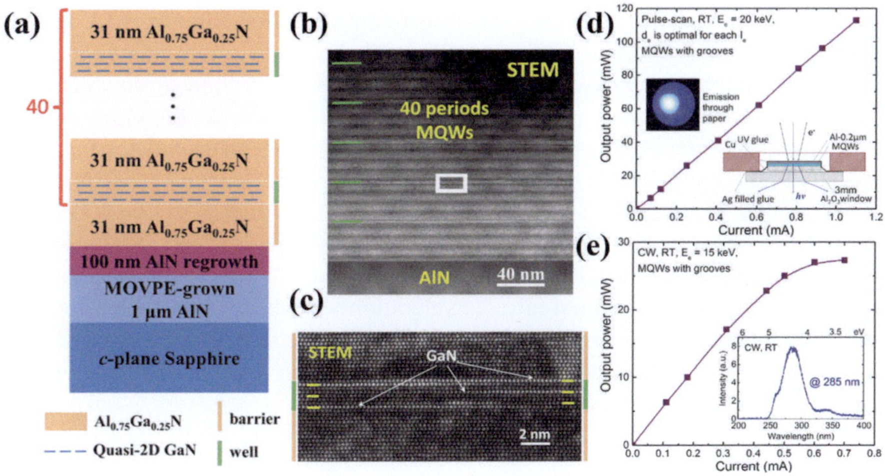

Fig. 8.12 **a** Schematic image of the designed quasi-2D GaN wells inserted in $Al_{0.75}Ga_{0.25}N$ matrix. **b** STEM-HAADF image of the MQWs/AlN structure. **c** STEM-HAADF image for the three periods of quasi-2D GaN layers inside the QWs. The dependence of output power on e-beam current in **d** pulse-scan mode and **e** CW pumping mode at RT. The pumping scheme of the MQWs structure by an e-beam was shown in the inset of (**a**) and the e-beam pumped emission spectrum in CW regime for the sample was shown in the inset of (**b**) [41]

O_2 and NO are acceptors, that is, they make AlN p-type due to the transfer of electron charge to the gas molecules. The resulting changes in the conductivity of AlN nanosheets can be electrically measured in specially designed channels on AlN, and the changes caused by the adsorption of different molecules can be detected.

In 2013, Rastegar et al. [42] studied the adsorption of NH_3 and NO_2 molecules on AlN nanosheets by DFT simulation. They found that the conductivity of AlN increased after NO_2 adsorption. This AlN sheet can selectively detect NO_2 molecules in the presence of NH_3, aiming to reveal the application prospects of gas sensors. The study revealed that the NH_3/AlN sheet composite had a stable configuration, and there was a strong interaction between the molecule and the AlN sheet. In the case of NO_2, two stable configurations were obtained. Due to its molecular structure, NH_3 may provide electrons to the AlN sheet. While NO_2 acted as a strong electron acceptor, the adsorption of the two molecules had different effects on the performance of the AlN nanosheet. Therefore, it is reasonable to assume that the charge transfer between the AlN nanosheet and NH_3 or NO_2 may change the electronic energy levels of the AlN sheet, and even change the band gap.

Theoretically, the band gap of hydrogenated 2D h-AlN decreased with the absorption of formaldehyde, which made hydrogenated 2D h-AlN suitable for formaldehyde detections [43]. The change in the band gap leaded to a difference in the electrical properties of 2D h-AlN with or without the gas absorption. According to the first principles calculation, the 2D h-AlN nanosheet was sensitive to CH_2O, CO_2, CO, H_2, O_2, NO, SO_2, and NO_2. [44] Charge density difference plots and the comparison of the band structures before and after the absorption were shown in Fig. 8.13. Among all the gas molecules, only CO_2 on the 2D AlN sheet had strong interaction with adsorption energy of 0.91 eV. While other gas molecules bound to the AlN sheet much weakly with the adsorption energy being less than 0.5 eV.

In addition, Yong et al. [45] also studied the properties of gas molecules (NO, NH_3, and NO_2) adsorbed on 2D GaN with a tetragonal structure (T-GaN) using first-principles methods. The recovery time of the T-GaN sensor at $T = 300$ K was estimated to be quite short for NO_2 and NO but very long for NH_3. Moreover, the magnetic properties of the T-GaN were changed obviously due to the adsorption of NO (or NO_2) molecule. Taking into consideration various aspects, such as adsorption energies, charge transfer, change of electric conductivity, and recovery time, it was concluded that T-GaN was a promising candidate for highly sensitive magnetic sensors for the detection of NO and NO_2.

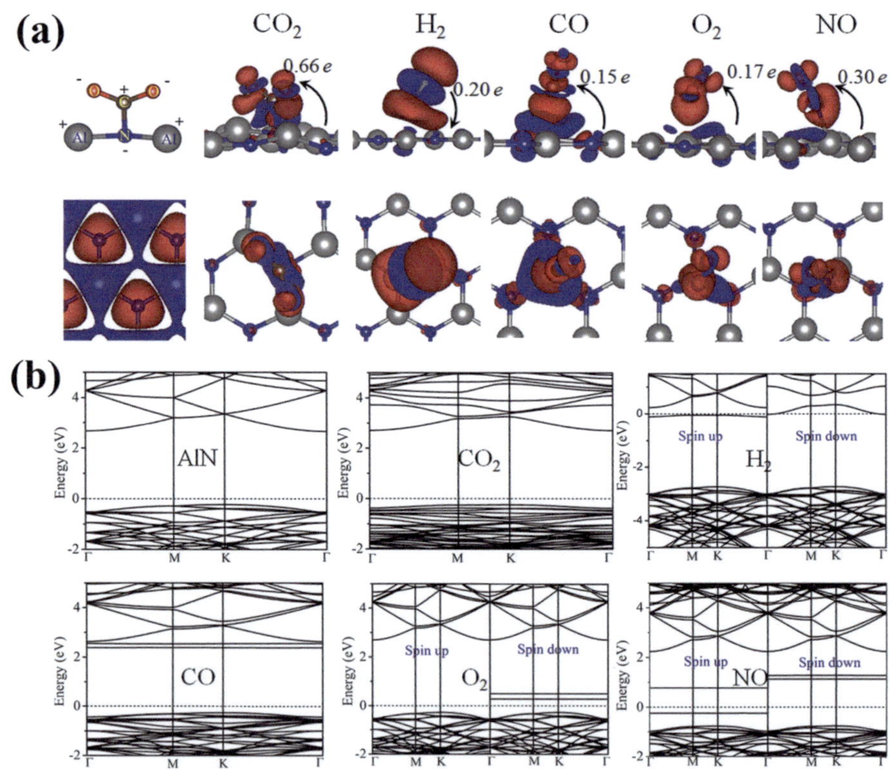

Fig. 8.13 **a** Charge density difference plots and **b** the band structures of the adsorption of CO_2, H_2, CO, O_2 and NO on 2D h-AlN [44]

8.3.3 Energy Storage and Conversion

In the preparation of Li- or Na–ion batteries, one of the biggest challenges is to find suitable anode materials. Xiaoming Zhang et al. proposed that 2D GaN was a good anode candidate material [46]. As the 2D GaN layer exhibited stable Li/Na adsorption and low theoretical voltage, it can also achieve high storage capacity and extremely fast Li/Na diffusion. The first-principles calculation of 2D GaN ML was carried out to study its feasibility as an electrode material. Through the calculation of three migration paths in Fig. 8.14a, they found that the diffusion barriers of Li and Na ions in the case of 2D GaN as an anode material were much lower than that of other 2D anode materials. The diffusion barrier of Li was 79 meV, and the diffusion barrier of Na was only 22 meV as shown in Fig. 8.14b. These data indicated that there was an extremely fast ion diffusion velocity on the 2D GaN layer. In addition, in their study on Li/Na storage capacity on 2D GaN, through the layer-by-layer calculation of adsorption energy, the average open circuit voltage (OCV) of Li was 0.060 V and that of Na was 0.050 V. In Fig. 8.14c, d, the maximum storage capacity of Li and Na was 938 mA h g^{-1} and 625 mA h g^{-1}, respectively. At the same time, it was also found that when the 2D GaN was adsorbed by Li and Na atoms, it would show metallization, which ensured the good electrical conductivity as an electrode. These results indicate that 2D GaN is an excellent electrode material with fast ion diffusion and high storage capacity, and will have great potential in the field of battery development.

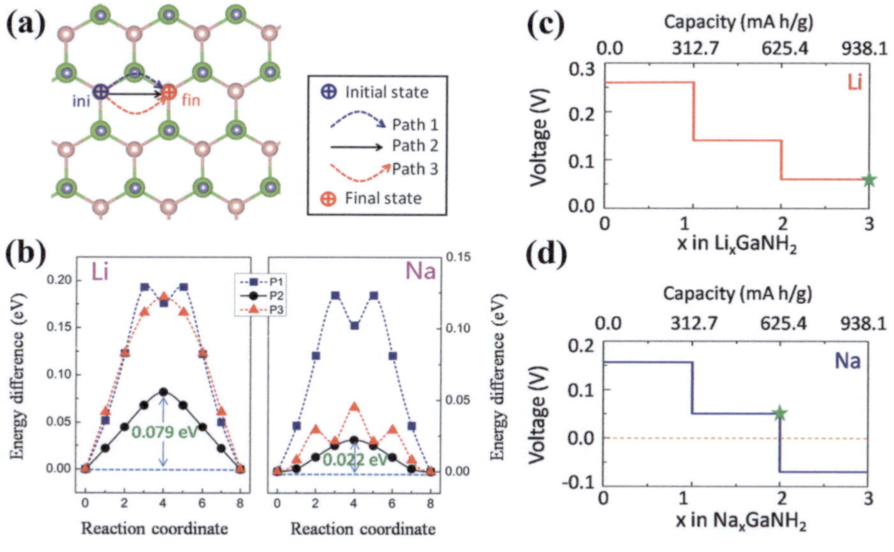

Fig. 8.14 **a** Possible Li/Na ion migration paths on the 2D GaN. **b** Corresponding diffusion barrier profiles for Li and Na. The calculated voltage profiles and storage capacities for **c** Li and **d** Na on the 2D GaN. The maximum Li and Na storage capacities are indicated by stars [46]

Furthermore, Amretashis Sengupta used DFT to simulate the Li and Na properties of 2D AlN sheets [47]. After the calculation, the maximum storage capacity of Li and Na adsorbed by 2D AlN was 500.8 mA h g^{-1} and 385.3 mA h g^{-1}. In most cases, Li atoms had stronger adsorption on 2D AlN. The maximum OCV was 1.49 V and 1.86 V, the average was 0.78 V and 0.94 V, and the minimum diffusion potential energy for different paths was 0.40 eV and 0.15 eV, respectively. These results indicated that 2D AlN can be used as an anode material in Li- or Na-ion batteries, and had a great potential in the application of supercapacitors and ion-air batteries. In the process of research, Sengupta also compared the transmission spectra of 2D AlN with and without Li/Na adsorption, and found that the addition of Li/Na improved the Fermi level of 2D AlN, and the adsorption change of Fermi energy would result in a significant change in the transmission spectrum. Therefore, the change in the transmission spectrum can be used to distinguish whether 2D AlN adsorbed Li or Na atoms. The electrical sensitivity and selectivity of 2D AlN make it useful for chemical/biosensor applications.

As a wide direct-gap semiconductor, GaN has emerged as a new-generation material for solar water splitting due to its unique structural, electrical, and optical properties. To date, a variety of 2D nanomaterials have been identified as promising candidates for photocatalytic or electrocatalytic applications, in association with their novel physical and chemical properties, including robust mechanical properties, high specific surface areas, and abundant active sites. In 2020, Lesheng Li et al. [48] systematically examined the prospect of using a single transition-metal -atom-doped g-GaN ML as an electrocatalyst for artificial nitrogen reduction via density functional theory calculations and the computational hydrogen electrode model. The Mo-doped g-GaN (Mo@g-GaN) ML was the only electrocatalyst predicted to be feasible for the nitrogen reduction reaction (NRR) with a low onset potential of − 0.33 V. In addition, Gani Purwiandono et al. [49] also prepared a GaN paste electrode from 2D hexagonal-shape GaN (HS-GaN) synthesized by a reaction in a molten salt using Ti plate as a substrate, and compared the photo-electro-chemical properties of the electrode with random-shaped GaN electrodes. They found that 2D HS-GaN generated higher photocurrent due to its more efficient water oxidation, which was about twice as high as that of random-shaped GaN, and the donor density of the 2D HS-GaN sample was one order of magnitude higher than that of the randomshaped. Thus, the precise growth control of 2D GaN crystals will facilitate new methods of developing high-performance photo-energy conversion materials.

8.3.4 Other Applications

In addition, Group III nitride crystals have many ideal dielectric properties, including high dielectric constant, low leakage current, and high breakdown voltage. 2D AlN is a prospective material for use as a dielectric related to electronic device applications [50], such as the gate dielectric in transistors. A major problem in this application is the presence of defects and traps at the semiconductor/dielectric interface, as well as the relatively high surface roughness of the dielectric layer, which plays a detrimental role in effective charge transfer. The most common forms of AlN crystals are 1D nanostructures and 3D bulk phases, while 2D AlN single crystals as gate dielectrics are still in their infancy. The development of this material is still affected by the non-ideal 2D form.

8.4 Summary

Owing to the wide band gap of 2D III-nitride materials, studies on 2D III-nitride materials have attracted considerable research interest. In this chapter, the recent advances in the properties, growth methods, and applications of 2D graphene-like III-nitride materials (h-GaN and h-AlN) are briefly summarized. 2D stable structures of GaN and AlN exhibit electronic and optical properties different from thin film nitrides and have important application prospects. However, the preparation of 2D GaN and AlN is currently difficult, and many properties remain to be explored. The main challenge at this stage of research on the 2D III-nitride materials is the growth of 2D III-nitride materials.

Currently, a variety of devices based on 2D GaN and 2D AlN have been fabricated, including FETs, LEDs, sensors, energy storage conversion, etc. In addition, they also have potential applications in catalysis and energy storage. 2D nitrides can open up new research avenues for realizing electronic and optoelectronic devices with ultra-wide bandgaps. Their excellent electronic performance and strength will have disruptive application effects, promoting the development of a new generation of lasers, electronic devices and sensors.

References

1. Zhou N, Yang R, Zhai T (2019) Two-dimensional non-layered materials. Mater Today Nano 8:100051
2. Wang ZX, Wang GD, Liu XT, Wang SZ, Wang TL, Zhang SY, Yu JX, Zhao G, Zhang L (2021) Two-dimensional wide band-gap nitride semiconductor GaN and AlN materials: properties, fabrication and applications. J Mater Chem C 9:17201–17232
3. Jiang HX, Lin JY (2017) Review-hexagonal boron nitride epilayers: growth, optical properties and device applications. ECS J Solid State Sci Technol 6:Q3012–Q3021
4. Jiang HX, Lin JY (2014) Hexagonal boron nitride for deep ultraviolet photonic devices. Semicond Sci Technol 29:084003
5. Li LH, Chen Y (2016) Atomically thin boron nitride: unique properties and applications. Adv Funct Mater 26:2594–2608
6. Sanders N, Bayerl D, Shi G, Mengle KA, Kioupakis E (2017) Electronic and optical properties of two-dimensional GaN from first-principles. Nano Lett 17:7345–7349
7. Gao YQ, Yu YL, Yang JK, Wang P, Duo YW, Yang JH, Huo ZQ, Ran JX, Wang JX, Wei ZM, Wei TB (2025) Polarization-sensitive solar-Blind ultraviolet photodetectors based on semipolar (11–22) AlGaN film. ACS Appl Mater Interfaces 17:9554–9562

8. Freeman CL, Claeyssens F, Allan NL, Harding JH (2006) Graphitic nanofilms as precursors to wurtzite films: theory. Phys Rev Lett 96:066102
9. Kecik D, Onen A, Konuk M, Gürbüz E, Ersan F, Cahangirov S, Aktürk E, Durgun E, Ciraci S (2018) Fundamentals, progress, and future directions of nitride-based semiconductors and their composites in two-dimensional limit: a first-principles perspective to recent synthesis. Appl Phys Rev 5:011105
10. Zhuang HL, Singh AK, Hennig RG (2013) Computational discovery of single-layer III-V materials. Phys Rev B 87:165415
11. Kolobov AV, Fons P, Tominaga J, Hyot B, Andre B (2016) Instability and spontaneous reconstruction of few-monolayer thick GaN graphitic structures. Nano Lett 16:4849–4856
12. Onen A, Kecik D, Durgun E, Ciraci S (2016) GaN: from three- to two-dimensional single-layer crystal and its multilayer van der Waals solids. Phys Rev B 93:085431
13. Yang KN, Zhang T, Wei B, Bai YJ, Jia SF, Cao GH, Jiang RH, Zhang CB, Gao EL, Chang XJ, Li JT, Li SM, Zhu DM, Tai RZ, Zhou H, Wang JB, Zeng MQ, Wang ZC, Fu L (2020) Ultrathin high-k antimony oxide single crystals. Nat Commun 11:2502
14. Şahin H, Cahangirov S, Topsakal M, Bekaroglu E, Akturk E, Senger RT, Ciraci S (2009) Monolayer honeycomb structures of group-IV elements and III-V binary compounds: first-principles calculations. Phys Rev B 80:155453
15. Durgun E, Tongay S, Ciraci S (2005) Silicon and III-V compound nanotubes: structural and electronic properties. Phys Rev B 72:075420
16. Wang Y, Shi S (2010) Structural and electronic properties of monolayer hydrogenated honeycomb III–V sheets from first-principles. Solid State Commun 150:1473–1478
17. Mei YF, Thurmer DJ, Deneke C, Kiravittaya S, Chen YF, Dadgar A, Bertram F, Bastek B, Krost A, Christen J, Reindl T, Stoffel M, Coric E, Schmidt OG (2009) Fabrication, self-assembly, and properties of ultrathin AlN/GaN porous crystalline nanomembranes: tubes, spirals, and curved sheets. ACS Nano 3:1663–1668
18. Qin ZZ, Qin GZ, Zuo X, Xiong QH, Hu M (2017) Orbitally driven low thermal conductivity of monolayer gallium nitride (GaN) with planar honeycomb structure: a comparative study. Nanoscale 9:4295–4309
19. Al Balushi ZY, Wang K, Ghosh RK, Vila RA, Eichfeld SM, Caldwell JD, Qin X, Lin YC, DeSario PA, Stone G, Subramanian S, Paul DF, Wallace RM, Datta S, Redwing JM, Robinson JA (2016) Two-dimensional gallium nitride realized via graphene encapsulation. Nat Mater 15:1166–1171
20. Ben JW, Liu XK, Wang C, Zhang YP, Shi ZM, Jia YP, Zhang SL, Zhang H, Yu WJ, Li DB, Sun XJ (2021) 2D III-nitride materials: properties, growth, and applications. Adv Mater 33:2006761
21. Zhang JM, Zhang J, Xu KW, Ji V (2015) Structural, electronic and magnetic properties of the Si chains doped zigzag AlN nanoribbons. Phys E 65:114–119
22. Sun TT, Wang YX, Du XJ, Ning ZR, Ke B (2014) First-principles study of a single Fe-chain doped zigzag AlN nanoribbons. Superlattices Microstruct 69:136–143
23. Sarma JVN, Chowdhury R, Jayaganthan R (2013) Molecular dynamics investigation of the thermomechanical behavior of monolayer GaN. J Appl Phys 113:243504
24. Peng Q, Chen XJ, Liu S, De S (2013) Mechanical stabilities and properties of graphene-like aluminum nitride predicted from first-principles calculations. RSC Adv 3:7083–7092
25. Akiyama T, Tsuboi Y, Nakamura K, Ito T (2019) An ab initio study for the formation of two-dimensional III-nitride compound ultrathin films: effects of Ag(111) substrate. J Cryst Growth 511:89–92
26. Behzad S (2018) Strain engineering of band dispersion and dielectric response of monolayer and bilayer AlN. J Comput Electron 17:514–520
27. Ghasemzadeh F, Kanjouri F (2018) Strain effect on the electronic properties of III-nitride nanosheets: *Ab-initio* study. Sci China: Technol Sci 61:535–541
28. Behera H, Mukhopadhyay G (2012) Strain tunable band-gaps of two-dimensional hexagonal BN and AlN: an FP-(L)APW+lo study. AIP Conf Proc 1447:273–274
29. Bacaksiz C, Sahin H, Ozaydin HD, Horzum S, Senger RT, Peeters FM (2015) Hexagonal AlN: dimensional-crossover-driven band-gap transition. Phys Rev B 91:085430

30. Keik D, Bacaksiz C, Senger RT, Durgun E (2015) Layer- and strain-dependent optoelectronic properties of hexagonal AlN. Phys Rev B 92:165408
31. Wang W, Li Y, Zheng Y, Li X, Huang L, Li G (2019) Lattice structure and bandgap control of 2D GaN grown on graphene/Si heterostructures. Small 15:1802995
32. Chen YX, Liu KL, Liu JX, Lv TR, Wei B, Zhang T, Zeng MQ, Chang ZC, Fu L (2018) Growth of 2D GaN single crystals on liquid metal. J Am Chem Soc 140:16392–16395
33. Zeng MQ, Zhang T, Tan LF, Fu L (2017) Liquid metal catalyst: Philosopher's stone of two-dimensional meterials. Acta Phys-Chim Sin 33:464–475
34. Li Q, Du BD, Gao JY, Xing BY, Wang DK, Ye JF, Liu J (2022) Room- temperature printing of ultrathin quasi-2D GaN semiconductor via liquid metal gallium surface confined nitridation reaction. Adv Mater Technol 7:2200733
35. Tsipas P, Kassavetis S, Tsoutsou D, Xenogiannopoulou E, Golias E, Giamini SA, Grazianetti C, Chiappe D, Molle A, Fanciulli M, Dimoulas A (2013) Evidence for Graphite-like hexagonal AlN nanosheets epitaxially grown on single crystal Ag(111). Appl Phys Lett 103:251605
36. Malin T, Mansurov V, Galitsyn Y, Zhuravlev K (2015) 2D AlN crystal phase formation on (0001) Al2O3 surface by ammonia MBE. Phys Status Solidi C 12:443–446
37. Chang H, Chen Z, Liu B, Yang S, Liang D, Dou Z, Zhang Y, Yan J, Liu Z, Zhang Z, Wang J, Li J, Liu Z, Gao P, Wei T (2020) Quasi-2D growth of aluminum nitride film on graphene for boosting deep ultraviolet light-emitting diodes. Adv Sci 7:2001272
38. Wang W, Zheng Y, Li X, Li Y, Zhao H, Huang L, Yang Z, Zhang X, Li G (2019) 2D AlN layers sandwiched between graphene and Si substrates. Adv Mater 31:1803448
39. Yang F, Jin L, Sun L, Ren X, Duan X, Cheng H, Xu Y, Zhang X, Lai Z, Chen W, Dong H, Hu W (2018) Free-standing 2D hexagonal aluminum nitride dielectric crystals for high-performance organic field-effect transistors. Adv Mater 30:1801891
40. Chen Q, Hu H, Chen X, Wang J (2011) Tailoring band gap in GaN sheet by chemical modification and electric field: Ab initio calculations. Appl Phys Lett 98:053102
41. Rong X, Wang X, Ivanov SV, Jiang X, Chen G, Wang P, Wang W, He C, Wang T, Schulz T, Albrecht M, Jmerik VN, Toropov AA, Ratnikov VV, Kozlovsky VI, Martovitsky VP, Jin P, Xu F, Yang X, Qin Z, Ge W, Shi J, Shen B (2016) High-output-power ultraviolet light source from quasi-2D GaN quantum structure. Adv Mater 28:7978–7983
42. Rastegar SF, Peyghan AA, Ghenaatian HR, Hadipour NL (2013) NO_2 detection by nanosized AlN sheet in the presence of NH_3: DFT studies. Appl Surf Sci 274:217–220
43. Noguerón A, Fernández-Escamilla N, Guerrero-Sánchez J, Takeuchi N (2019) Formaldehyde adsorption on a hydrogenated aluminum nitride monolayer: a self-propagated reaction. Comput Theor Chem 1159:18–22
44. Wang Y, Song N, Song X, Zhang T, Yang D, Li M (2018) A first-principles study of gas adsorption on monolayer AlN sheet. Vacuum 147:18–23
45. Yong XL, Su XY, Cui HL, Zhou QY, Kuang YM, Li XH (2017) Two-dimensional tetragonal GaN as potential molecule sensors for NO and NO_2 detection: a first-principle study. ACS Omega 2(12):8888–8895
46. Zhang X, Jin L, Dai X, Chen G, Liu G (2018) Two-dimensional GaN: an excellent electrode material providing fast ion diffusion and high storage capacity for Li-ion and Na-ion batteries. ACS Appl Mater Interfaces 10:38978–38984
47. Sengupta A (2018) Lithium and sodium adsorption properties of two-dimensional aluminum nitride. Appl Surf Sci 451:141–147
48. Li LS, Martirez JMP, Carter EA (2020) Predictioin of highly selective electrocatalytic nitrogen reduction at low overpotential on a Mo-doped g-GaN monolayer. ACS Catal 10:12841–12857
49. Purwiandono G, Manseki K, Sugiura T (2020) Photo-electrochemical property of 2D hexagonal-shape GaN nanoplates synthesized using solid nitrogen source in molten salt. J Photochem Photobiol A 394:112499
50. Beshkova M, Yakimova R (2020) Properties and potential applications of two-dimensional AlN. Vacuum 176:109231

MIX
Papier aus verantwortungsvollen Quellen
Paper from responsible sources
FSC® C105338

If you have any concerns about our products,
you can contact us on
ProductSafety@springernature.com

In case Publisher is established outside the EU,
the EU authorized representative is:
Springer Nature Customer Service Center GmbH
Europaplatz 3, 69115 Heidelberg, Germany

Printed by Libri Plureos GmbH
in Hamburg, Germany